Infinitesimal Differences

Infinitesimal Differences

Controversies between Leibniz and his Contemporaries

Edited by
Ursula Goldenbaum and Douglas Jesseph

Walter de Gruyter · Berlin · New York

♾ Printed on acid-free paper which falls within the guidelines,
of the ANSI to ensure permanence and durability.

ISBN 978-3-11-020216-8

Bibliographic information published by the Deutsche Nationalbibliografie

The Deutsche Nationalbibliothek lists this publication in the Deutsche Nationalbibliografie;
detailed bibliographic data are available in the Internet at http://dnb.d-nb.de.

Printed in Germany
Cover design: Martin Zech, Bremen
Typesetting: Dörlemann Satz GmbH & Co. KG, Lemförde
Printing and binding: Hubert & Co. GmbH & Co. KG, Göttingen

Table of Contents

Introduction

This volume had its beginnings in a conference entitled *The Metaphysical and Mathematical Discussion of the Status of Infinitesimals in Leibniz's Time* held in April 2006 at Emory University in celebration of the 50th anniversary of the university's graduate program in philosophy. Leroy E. Loemker, who initiated the graduate program as the first chair of the new Department of Philosophy at Emory, is well known as the father of North American Leibniz scholarship. Thus, the conference was dedicated as well to Loemker and his outstanding work on Leibniz, most notably his volume of Leibniz's *Philosophical Papers and Letters*, which remains one of the central works of Leibnizian scholarship in English.

The conference centered on a topic of interest for many scholars in philosophy as well as the history of mathematics, and it gave rise to many lively and interesting discussions about the nature and status of infinitesimals. Participants also had much to say about the notion of fiction, and especially the concept of a "well-founded fiction" in Leibniz's system. However, as we can now see on the basis of the largely revised papers, this conference also initiated a new effort to work out a clearer and more comprehensive understanding of these questions, some focusing particularly on *methodological* approaches to the infinitesimals in mathematics, physics and metaphysics. As a result, this volume offers a tightly focused collection of papers that address the metaphysical, physical, and mathematical treatment of infinitely small magnitudes in Leibniz's thought and that of his contemporaries, whether in the foundations of the *calculus differentialis*, the physics of forces, the theory of continuity, or the metaphysics of motion.

Although the central focus of the volume is on the development of Leibniz's calculus, the contributions provide a consistent and comprehensive overview of seventeenth and early eighteenth century discussions of the infinitesimal. In addition to addressing the role of infinitesimals in Leibniz's thought, contributors also consider the approaches of his predecessors, contemporaries, and immediate successors as Bonaventura Cavalieri, Evangelista Torricelli, Gilles Personne de Roberval, Thomas Hobbes, John Wallis, Isaac Newton, Blaise Pascal, Christiaan Huygens, Johann Bernoulli, Guillaume de L'Hôpital, Jacob Hermann, and Bernard Nieuwentijt.

The resulting collection therefore offers insight into the origins of Leibniz's conception of the infinite (and particularly the infinitely small), as well as the role this conception plays in different aspects of his mature thought on mathematics, physics, and metaphysics.

Leibniz mastered the mathematics of his day and developed his own calculus over the short span of a few years. But despite the success of his calculus in solving outstanding mathematical problems, the apparent ambiguity of Leibniz's conception of infinitesimals as fictions led to controversy at the end of the 17th century. Although urged to explain his approach more explicitly, Leibniz was generally reluctant to present the foundations of his new method. Moreover, he had offered very different accounts of the infinitesimal to different correspondents, further complicating a univocal understanding of his approach to the calculus. Even without an explicit statement of foundations, however, it is clear that Leibniz's mature view never characterized infinitesimals as real quantities, although he considered the prospects of a "realist" approach to infinitesimals in his earlier years. Although the calculus was undoubtedly successful in mathematical practice, it remained disputed precisely because its procedures seemed to lack an adequate metaphysical or methodological justification. In addition, Leibniz freely employed the language of infinitesimal quantities in the foundations of his dynamics and theory of forces, so that disputes over the very nature of infinitesimals naturally implicate the foundations of the Leibnizian science of bodies.

The fourteen essays collected here enhance and develop current scholarly understanding of the different conceptual and metaphysical issues raised by the mathematics of infinitesimals. Some essays are concerned principally with the historical origins of the mathematics of the infinitesimal, while others focus on the theoretical foundation of the calculus or on Leibniz's mature "fictionalism" about the infinite. In addition, a number of contributors seek to clarify the physics of forces Leibniz expressed in the language of the calculus. *Richard Arthur's* paper compares the Leibnizian doctrine of the infinitesimal and Newton's method of prime and ultimate ratios. He argues that these two approaches are not nearly as different as has commonly been supposed, and that both are motivated by surprisingly similar concerns about the rigorous development of a theory of continuously varying quantities. *Philip Beeley's* essay discusses John Wallis' motives for reforming Cavalieri's geometry of continua, known as indivisibles. These had already been transformed into infinitely small entities through authors such as Torricelli, Roberval and Pascal. Beeley argues that Wallis sought for the first time to combine the concepts of "infinitesimals" and arithmetical limits, when coming up with their arithmetization. Beeley also gives an account of some of the debates which ensued with the likes of Hobbes and Fermat.

Ursula Goldenbaum argues on the basis of newly discovered marginalia of Leibniz in Hobbes' *Opera philosophica* (1668) that Leibniz embraced Hobbes' *conatus* while reading *De homine* in the end of 1669. Leibniz's great expectation toward Hobbes' theory of sensation, due to his own projected philosophy of mind, spurred him to study the conatus conception of Hobbes in *De corpore* and consequently the mathematics of indivisibles. *Siegmund Probst* presents and analyzes newly discovered material concerning Leibniz's use of indivisibles and infinitesimals in his earliest mathematical writings, shedding some light on his unknown mathematical studies of Hobbes. In particular, he draws attention to mathematical manuscripts of Leibniz that "illustrate how Leibniz operated with concepts such as indivisibles and infinitesimals," in the early 1670s.

Samuel Levey's paper analyzes the reasons for Leibniz's ultimate abandoning of his earlier commitment to actual infinitesimals in 1676. He then takes up the question of how Leibniz's fictionalism about infinitesimals should be understood, concluding that there is no single "fictionalist" treatment to which Leibniz was invariably committed, although they all can be styled "Archimedean" in their reliance on classical exhaustion techniques. *O. Bradley Bassler* distinguishes Leibniz's metaphysical concerns with infinitesimals (which concern their fictional status) from his mathematical treatment of infinitesimals as differentials. Bassler argues that the central technical issue surrounding the status of differentials concerns the specification of the "progression of variables." He then suggests some ways in which Leibniz's metaphysical and mathematical approaches to infinitesimals can be related.

Emily Grosholz emphasizes Leibniz's "productively ambiguous notation" as crucial for his development of the calculus. Leibniz's ambiguous notation, connected with the law of continuity, allowed for yoking together very unlike things and offers a means of making them mutually intelligible. Thus Leibniz's development of the infinitesimal calculus and his investigations of transcendental curves can be read as instances of ambiguity which, far from hindering understanding, makes novel mathematical objects comprehensible. *Eberhard Knobloch's* contribution focuses on Leibniz' claim for the generality of his calculus. He investigates Leibniz's declared debt to the ancients, particularly to Archimedes' emphasis on geometrical rigor. Although Leibniz made his great mathematical progress by studying the work of most recent mathematicians in Paris, Knobloch shows how Leibniz avoided "the danger" of the method of indivisibles by a conscious turn to Archimedean methods. In addition, Knobloch draws an illuminating contrast between the Leibnizian theory of infinitesimals and the more robust (but ultimately incoherent) realism about infinitesimals embraced by Leonhard Euler. *Herbert Breger's* essay focuses on Leibniz' mathematical development after his departure from Paris. He emphasizes the strong influence of Pascal and Huygens on Leibniz's approach and gives an instructive survey of their methods and arguments. The result is that, given this background to the calculus, there was in fact no genuine "foundational problem" to be addressed. Breger argues that "what was really new and what posed the actual problem of understanding the new method of calculation was the higher level of abstraction."

Some of Leibniz's contemporaries objected that his methods violated standards of mathematical rigor, and the resulting controversies are important in understanding the reception of the calculus. Two papers in this collection are directed toward these controversies. *Fritz Nagel's* contribution investigates into the conception of the infinitesimal put forward by Hermann, which arose in response to the criticisms advanced by Nieuwentijt against Leibniz in the 1690s. Nagel notes that Herman's approach, endorsing Leibniz' position, has a significant degree of methodological and technical sophistication, and understanding it can shed some considerable light on the foundations of the calculus at the close of the seventeenth century. *Douglas Jesseph's* essay deals with both early and late Leibnizian writings on the calculus. He argues that some of the fundamental notions in the *calculus differentialis* can be found in Hobbes's concept of *conatus*. Jesseph then interprets the fictionalism espoused by Leibniz in response to criticisms as a further development of some of the key concepts that he had first encountered decades earlier in his reading of Hobbes.

The role of infinitesimals in Leibnizian physics is the focus of three of the contributions to this volume. *François Duchesneau's* discusses the often mentioned ambiguities of Leibnizian scientific statements, arguing that such ambiguous analogies for Leibniz, when duly controlled, could become crucial means for promoting the art of discovery (*ars inveniendi*). Duchesneau shows how Leibniz' scientific methodology itself favors hypothetical constructions. With hypotheses, truths of reason may be applied to the analysis of contingent truths expressing the connection of natural phenomena. Along this line, a condition of valid hypothesizing consists in the framing of relevant mathematical models within science. *Donald Rutherford* focuses on the notion of force and the connection between the physical theory of forces and the calculus. His essay aims to reconcile two Leibnizian claims: first, that force as the only "real and absolute" property of bodies is an infinitesimal element of action which produces continuous change over time; and second, that the infinitesimal quantities which model forces are mere fictions rather than real entities. This reconciliation is undertaken by seeing that a substance's transition from state to state is to be understood in terms of internal forces, which Leibniz thinks can best be modeled on the internal dynamics of the soul. *Daniel Garber's* essay is also concerned with this tension between the physical and mathematical understanding of infinitesimals, notably the notion of "dead force" in Leibniz's mechanics, and the connection between it and the notion of an infinitesimal magnitude. Garber argues that Leibniz distinguishes mathematics (and such fictions as infinitesimals) from the physical world in a way that allows physically real forces to be modeled or represented by mathematical devices that are not, strictly speaking, real entities.

We are grateful to the Graduate School of *Emory* having supported and generously sponsored this conference and to the colleagues of the Philosophy Department who encouraged us to organize this conference. We also thank the *Gottfried-Wilhelm-Leibniz Gesellschaft* at Hannover and the *North American Leibniz Society* for their official support and promotion of the conference. We are particularly grateful to Gertrud Grünkorn at the *de Gruyter Publishing House* at Berlin as well as to Andreas Vollmer for their supportive cooperation and the careful work on this volume, whose technical content makes it rather difficult. We would also like to thank Matt Traut, graduate student at Emory, and Stephen P. Farrelly, former graduate student at Emory (and now Assistant Professor at the Department of Philosophy at the University of Arkansas at Little Rock) for their great support in revising the papers for the publisher. Last but not least we are very grateful for the reliable cooperation with all the authors of this volume whose readiness to improve their papers mirrored the cheerful and enthusiastic atmosphere of our conference.

March 2008 Ursula Goldenbaum and Douglas Jesseph

Richard T. W. Arthur

Leery Bedfellows: Newton and Leibniz on the Status of Infinitesimals[1]

1. Newtonian and Leibnizian Foundations: The Standard Contrast

As is well known, Newton did not welcome Leibniz's efforts at establishing a differential calculus: his attitude, one might say, ranged between deep suspicion, disdain and utter hostility. In his eyes, Leibniz's differential calculus was at best a sample of the new method of analysis, an unrigorous symbolic method of discovery that could not meet the standard of rigorous proof required in geometry; and at worst, not just a plagiarism of his own work, but a dressing up and masking in Leibniz's fancy new symbols of the deep truths of his method of fluxions, which did not depend on the supposition of infinitesimals but was instead founded directly in the "real geneses of things." Leibniz, for his part, while accepting many of Newton's results, harbored doubts about Newton's understanding of orders of the infinitely small, which to his way of thinking was betrayed by the unfoundedness of Newton's composition of non-uniform with uniform motions in the limit.

There are some profound differences here in the respective thinkers' philosophies of mathematics, involving differing conceptions of proof, of the utility of symbolism, and in the conceptions of how mathematics is related to the physical world. I do not want to understate them. Nevertheless, I shall contend here, there is a very real consilience between Newton's and Leibniz's conceptions of infinitesimals, and even in the foundations they provide for the method of fluxions and for the differential calculus.

Newton's own evaluation of the difference in their methods was given by him in the supposedly "neutral" report he submitted anonymously to

[1] I would like to thank Sam Levey and Niccolò Guiciardini for their helpful feedback on earlier drafts.

the Royal Society in 1715, *Account of the Commercium Epistolicum* (in MPN VIII). There he depicted his method as proceeding "as much as possible" by finite quantities, and as founded on these and the continually increasing quantities occurring in nature, in contrast to Leibniz's, founded on indivisibles that are inadmissible in geometry and non-existent in nature:

> We have no Ideas of infinitely little, & therefore Mr. Newton introduced Fluxions into his Method that it might proceed by finite Quantities as much as possible. It is more Natural & Geometrical because founded upon the *primae quantitatum nascientum rationes* w ch have a Being in Geometry, whilst *Indivisibles* upon which the Differential Method is founded have no Being either in Geometry or in Nature. There are *rationes primæ quantitatum nascentium* but not *quantitates primæ nascentes*. Nature generates Quantities by continual Flux or Increase, & the ancient Geometers admitted such a Generation of Areas & Solids [...]. But the summing up of Indivisibles to compose an Area or Solid was never yet admitted into Geometry. (MPN VIII, 597–8)

This has been an influential account. Although it has long been recognized that Leibniz's differential calculus is a good deal more general than the Cavalierian geometry of indivisibles, and that Newton's characterizing of it as founded on indivisibles must be interpreted accordingly, the idea that Leibniz's methods were committed to the existence of infinitesimals has stuck. As a result, his official position that they are to be taken as fictions has been regarded as a not very successful attempt to distance himself from the foundational criticisms brought to bear by Nieuwentijt, Rolle, and the Newtonians, when in fact his method is based upon infinite sums and infinitely small differences, and thus firmly committed to infinities and infinitesimals. Newton, on the other hand, has been seen as moving from an early purely analytic method depending on a free use of infinitesimals to a mature view, represented in his *Method of First and Ultimate Ratios* (MFUR) published in the *Principia*, where (officially, at least) there are only limiting ratios of nascent or evanescent quantities, and never infinitely small quantities standing alone.

Newton's *Account of the Commercium Epistolicum* is a late text in his mathematical development, occurring as the culmination of a process of distancing himself from Analysis. By the 1680s he had turned away from the "moderns" in favor of Pappus and Apollonius, and an insistence on geometric demonstration. But the contrast between an early analytic Newton and the later conservative geometrician should not be overemphasized. The conception of fluxions or velocities by means of which Newton articulated what we call the *Fundamental Theorem of the Calculus* is intimately bound up with the kinematic conception of curves that he inherited from

Barrow and Hobbes. Thus although Newton's first formulations of his theory of fluxions are analytic in the sense that they are couched in terms of equations and algebraic variables, his kinematic understanding of curves and surfaces already implicitly involves a notion of the quantities represented by the variables as geometric, and as generated in time.

I shall argue, accordingly, that there is not such a huge gulf between Newton's analytic method of fluxions and the synthetic methods he later developed. Moreover, I contend, when Newton comes to secure the foundations of his synthetic method in the *Method Of First and Ultimate Ratios*, he appeals to Lemma 1, which is a synthetic version of the *Archimedean axiom*: "Quantities, and also ratios of quantities, which in any finite time constantly tend to equality, and which before the end of that time approach so close to one another that their difference is less than any given quantity, become ultimately equal" (Newton, 1999, 433). The axiom then serves to justify Newton's appeals to infinitesimal moments in supposedly geometric proofs such as that of Proposition 1 of the *Principia*, since these moments can be understood as finite but arbitrarily small geometric quantities in accordance with the *Archimedean axiom*. Furthermore, although Newton himself is careful to apply Lemma 1 only to *ratios of quantities*, the lemma as stated by him also applies directly to *quantities*; and Leibniz will appeal to a very similar principle applied to quantities as the foundation of his own method. In fact, the principle Leibniz appeals to, which takes differences smaller than any assignable to be null, is stated independently by Newton in his analytic method of fluxions (1971), and is a straightforward application of the *Archimedean axiom*.

Contrary to the standard depiction of their methods, then, there is a great similarity in the foundations of Newton's and Leibniz's approaches to the calculus. In fact, as I show by a detailed analyses of Newton's proof of Lemma 3 of his MFUR, and of Leibniz's proof of his Proposition 6 of *De quadratura arithmetica* (1676; DQA), their (contemporary and independent) attempts to provide rigorous foundations for their infinitesimalist methods by an appeal to the *Archimedean axiom* are in detailed correspondence, and perfectly rigorous. The rigor of Leibniz's approach to proving proposition 6 has already been stressed by Eberhard Knobloch (Knobloch, 2002). Here I extend that analysis to show its compatibility with the *syncategorematic interpretation of infinitesimals* attributed to Leibniz by Hidé Ishiguro.

2. Newton's Moments and Fluxions

The paper that is generally taken as containing Newton's first full statement of his analytic method of fluxions is *To Resolve Problems by Motion*, written in October 1666 as the culmination of several redraftings (MPN I, 400–448). The commitment to the kinematical representation of curves is evident in its title, and this is so also for the earlier drafts out of which it develops: two drafts of *How to draw tangents to Mechanicall lines* [30? October 1665 and 8 November 1665, resp.], a third draft titled *To find ye velocitys of bodys by ye lines they describe*, [November 13th 1665], and a fourth titled *To resolve these & such like Problems these following propositions may bee very usefull*, [May 14, 1666].[2] Thus Newton's recipe in Proposition 7 for what we, after Leibniz, call differentiation, is couched by him in terms of the velocities of bodies:

> [Prop.] 7. Haveing an Equation expressing y^e relation twixt two or more lines x, y, z &c: described in y^e same time by two or more moveing bodies A, B, C, &c [Fig. 1]: the relation of their velocitys *p*, *q*, *r*, &c may bee thus found, viz:

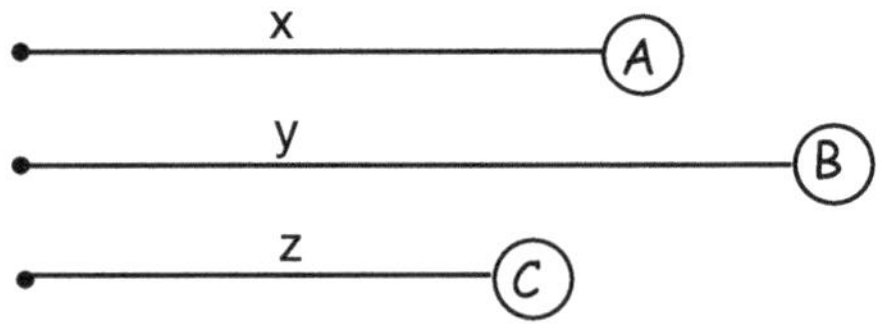

Figure 1.

> Set all y^e termes on one side of y^e Equation that they may become equall to nothing. And first multiply each terme by so many times p/x as x hath dimensions in y^t terme. Secondly multiply each terme by so many times q/y as y hath dimensions in it. Thirdly (if there be 3 unknowne quantitys) multiply each terme by so many times r/z as z hath dimensions in y^t terme, (& if there bee still more unknowne quantitys doe like to every unknowne quantity). The summe of all these products shall be equall to nothing. wch Equation gives ye relation of ye velocitys *p*, *q*, *r*, &c. (MPN I, 402)

The first thing to notice about this algorithm is that it is not purely analytic: the equations are given a geometrical interpretation in terms of lines traced by moving bodies. Second, what results from the algorithm is not a veloc-

2 These drafts are given in MPN I, 369–377, 377–382, 382–389, and 390–392, resp. The last draft was subsequently cancelled and rewritten as *To resolve Problems by motion ye 6 following prop. are necessary and sufficient*, dated May 16, 1666 (MPN I, 392–399).

ity but the *ratio* of two velocities, and these velocities (say, p and q) are the instantaneous velocities of two bodies at the beginning of the moment o for which they are assumed to travel with that velocity.

A very simple example of applying this algorithm is provided by the result Newton quotes in his demonstration of Proposition 1 of this tract – this being perhaps the very first application of the method of fluxions in physics.[3] Proposition 1 is a statement of the resolution of velocities, and its demonstration depends on finding the relation between the velocities of the body A in two directions, towards d and towards f, as it travels along the line gc below, with $df \perp ac$, at the very point a when it reaches the perimeter of the circle. Letting $df = a$, $fg = x$, and $dg = y$, we have (since Δadf is a right triangle)

$$a^2 + x^2 - y^2 = 0.$$

According to Newton's algorithm given in Proposition 7 above, we must multiply each term in x in the equation by $2p/x$ and each term in y by $2q/y$, yielding

$$2xp - 2yq = 0.$$

This result is quoted by Newton in his demonstration as follows:

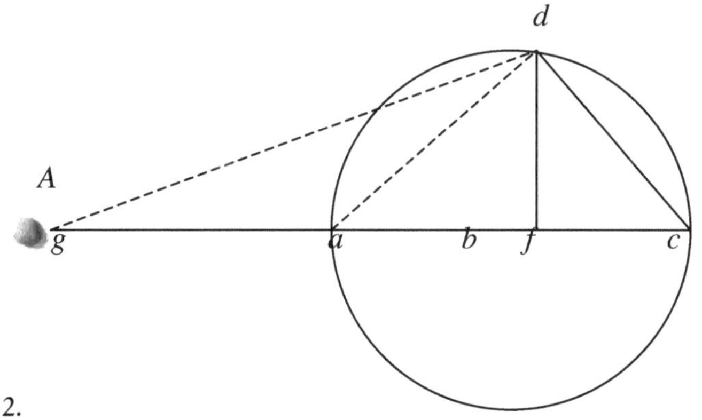

Figure 2.

Now by Prop 7th, may y^e^ proportion of (p) y^e^ velocity of y^t^ body towards f; to (q) its velocity towards d bee found viz: $2px - 2pq = 0$. Or $x{:}y{::}q{:}p$. That is $gf : gd ::$ its velocity to d : its velocity towards f or c. & when y^e^ body A is at a, y^t^ is when y^e^ points g & a are coincident then is $ac{:}ad :: ad{:}af ::$ velocity to c: velocity to d. (MPN I, 415)

[3] Newton first gives the demonstration of Proposition 1 immediately after stating all 8 propositions (MPN I, 415), but as Whiteside notes, Newton alludes to the fact that it can be so demonstrated in the draft of May 14th, 1666 (MPN I, 390).

Or, as we would say in more Leibnizian terms, differentiating $a^2 + x^2 - y^2 = 0$ yields $2xp - 2yq = 0$, with p and q the derivatives of x and y respectively. Thus the velocities p and q are in the inverse ratio of x and y. Now when the body A reaches the point a we have $x = af$, $y = ad$, $q = v_{ad}$ and $p = v_{ac}$, yielding

$$v_{ac} : v_{ad} = ad : af$$

and since by similar triangles $ad{:}af = ac{:}ad$, we obtain finally

$$v_{ac} : v_{ad} = ac : ad \qquad or \qquad v_{ad} : v_{ac} = ad : ac$$

which, in modern notation, is the correct formula for the resolution of velocities in an oblique direction:

$$v_{ad} = v_{ac} \cos \phi, \text{ where } \phi = \angle \mathrm{dac}$$

Of interest to us here is the justification Newton gives in 1666 for Proposition 7. The demonstration he provides is by reference to a specific equation, $x^3 - abx + a^3 - dy^2 = 0$. There is no loss in generality in our substituting for it the above equation for Proposition 1, $a^2 + x^2 - y^2 = 0$. Newton first supposes two bodies A and B moving uniformly, the one from a to c, d, e, f, the other from b to g, h, k, l, in the same time. Then the pairs of lines ac and bg, cd and gh, de and hk etc. are "as their velocitys" p and q (MPN I, 414):

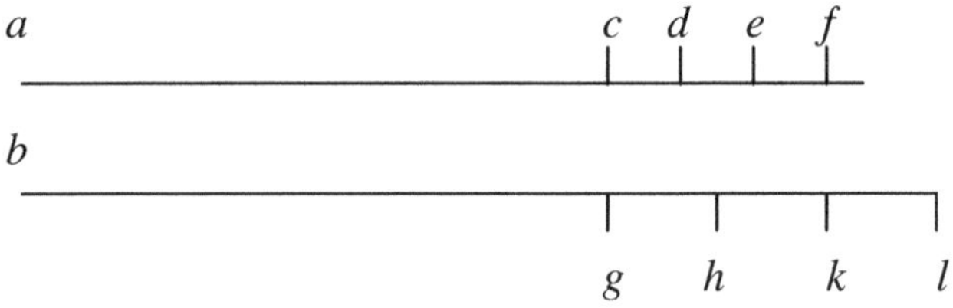

Figure 3.

He then reasons that:

> And though they move not uniformely, yet are y^{e} infinitely little lines wch each moment they describe, as their velocitys wch they have while they describe y^{m}. As if y^{e} body *A* wth y^{e} velocity *p* describe y^{e} infinitely little line ($cd =$) p x o in one moment, in y^{t} moment y^{e} body *B* wth y^{e} velocity *q* will describe y^{e} line ($gh =$) q x o. For $p{:}q :: po{:}qo$. Soe y^{t} if y^{e} described lines bee ($ac =$) x, & ($bg =$) y, in one moment, they will bee ($ad =$) $x + po$, & ($bh =$) $y + qo$ in y^{e} next. (MPN I, 414)

Now he claims that "I may substitute $x + po$ & $y + qo$ into y^{e} place of x & y; because (by y^{e} lemma) they as well as x & y, doe signify y^{e} lines described by y^{e} bodys A & B" (414). Thus for the equation $a^2 + x^2 - y^2 = 0$ we get

$$a^2 + x^2 + 2pox + p^2o^2 - y^2 - 2qoy - q^2o^2 = 0$$

Subtracting the original equation gives

$$2pox + p^2o^2 - 2qoy - q^2o^2 = 0$$

"Or dividing it by o tis [$2px + p^2o - 2qy - q^2o = 0$]. Also those termes are infinitely little in wch o is. Therefore omitting them there rests [$2xp - 2yq = 0$]. The like may bee done in all other equations" (MPN I, 414–15).

Here Newton's division by o prior to omitting terms in o because they are "infinitely little" is, of course, lacking in rigor. Either, one may object, adding po to x takes body A to "y^e^ next" point on the line representing its path, and one is committed to composing that line out of successive infinitesimal linelets (and thus succumbing to the paradoxes of the continuum); or indeed, $x + po$ does not at all differ from x, in which case division by o is completely illegitimate. And yet Newton's algorithm is framed in terms of *ratios* of quantities and their velocities in the moment o. Of course, there is no way to represent an instantaneous velocity geometrically save by showing the line segment (*cd* in figure 3) that a body would cover if it continued with that velocity for a time o. From this point of view, the moment o is more nearly a device enabling instantaneous velocities to be geometrically represented: po is the distance the body A would have covered if it had proceeded with the velocity p for some time o. The ratio $po{:}qo$ is of course equal to the ratio of p and q for any finite o. Moreover, it is implicit in the kinematic representation that the velocities p and q are the velocities at the very beginning of the moment o, so that the term for $po{:}qo$ calculated by Newton's algorithm, which will still generally contain terms in o, will be closer to $p{:}q$ the closer o is to 0. The justification for neglecting the remaining terms in o is therefore not so much that they are conceived as "infinitely little" in the sense of absolutely infinitely small, but in the sense that the ratio $p{:}q = po{:}qo$ represents the ratio of p and q right at the beginning of the moment, so that the smaller o is made, the smaller will be the terms still containing o, and the more nearly will the resulting expression represent the ratio.

Thus in the context of a kinematic and geometric interpretation of the quantities involved, Newton's early appeal to the infinitely small cannot simply be taken as committing him to a composition of quantities out of infinitesimals. In fact, his procedure already implicitly involves a kind of limiting process: to find the ratio of the velocities precisely at the beginning of the moment o (e.g. at the instant the moving body A reaches the point a in the above diagram), o must be shrunk to zero, so that the extra terms in the expression of this ratio still depending on the quantity o will therefore also vanish, with the resulting expression yielding the "first ratio" of these velocities.

Newton himself recognized this soon enough, and proceeded to make the limit concept implicit in the kinematical representation the foundation of the synthetic method of fluxions. He drew up these early results, as well as those outlined in his *De Analysi per æquationes numero terminorum infinitas* (1669; publ. 1711), into a formal Latin treatise intended for publication, the *Tractatus de methodis serierum et fluxionum* (1671; publ. 1736; MPN III, 32–328), or *Treatise on Fluxions* for short, where the terminology of "fluxions" was first introduced. But he remained unsatisfied with the foundations of his methods, and in an Addendum on *The Theory of Geometrical Fluxions* made just after completing the latter, he developed a wholly synthetic approach, "based on the genesis of surfaces by their motion and flow" (MPN III, 328–31; Guicciardini, 2003, 315). Axiom 4 of this Addendum was "Contemporaneous moments are as their fluxions" (MPN III, 330), or more perspicuously perhaps, "Fluxions are as the contemporaneous moments generated by those fluxions" (draft). Whiteside observes: "This fundamental observation opens the way to subsuming limit-increment arguments as fluxional ones, and conversely so" (MPN III, 330, fn 7).

As Guicciardini has noted (Guicciardini, 2002, 414–17), these foundations are synthetic in two distinct senses: they are based on explicit axioms from which propositions are derived, "synthesis" as opposed to "analysis"; and the quantities are not the symbols but fluent geometrical figures, synthetic in the sense of flowing, increasing, staying constant, or decreasing continuously in time. The emphasis on synthesis (in this dual sense) is a symptom of Newton's progressive disenchantment with analysis in the 1670s, and a growing respect for the geometry of the ancients. This process is taken further in *Geometria curvilinea*, written some time between 1671 and 1684, where Newton stresses the generation of geometric quantities in time:

> Those who have measured out curvilinear figures have usually viewed them as consisting of infinitely many infinitely small parts. But I will consider them as generated by growing, arguing that they are greater, equal or smaller according as they grow more swiftly, equally swiftly or more slowly from the beginning. And this swiftness of growth I shall call the fluxion of a quantity. (MPN IV, 422–23)

This interpretation of his mathematics explains the contrast Newton draws between the ontological foundation of his methods ("This Method is derived immediately from Nature her self") and the lack of such a foundation in the analysis of Leibniz. It is emphasized even more strongly in the *De quadratura curvarum* of 1693, where Newton writes:

> I don't here consider Mathematical Quantities as consisting of *indivisibles*, whether least possible parts or infinitely small ones, but as described by a continual motion. Lines are described, and by describing are generated, not by any apposition of Parts, but by the continuous motion of Points, Surfaces by the motions of Lines, Solids by the motion of Surfaces, Angles by the Rotation of their Legs, Time by a continual Flux, and so on in all the rest. These *Geneses* are founded upon Nature, and are every Day seen in the motions of Bodies. (Newton, 1964, 141)

In these passages Newton not only claims that geometric quantities are founded *in rerum natura*, he also explicitly repudiates their composition out of infinitely small parts (infinitely small quantities have "no Being either in Geometry or in Nature"). As he had come to recognize, the moments of quantities do not have to be supposed as infinitely small quantities, falling outside the scope of geometry based on the *Archimedean axiom*, but can instead stand for finite quantities that can be taken as small as desired. This is the foundation of his synthetic method of fluxions Newton presents in the *Geometria curvilinea*, and which he will publish in the *Principia* under the new moniker the *Method of First and Ultimate Ratios*. Although infinitely small quantities still occur in Newton's mature work, they are interpreted as standing for finite but small quantities that are on the point of vanishing, with the ratio between two such quantities remaining finite in this temporal limit.

An example of this synthetic method of fluxions, I claim, is provided by Newton's demonstration of Proposition 1, in Book 1 of the *Principia*. In fact, this proposition provides a particularly good specimen of the advantages of the synthetic method of fluxions. For not only is the proof extremely economical compared to any analytic derivation of Kepler's *Area Law*, it also depends on no assumptions about the nature of the force except that it is continuous and centrally directed.[4] Newton's demonstration goes as follows:

> Let the time be divided into equal parts, and in the first part of the time let a body by its inherent force describe the straight line *AB*. In the second part of the time, if nothing hindered it, this body would (by law 1) go straight on to *c*, describing line *Bc* equal to *AB*, so that – when radii *AS*, *BS* and *cS* are drawn to the centre – the equal areas *ASB* and *BSc* would be described. But when the body comes to *B*, let a centripetal force act with a single but great impulse and make

[4] Also, of course, as explained by Nauenberg, 2003, 450, the curvature of the curve must remain finite, and the radius vector cannot become tangential to it.

> the body deviate from the straight line *Bc* and proceed in the straight line *BC*. (Newton, 1999, 444)

Newton now completes the parallelogram *VBcC* to compute where the body would end up under the joint action of the inertial force and the force impressed at *B* by applying the parallelogram law (corollary 1), and uses elementary geometry to prove the equality of the triangles *SAB* and *SBC*. The motion along *BC* will now be the new inertial motion, and the same reasoning can be applied to triangles *SBC* and *SCD*, etc.

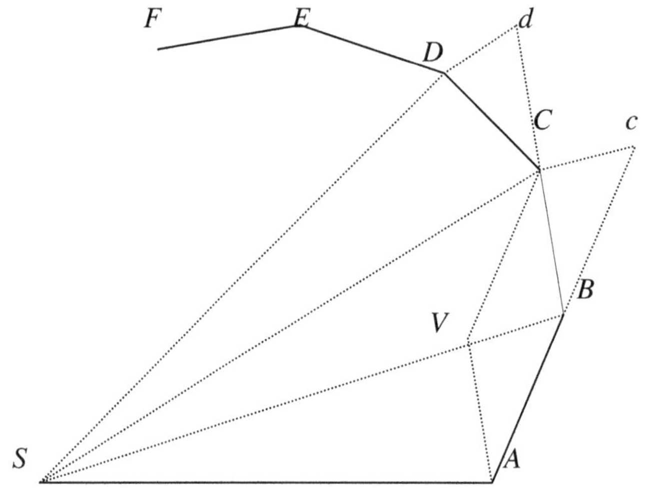

Figure 4.

> Now let the number of triangles be increased and their width decreased indefinitely, and their ultimate perimeter *ADF* will (by lem. 3, corol. 4) be a curved line; and thus the centripetal force by which the body is continually drawn back from the tangent of this curve will act uninterruptedly, while any areas described, *SADS* and *SAFS*, which are always proportional to the times of description, will be proportional to the times in this case. *Q.E.D.* (Newton, 1999, 444)

Crucial in this proof is the appeal to Lemma 3, Corollary 4: "And therefore these ultimate figures (with respect to their perimeters *acE*) are not rectilinear, but curvilinear limits of rectilinear figures" (Newton, 1999, 434).[5]

[5] Michael Nauenberg (Nauenberg, 2003, 444ff.) was the first to draw attention to the importance of this lemma in Newton's justification of Proposition 1. A minor oddity of this appeal to Lemma 3 is that the figure for Lemma 3 involves curvilinear limits of rectangles under the curve, rather than curvilinear limits of the triangles subtended under it in Proposition 1. But this does not undermine the appeal to this Lemma, since in principle the same arguments can be run for triangular rather than rectangular elements.

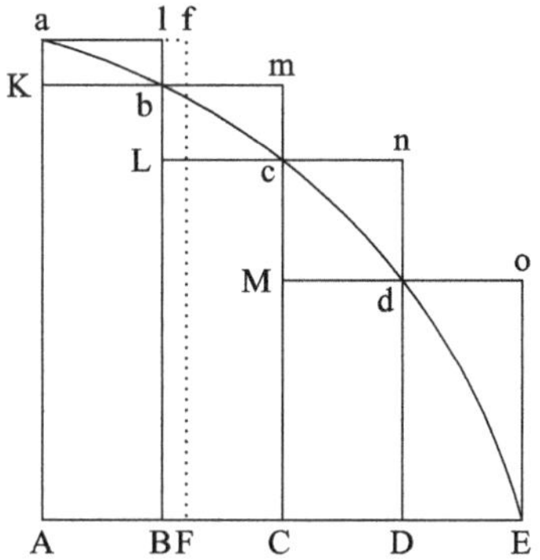

Figure 5

Lemma 3 itself is: "*the ultimate ratios [which the inscribed figure* AK*b*L*c*M*d*D, *the circumscribed figure* A*albmcndo*E, *and the curvilinear figure* A*abcd*E *have to one another] are also ratios of equality when the widths* AB, BC, CD, [...] *of the parallelograms are unequal and are all diminished indefinitely*" (Newton, 1999, 433). Newton uses this result to argue in Corollary 1 that "the ultimate sum of the vanishing parallelograms coincides with the curvilinear figure in its every part," in Corollaries 2 and 3 that the figure comprehended by the chords or the tangents of the vanishing arcs "coincides ultimately with the curvilinear figure," and in Corollary 4 that "therefore these figures (with respect to their perimeters *ac*E) are not rectilinear, but curvilinear limits of rectilinear figures" (Newton, 1999, 434). Thus by a similar argument the triangles in Figure 4 are not rectilinear, but curvilinear limits of rectilinear figures, the ratios between any two of which are equal.

Let us now turn to Newton's justification of this Lemma. He demonstrates it by reference to the same figure used for all the first four Lemmas. Having proved Lemma 2 on the supposition of equal intervals AB, BC, DE, etc., he now supposes them unequal, and lets "AF be equal to the greatest width" of any of the rectangles. Hence FA*af* is at least as wide as any of the rectangles, and its total height will be the sum of the heights of the differences between the circumscribed and inscribed figures. "This parallelogram will therefore be greater than the difference of the inscribed and circumscribed figures; but if its width AF is diminished indefinitely, it will become less than any given rectangle. Q.E.D." (Newton, 1999, 434)

The last step of this proof is an application of Lemma 1 of the *Method of First and Ultimate Ratios*, which I quote here in its original wording from the First Edition:

> *Quantities, and also ratios of quantities, which in a given time constantly tend to equality, and which before the end of that time approach so close to one another that*

> *their difference is less than any given quantity, become ultimately equal.* (Newton, 1999, 434)

Here one might object that an infinitesimal is precisely a quantity that is "less than any given quantity," so that if there exist non-zero infinitesimals then such a difference will be non-zero. In default of some further premise, the argument therefore seems to assume what it needs to prove. The missing premise is that in order for the quantities to count as geometrical quantities, they must obey the *Archimedean axiom*:

> If a and b are two line segments (or other continuous geometric quantities) with $a < b$, we can always find a (finite) number n such that $na > b$.

This axiom rules out the existence of an actual infinitesimal quantity, such as the "difference less than any given quantity" mentioned in Lemma 1. As Newton argues in his demonstration of the Lemma:

> If you deny this, let their ultimate difference be D. Then they cannot approach so close to equality that their difference is less than the given difference D, contrary to the hypothesis. (Newton, 1999, 433)

The "hypothesis" in question here is that they can always "approach so close to one another that their difference is less than any given quantity." This is simply an expression in synthetic form of the *Archimedean axiom*: given two quantities whose difference D is less than some quantity a, we can always find a number n such that $n\mathrm{D} > a$, so that $c = a/n < \mathrm{D}$.

In fact, if we explore the origins of Lemma 1 of the MFUR we can trace a direct line of descent from the "Treatise on Fluxions." Two paragraphs of this are rewritten into the "Addendum on Geometrical Fluxions," the latter is reworked into the *Geometria curvilinea*, and it is from this that the *Method of First and Ultimate Ratios* is derived. The first of the two paragraphs of the *Treatise on Fluxions* runs:

> This method for proving that curves are equal or have a given ratio by the equality or given ratio of their moments, I have used because it has an affinity with methods usually employed in these cases; but a method based on the genesis of surfaces from the motion of their flowing seems more natural [...]. (MPN III, 282)

This is transcribed to the Addendum, with the addition "[...], one which will prove to be more perspicuous and elegant if certain foundations are laid out in the style of the synthetic method; such as the following" (MPN III, 328–330), and this introduces the axioms and theorems that constitute the synthetic method. But the previous method referred to in this paragraph, that of proving "through the equality of moments," is described in the immediately preceding paragraph of the Treatise as follows:

> In demonstrations of this sort it should be observed that I take those quantities to be equal whose ratio is one of equality. And a ratio of equality is to be regarded as one which differs less from equality than any unequal ratio that can be assigned. Thus in the preceding demonstration I set the rectangle *Ep* × *Ac*, that is, *Feqf*, equal to the space *FEef* since (because their difference *Eqe* is infinitely smaller than them, i.e. nothing with respect to them), they have no ratio of inequality. And for the same reason, I set *Dp* × *HI* = *HIih*, and likewise in the others. (MPN III, 282)

The principle appealed to here is this:

> If an inequality is such that its difference from a strict equality can be made smaller than any that can be assigned, it can be taken for an equality.

Let us call this the *Principle of Unassignable Difference*. This principle, clearly, is the analytic equivalent of the chief synthetic axiom, Lemma 1 of the *Method of First and Ultimate Ratios*. And like that Lemma, it derives its warrant from the *Archimedean axiom*. This common warrant underwrites the equivalence between the analytic and synthetic methods of fluxions, allowing the translatability of statements given in terms of "indivisibles" (i.e. infinitesimals) into fluxional terminology, thus justifying Newton's claim in the *Principia* that having reduced the propositions there to the limits of the sums and ratios of First and Ultimate ratios of nascent and evanescent quantities, he had thereby "performed the same thing as by the method of indivisibles." He continues:

> Accordingly, whenever in what follows [...] I use little curved lines in place of straight ones, I wish it always to be understood that I mean not indivisibles but evanescent divisible quantities, and not the sums and ratios of determinate parts, but the limits of such sums and ratios; and that the force of such demonstrations always rests on the method of the preceding lemmas. (Newton, 1999, 441–2; trans. slightly modified)

3. Leibniz's Syncategorematic Infinitesimals

Now let us turn to Leibniz. During the same period (1671–1684) in which Newton was perfecting his synthetic interpretation of the results he had obtained in 1666, Leibniz was independently developing the algorithms and techniques he was to present as the differential and integral calculus. In his approach to the development and application of his calculus, Leibniz often stressed the pragmatic utility of his techniques, and how they could be exploited by mathematicians without their having to trouble themselves with foundational problems. These comments, together with the lack of clarity regarding foundations in his early publications, and his late pronouncements on the nature of infinitesimals precipitated by the controversies involving Rolle, Nieuwentijt and Varignon, have conspired to produce the impression that Leibniz developed his calculus without much attention to its foundations.

But this impression is entirely mistaken. For just as Newton had attempted to strengthen the foundations of his methods in his Latin treatise *De methodis serierum et fluxorum* in 1671, and again in *Geometria curvilinea* not long afterwards, so in 1675–76 Leibniz had also written a comprehensive Latin treatise on his infinitesimal methods, *De quadratura arithmetica*, which has only recently been edited and published by Eberhard Knobloch (DQA); and in this treatise, as Knobloch has shown, "Leibniz laid the rigorous foundation of the theory of infinitely small and infinite quantities" (Knobloch, 2002, 59). I have argued elsewhere (Arthur, 2008a) that Knobloch's interpretation of Leibniz's foundational work is fully in keeping with Hidé Ishiguro's attribution to Leibniz of an interpretation of infinitesimals as "syncategorematic." That is, as I have tried to show, Leibniz's mature interpretation of infinitesimals as "fictions" has a precise mathematical content, perfectly consistent with his philosophy of the infinite and solution to the continuum problem (Arthur, 2001, 2008b). Moreover, I shall argue here, this content is given by the foundation of the method on the *Archimedean axiom*. Thus Leibniz's justification of his infinitesimal methods will be seen to be in surprising conformity with Newton's.

As regards foundations, the nub of the *De quadratura arithmetica* occurs in Proposition 6 (DQA, 28–36), as Eberhard Knobloch has explained. Leibniz himself describes it as

> spinosissima in qua morose demonstratur certa quaedam spatia rectilinea gradiformia itemque polygona eousque continuari posse, ut inter se vel a curvis differant quantitate minore quavis data, quod ab aliis plerumque assumi solet. Praeteriri initio ejus lectio potest, servit tamen ad fundamenta totius Methodi indivisibilium firmissime jacienda.[6] (DQA, 24).

[6] "[...] most thorny; in it, it is demonstrated in fastidious detail that the construction of certain rectilinear and polygonal step spaces can be pursued to such a degree that they differ from one another or from curves by a quantity smaller than any given, which is something that is most often [simply] assumed by other authors. Even though one can skip over it at

The "thorniness" is evident from *Figure 6* (fig. 3 in the DQA):

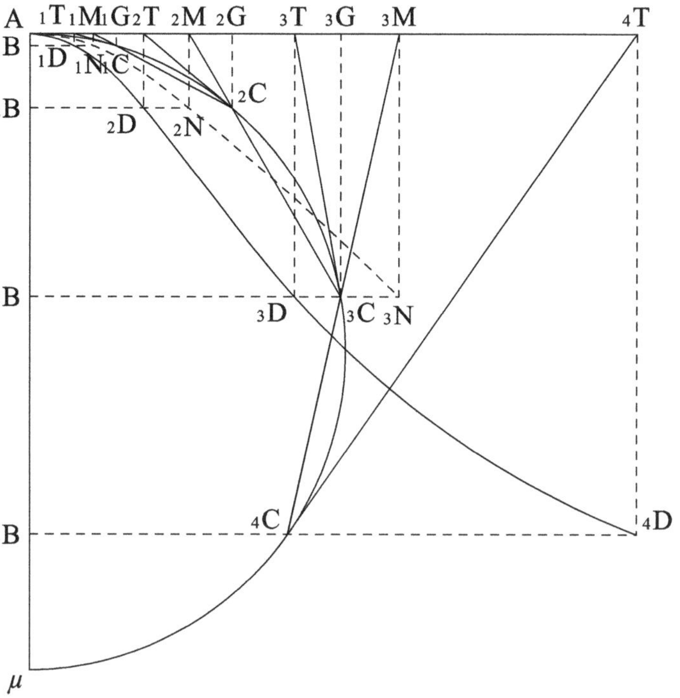

Figure 6.

In this figure, the x-axis is vertical, and the y-axis is the horizontal axis across the top. The curve considered here is a circular arc C, the tangents to which at successive points on this curve ($_1$C, $_2$C, $_3$C, $_4$C) cut the y-axis at the points $_1$T, $_2$T, $_3$T, $_4$T. Now a second, auxiliary curve D is defined by the points of intersection of these tangents to C with the ordinates $_1$B, $_2$B, $_3$B, $_4$B, yielding the points $_1$D, $_2$D, $_3$D, $_4$D, on this new curve. The secants joining successive pairs of points on the original curve, $_1$C$_2$C, etc., are extended to cut the y-axis in the points $_1$M, $_2$M, $_3$M. The points of intersection of the perpendiculars from these points M down through the ordinates B of the original curve define another set of points $_1$N, $_2$N, $_3$N. Provided certain conditions are satisfied – continuity, no point of inflection, no point with a vertical tangent – this construction is always possible, and as Knobloch

first reading, it serves to lay the foundations for the whole method of indivisibles in the soundest possible way".

comments, “once the second curve has been constructed, the first curve can be omitted.”[7]

Following Knobloch, we will now give a simplified figure depicting a portion of the area under the curve D between the ordinates ${}_1B$ and ${}_3B$:

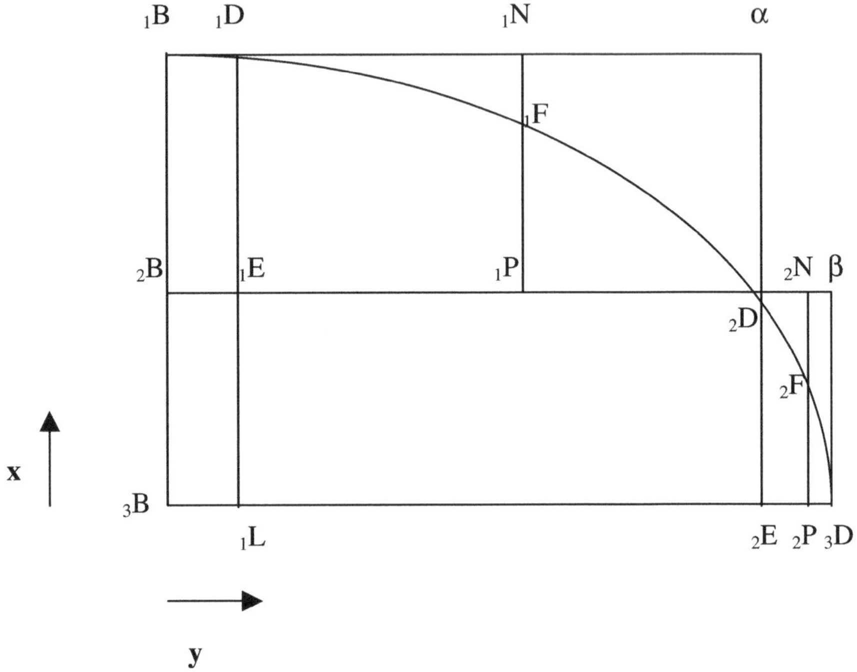

Figure 7.

The demonstration of Proposition 6 then proceeds in eight numbered stages. First Leibniz partitioned the interval containing the area under the curve D is into a finite number of unequal subintervals (in the above figure there are two, ${}_1B_2B$ and ${}_2B_3B$). The rectangles bounded by the ordinates, the x-axis to the left, and the normals through N to the right, here ${}_1B_1N_1P_2B$ and ${}_2B_2N_2P_3B$, he called elementary rectangles; the rectangles overlapping these bounded by successive points on the curve, here ${}_1D\alpha_2D_1E$ and ${}_2D\beta_3D_2E$, he called *complementary rectangles.*

In stage 2, he computed the (absolute value of the) difference between the area under the mixtilinear figures ${}_1B_1D_2D_2B$ and ${}_2B_2D_3D_3B$, and their corresponding elementary rectangles ${}_1B_1N_1P_2B$ and ${}_2B_2N_2P_3B$. In each case this difference is less than the corresponding complementary rectangle:

[7] See Knobloch, 2002, 63, for a discussion of these conditions.

$|{}_1B_1D_2D_2B - {}_1B_1N_1P_2B| < {}_1D\alpha_2D_1E$, etc. This is proved in stage 3 by subtracting from each their common part, ${}_1B_1D_1F_1P_2B$, etc., leaving a difference of two trilinear areas. Even the sum of these two areas is less than the complementary rectangle, so their difference certainly is. Thus $|{}_1B_1D_2D_2B - {}_1B_1N_1P_2B| = |{}_1D_1N_1F - {}_1F_2D_1P| < {}_1D\alpha_2D_1E$, etc. In step 4, it is shown that this inequality holds for all such differences between curvilinear areas and their corresponding elementary rectangles. As Knobloch has shown, Leibniz is here implicitly appealing to the triangle inequality $||A| - |B|| \leq |A - B|$ (Knobloch, 2002, 65).

Therefore (stages 5 and 6) the absolute value of the difference between the sum M of all the mixtilinear areas (the area under the curve, called by Leibniz the "total Quadrilineal") and the sum E of all the elementary rectangles approximating the area under the curve (the Riemannian sum, called by Leibniz the "step space [*spatium gradiforme*]") is less than the sum C of all the complementary rectangles: $|M - E| < C$. But the sum C of all the complementary rectangles ${}_1D\alpha_2D_1E$, ${}_2D\beta_3D_2E$, etc. would be less than the sum of all their bases times their common height, if all the ordinates were equally spaced. Since by hypothesis they are not, let the greatest height (say, the difference between successive ordinates ${}_3B$ and ${}_4B$) be h_m. The sum of all the bases is the difference between the greatest and smallest ordinate, ${}_1L_3D$. Therefore C is smaller than the rectangle equal to the product ${}_3B_4B \times {}_1L_3D$, i.e. $C < {}_1L_3D\ h_m$. Hence, since $|M - E| < C$, we have

$$M - E < {}_1L_3D \cdot h_m,$$

where h_m is the greatest height of any of the elementary rectangles.

But (stage 7) the abscissa representing this greatest height, "tametsi caeteris majus, aut certe non minus sit assumtum intervallis, tamen assignata quantitate minus assumi potest; nam ipso sumto utcunque parvo caetera sumi possunt adhuc minora".[8] (DQA, 31–32) Therefore "sequetur *ut rectangulum ψ ₄D ₁L*, altitudinem habens quae data recta minor sumi posit, etiam *data aliqa superficie reddi posse minus.*"[9] (DQA, ibid.).

It therefore follows (stage 8) that *"Differentia hujus Quadrilinei*, (de quo et proposition loquitur) *et spatii gradiformis data quantitate minor reddi po-*

8 "[…] even though it is greater than, or at any rate not less than, any of the other intervals assumed, can nevertheless be assumed smaller than any assigned quantity; for however small it is assumed to be, others can be assumed still smaller."

9 "[…] it will follow *that the rectangle ψ 4D 1L [3B4B1L3D]*, having a height which can be assumed smaller than any given line, also can be made smaller than any given surface".

test"[10] (DQA, 32). That is, the difference between the Riemannian sum and the area under the curve is smaller than any assignable, and therefore null.

As Leibniz points out, the prolixity of this proof is due in part to the fact that it is considerably more general than the "communi methodo indivisibilium"[11] (DQA, 32), where one is "securitatis causa cognimur, ut Cavalierius, ad ordinatas parallelas methodum restringere, et aequalia semper duarum proximarum ordinatarum intervalla postulare"[12] (DQA, 69). In that case the points N and the points D coincide and "longe facilior fuisset demonstratio"[13] (DQA, 32), as he proceeds to show.

Several things about this demonstration are worthy of note. As Leibniz observes in the Scholium to Proposition 7: "Demonstratio illud habet singulare, quod rem non per inscripta ac circumscripta simul, sed per sola inscripta absolvit."[14] (DQA, 35) More accurately though, the step figure is, as Knobloch says, "something in between" an inscribed and a circumscribed one (Knobloch, 2002, 63). Leibniz's method, in fact, is extremely general and rigorous; the same construction of elementary and complementary rectangles could be constructed for any curve whatsoever satisfying the three conditions outlined. It amounts in modern terms to a demonstration of "the integrability of a huge class of functions by means of Riemannian sums which depend on intermediate values of the partial integration intervals" (Knobloch, 2002, 63).

Second, it is strictly finitist. As Leibniz observes, the traditional Archimedean method of demonstration was by a double *reductio ad absurdum*. But his preference is instead to proceed by a direct *reductio* to prove that "inter duas quantitates nullam esse differentiam".[15] (DQA, 35) As he explains in the continuation of the Scholium to Prop. 7,

> Equidem fateor nullam hactenus mihi notam esse viam, qua vel unica quadratura perfecte demonstrari possit sine deductione ad absurdum; imo rationes habeo, cur verear ut id fieri possit per naturam rerum sine quantitatibus fictitiis, infinitis

---

10 "[...] the difference between this Quadrilineal (which is the subject of this proposition) and the step space [i.e. M – E] can be made smaller than any given quantity".

11 "[...] common method of indivisibles".

12 "[...] compelled for safety's sake, as was Cavalieri, to restrict the method to parallel ordinates, and to suppose that the intervals between any two successive ordinates are always equal".

13 "[...] the demonstration is far easier".

14 "[...] the demonstration has the singular feature that the result is achieved not by inscribed and circumscribed figures taken together, but by inscribed ones alone".

15 "[...] the difference between two quantities is nothing".

> scilicet vel infinite parvis assumtis: ex omnibus tamen ad absurdum deductionibus nullam esse credo simplicem magis et naturalem, ac directae demonstratione propiorem, quam quae non solum simpliciter ostendit, inter duas quantitates nullam esse differentiam, adeoque esse aequales, (cum alioquin alteram altera neque majorem neque minorem esse ratiocinatione duplici probari soleat) sed et quae uno tantum termino medio, inscripto scilicet circumscripto, non vero utroque simul, utitur.[16] (DQA, 35)

We see here a distinction between the method of integration using infinitely many infinitely small elements, which Leibniz characterizes as fictitious, and the direct *reductio ad absurdum method* just exploited in the demonstration above. As we saw there, this involves an inference from the fact that a difference between two quantities can be made smaller than any that can be assigned, to their difference being null. This is a *reductio* in the sense that whatever minimum difference one supposes there to be, one can prove that the difference is smaller. As we have seen, that is the very same reasoning Newton appeals to in his *Principia* to demonstrate Lemma 1 of his *Method of First and Ultimate Ratios.*

Third, Leibniz's demonstration of Proposition 6, just like Newton's Lemmas 1–4, licenses his infinitesimal techniques in quadratures, "servit tamen ad fundamenta totius Methodi indivisibilium firmissime jacienda."[17] (DQA, 24). The term "indivisible" here needs to be taken with a pinch of salt: Leibniz is clear that "plurimum interest inter indivisibile et infinite parvum",[18] and that "Fallax est indivisibilium Geometria, nisi de infinite parvis explicetur; neque enim puncta vere indivisibilia tuto adhibentur, sed lineis utendem est, infinite quidem parvis, lineis tamen, ac proinde divisibilius."[19]

---

16 "For my part I confess that there is no way that I know of up till now by which even a single quadrature can be perfectly demonstrated without an inference ad absurdum. Indeed, I have reasons for doubting that this would be possible through natural means without assuming fictitious quantities, namely, infinite and infinitely small ones; but of all inferences ad absurdum I believe none to be simpler and more natural, and more proper for a direct demonstration, than that which not only simply shows that the difference between two quantities is nothing, so that they are then equal (whereas otherwise it is usually proved by a double reductio that one is neither greater nor smaller than the other), but which also uses only one middle term, namely either inscribed or circumscribed, rather than both together."

17 "[...] laying the foundations of the whole method of indivisibles in the soundest possible way".

18 "[...] there is a profound difference between the indivisible and the infinitely small".

19 "The Geometry of Indivisibles is fallacious unless it is explicated by means of the infinitely small; for truly indivisible points may not safely be applied, and instead it is necessary to use lines which, although infinitely small, are nevertheless lines, and therefore divisible."

(Scholium to Proposition 11, DQA, 133)[20] In Proposition 7, explaining that "Per Summam Rectarum ad quondam axem applicatarum"[21] (DQA, 39) he means "figurae perpetua applicatione factae aream",[22] he comments:

> Quicquid enim de tali summa demonstrari poterit, sumto intervallo, utcunque parvo, id quoque de areae curvilineae ${}_0C_0B_3B_3C_0C$ magnitudine demonstratum erit, cum summa ista (intervallo satis exiguo sumto) talis esse posit, ut ab ista summa rectangulorum differentiam habeat data quavis minorem. Et proinde si quis assertiones nostras neget facile convinci posit ostendendo errorem quovis assignabili esse minorem, adeoque nullum.[23] (DQA, 39)

This is precisely the same as the principle appealed to by Newton to found his analytic method of fluxions, which I called above the *Principle of Unassignable Difference*; it is simply an application of the *Archimedean axiom.*

Fourth, Leibniz is explicit that the equivalence between a proof effected by infinitesimals and the corresponding rigorous kind of proof from first principles given in Proposition 6, means that infinitesimals can always be taken as a kind of shorthand for the arbitrarily small finite lines occurring in the latter. Acknowledging his free use of infinite and infinitely small quantities in proving his results concerning the circle, the ellipse and the infinite hyperboloid, Leibniz writes in the Scholium to Proposition 23:

> Quae de infinitis atque infinite parvis huc usque diximus, obscura quibusdam videbuntur, ut omnia nova; sed mediocri meditatione ab unoquoque facile percipientur: qui vero perceperit, fructum agnoscet. Nec refert an tales quantitates sint in rerum natura, sufficit enim fictione introduci, cum loquendi cogitandique, ac proinde inveniendi pariter ac demonstrandi compendia praebeant, ne semper inscriptis vel circumscriptis uti, et ad absurdum ducere, et errorem assignabili quovis minorem ostendere necesse sit. Quod tamen ad modum eorum quae *prop. 6. 7. 8.* diximus facile fieri posse constat. Imo si quidem possibile est directas de his rebus exhiberi demonstrationes, ausim asserere, non posse eas dari, nisi

---

20 This Scholium to Proposition 11 is recorded as deleted by Knobloch (DQA, 132–33), but is included in the main text without comment in the edition of Parmentier (Leibniz, 2004, 96–101).

21 "[...] by the sum of the straight lines applied to a certain axis".

22 "[...] the area of the figure formed by this continued application".

23 "For whatever properties of such a sum could be demonstrated by taking the interval arbitrarily small, will also be demonstrated of the curvilinear area ${}_0C\ {}_0B_3B_3C_0C$, since, if the interval is taken sufficiently small, this sum could be such that its difference from the sum of the rectangles will be smaller than any given. And so anyone contradicting our assertion could easily be convinced by showing that the error is smaller than any assignable, and therefore null."

> his quantitatibus fictitiis, infinite parvis, aut infinitis, admissis, adde supra prop. 7. schol.[24] (DQA, 69)

An infinitesimal, therefore, is simply a shorthand for a quantity that may be taken as small as desired; likewise an infinite quantity is a quantity "qualibet a nobis assignabili, numerisve designabili, majorem."[25] (DQA, 133; Leibniz, 2004, 98) Both are, with respect to geometry, *fictions*. On whether they can be found in nature, Leibniz is here agnostic; but "Geometrae sufficit, quid ex ipsis positis sequatur, demonstrare (Scholium to Prop. 11; DQA, 133; Leibniz, 2004, 98).[26]

This interpretation, as I have argued elsewhere (Arthur, 2001, 2008a, b), is completely in accord with the insightful presentation of Leibniz's mature interpretation of infinitesimals given by Hidé Ishiguro in the second edition of her *Leibniz's Philosophy of Logic and Language* (1990). According to Ishiguro, Leibniz held, analogously to Russell's position regarding definite descriptions,

> that one can have a rigorous language of infinity and infinitesimal while at the same time considering these expressions as being syncategorematic (in the sense of the Scholastics), i.e. regarding the words as not designating entities but as being well defined in the proposition in which they occur (Ishiguro, 1990, 82).

As she goes on to argue, "Leibniz denied that infinitesimals were fixed magnitudes, and claimed that [in our apparent references to them] we were asserting the existence of variable finite magnitudes that we could choose as small as we wished" (Ishiguro, 1990, 92). This is indeed the case, as we have seen.

---

24 "The things we have said up to now about infinite and infinitely small quantities will appear obscure to some, as does anything new; nevertheless, with a little reflection they will be easily comprehended by everyone, and whoever comprehends them will recognize their fruitfulness. Nor does it matter whether there are such quantities in the nature of things, for it suffices that they be introduced by a fiction, since they allow economies of speech and thought in discovery as well as in demonstration. Nor is it necessary always to use inscribed or circumscribed figures, and to infer by reductio ad absurdum, and to show that the error is smaller than any assignable; although what we have said in Props. 6, 7 & 8 establishes that it can easily be done by those means. Moreover, if indeed it is possible to produce direct demonstrations of these things, I do not hesitate to assert that they cannot be given except by admitting these fictitious quantities, infinitely small or infinitely large (see above, Scholium to Prop 7)."

25 "[...] greater than any assignable by us, or greater than any number that can be designated".

26 "[...] for Geometry it suffices to demonstrate what follows from their supposition".

There is, of course, much more to say on Leibniz's syncategorematic interpretation, in particular, concerning the philosophical status of infinitesimals as fictions. Other contributors to this volume will have more to say here on such issues. But I think it will be instructive for me to close by showing how Leibniz's use of infinities and infinitesimals can be justified by the Archimedean foundation he shared with Newton. Eberhard Knobloch has identified twelve rules occurring in his treatise that may be said to constitute Leibniz's "arithmetic of the infinite" (Knobloch, 2002, 67–8). In the interests of space I shall just consider a small sample. The first of these rules is "Finite + infinite = infinite". Rule 2.1 is "Finite ± infinitely small = finite," and Rule 2.2 is "If $x = y$ + infinitely small, then $x - y \approx 0$ (is unassignable)" where x and y are finite quantities.

Let us take 2.2 first. If $x = y + dy$, where dy is an arbitrarily small finite variable quantity, and D is any pre-assigned difference between x and y, no matter how small, then dy may always be taken so small that $dy < \mathrm{D}$. In particular, if D is supposed to be some fixed ultimate difference between them, then dy can be supposed smaller: so long as D and dy are quantities obeying the *Archimedean axiom*, the variability of dy means that it can always take a value such that $dy < \mathrm{D}$ for any assigned D. Therefore, since the difference between x and y is smaller than any assignable, it is unassignable, and effectively null. The same reasoning justifies 2.1.

Leibniz gives such an argument explicitly in a short paper dated 26 March, 1676:

> Videndum exacte an demonstrari possit in quadraturis, quod differentia non tamen sit infinite parva, sed quod omnino nulla, quod ostendetur, si constet eousque inflecti semper posse polygonum, ut differentia assumta etiam infinite parva minor fiat error. Quo posito sequitur non tantum errorem non esse infinitum parvum, sed omnino esse nullum. Quippe cum nullus assumi possit.[27] (A VI, 3, 434)

Notable here is his claim that this argument works even if the difference D is assumed infinitely small; it does so, of course, only if the variable dy obeys the *Archimedean axiom*.[28]

---

27 "We need to see exactly whether it can be demonstrated in quadratures that a difference is not after all infinitely small, but nothing at all. And this will be shown if it is established that a polygon can always be inflected to such a degree that even when the difference is assumed infinitely small, the error will be smaller. Granting this, it follows not only that the error is not infinitely small, but that it is nothing at all – since, of course, none can be assumed." (DLC, 64–65)

28 As Sam Levey has pointed out to me, this will also entail that the n in the Archimedean axiom would have to be allowed to range over infinite numbers. In that case, by the same

To prove Knobloch's Rule 1, suppose *dz* is another arbitrarily small finite variable quantity such that the ratio *dy*:*dz* remains finite as *dz* is made arbitrarily small. Now again suppose $x = y + dy$, and divide all through by *dz*, and let *dz* become arbitrarily small. As it does so, x/dz and y/dz will each become arbitrarily large; indeed, no matter how large a quantity Q is given, *dz* can be taken sufficiently small that x/dz and y/dz will each exceed it. Thus x/dz and y/dz will each be greater than any given quantity Q, and thus infinite by Leibniz's definition, while *dy*:*dz* remains finite, yielding rule 1, Finite + infinite = infinite. Similar justifications can be given for Knobloch's other rules.

This is, of course, only a start to providing a satisfactory foundation for the infinitesimal methods used by Leibniz and Newton. In particular, it needs to be extended to the limit approach to tangents and curvature dealt with by Newton in his Lemmas, and also to issues surrounding higher-order infinitesimals. It is in fact possible to give a successful account of second-order infinitesimals on Leibniz's syncategorematic interpretation, as I have argued elsewhere (Arthur, 2008a). But this beginning must suffice for present purposes.

4. Comparison: A Consilience of Foundations

In the foregoing discussion we have seen a consilience in the foundational writings of Newton and Leibniz that is quite remarkable. Not only does each thinker appeal to the *Archimedean axiom* in the form of the *Principle of Unassignable Difference* (or its synthetic counterpart, Lemma 1) to justify methods that apparently appeal to infinitely small differences or moments of quantities, each gives an explicit foundation for the "Method of Indivisibles" in essentially identical terms by a method which is by all relevant standards completely rigorous, being effectively equivalent to what is now known as Riemannian Integration.

Here I have only described this consilience; I have not sought to explain it. I surmise that the explanation lies in the common sources Newton and Leibniz had for their mathematics; Niccolò Guicciardini (private com-

---

reasoning as I gave in explaining Newton's proof of Lemma 1, if D is given (fixed), even if infinitely small, then we can find a quantity $c = a/n$ still smaller (and also infinitely small), provided we allow quantities to approach as close to zero as desired. But clearly such an extension of the Archimedean axiom needs more discussion than I can give it here; see Levey's paper in this volume.

munication) has suggested to me that the Archimedean foundation is perhaps due particularly to what the two rivals (and also Wallis) found in the work of Pascal and Barrow; but that is a topic that will have to wait for another time.

Philip Beeley

Infinity, Infinitesimals, and the Reform of Cavalieri: John Wallis and his Critics

1. Introduction

Sometimes circumstances can throw together people of quite different political and religious persuasion into a common camp. Such was the case in revolutionary England in the mid seventeenth century. After the end of the Civil Wars, the establishment of a republican constitution and the purging of most college heads and fellows at the universities for actual or at least alleged royalist sympathies, university teaching itself came under the scrutiny of Puritan reformers such as John Webster and William Dell (Cf. Debus, 1970, 37–51). Proposing the introduction of a curriculum based on a concept of new learning which combined hermeticism with Baconian principles, Webster attacked the universities in his programmatic treatise *Academiarum examen* of 1654 as having been prevented from embracing the new spirit of natural philosophy by their persistent adherence to the teachings of Aristotle. In particular, recent developments in mathematics had, according to Webster, been neglected by the universities of Oxford and Cambridge. Advances in the subject had not been made in these ancient institutions, but rather by private individuals elsewhere:

> [...] but that some private spirits have made some progress therein, as Napier, Briggs, Mr. Oughtredge, and some others, it had lain as a fair garden unweeded or cultivated, so little have the Schools done to advance learning, or promote Sciences. (Webster, 1654, 41)

Webster's remarks echoed those made by Thomas Hobbes three years earlier in his *Leviathan* (1651). With a very different motivation to that of the enthusiastic reformer Webster, the philosopher Hobbes directed his venom at the universities as having been the ideological sources of civil war, as being upholders of the authority of the pernicious doctrines of Aristotle, and as being educational institutions to which the modern mathematical sciences had scarcely found admittance:

And for Geometry, till of very late times it had no place at all; as being subservient to nothing but rigide Truth. And if any man by the ingenuity of his own nature, had attained to any degree of perfection therein, he was commonly thought a Magician, and his Art Diabolical. (Hobbes, 1651, IV, § 46, 370 = EW III, 670–671)[1]

The universities were quick to respond to these attacks, and particularly to that of Webster, who was perceived to be a real danger on account of his close ties to Oliver Cromwell. The principal task of reply fell upon Seth Ward, Savilian professor of astronomy in the University of Oxford since 1649, when he had been intruded to succeed his expelled predecessor John Greaves (Cf. Flood and Fauvel 2000, 98; Jesseph 1999, 67–72). In his detailed reply to Webster, entitled *Vindiciae academiarum* (1654), Ward described the mathematics and astronomy teaching provided at the university and at the same time alluded to the advances made possible by the new political order:

Arithmetick and Geometry are sincerely & profoundly taught, Analyticall Algebra, the Solution and Application of Æquations, containing the whole mystery of both those sciences, being faithfully expounded in the Schooles by the professor of Geometry, and in many severall Colledges by particular Tutors [...] These Arts he mentions, are not only understood and taught here, but have lately received reall and considerable advances (I mean since the Universities came into those hands wherein now it is) particularly Arithmetick, and Geometry, in the promotion of the Doctrine of Indivisibilia, and the discovery of the naturall rise and management of Conic Sections and other solid places. (Ward, 1654, 28–29)

The professor of geometry to whom Ward refers was none other than John Wallis, Savilian professor of geometry in the University of Oxford, who like Ward had been intruded in this post in 1649. Despite having very different credentials – Ward an Anglican with royalist sympathies, Wallis a Presbyterian who had actively served the parliamentary cause during the Civil Wars after the discovery of his skill in the art of deciphering – both men profited decisively from the revolution (Cf. Flood and Fauvel, 2000, 97). But their appointments could hardly have been more propitious: through Wallis and Ward Oxford became one of the most important centers of the mathematical sciences in Europe during the following years.

As Ward makes clear, mathematics teaching had a well-established tradition in the universities at least since the time of Sir Henry Savile, who had

[1] As Jesseph points out, Hobbes in his program for reforming the universities places the content of the curriculum, including that of mathematics, under the authority of the monarch. Cf. Jesseph, 1999, 59–60.

established the two professorships named after him in 1619 (Cf. Webster, 1975, 122–124; Feingold, 1997, 371–374). What was taught was largely laid down by the classically-orientated Savilian statutes. But contrary to what Webster had asserted, that advances had taken place outside the walls of the universities, Ward was able to point out that already Henry Briggs, the first incumbent of the geometry chair, had contributed significantly to the advancement of mathematics through the development of logarithmic tables (Ward, 1654, 28), while the present incumbent, Wallis, who had yet to publish any mathematical work, let alone his most recent, was busy developing the method of indivisibles and working out ways of freeing curves, traditionally based on conic sections, from their geometric background.

2. The Rise of the Geometry of Indivisibles in the First Half of the Seventeenth Century

Ward could scarcely have chosen a more appropriate means of exemplifying the up-to-date nature of Wallis' work than by referring to the method of indivisibles, which no self-respecting mathematician at the time could dare to ignore and which for many years would continue to play a decisive role in advancing techniques for finding the areas enclosed by curved lines (quadratures), the volumes enclosed by curved figures (cubatures), as well as for determining the centers of gravity of surfaces and bodies. Most contemporary authors ascribed the method to Bonaventura Cavalieri, although few had been able to obtain copies of his books and even fewer had had the patience to work through them. As François de Gandt has recently pointed out, the reference to Cavalieri in connection with the method of indivisibles soon acquired an almost obligatory character which bore little reflection on true lines of intellectual dependence (Cf. De Gandt, 1992b, 104). In fact, techniques for employing indivisibles for the measurement of areas and volumes were in the air already before Cavalieri published his *Geometria indivisibilibus continuorum nova quadam ratione promota* in 1635. Paul Guldin famously accused Cavalieri of having appropriated his technique from Kepler[2] and it is fairly clear that Pierre de Fermat and Gilles Personne de Roberval developed similar techniques independently of

[2] Leibniz writes to Johann Bernoulli on October [12]/23, 1716: "Et notatum jam est a Guldino aliisque, Keplerum in libro de Dolio Austriaco ipsi Cavallerio ad hanc Geometriam, quam indivisibilium vocat, viam aperuisse." (GM III, 971); see also Leibniz, Historia et origo calculi differentialis (GM V, 393); cf. Festa, 1992, 199–200.

Cavalieri but did not publish results obtained through them (De Gandt, 1992b, 104).

William Oughtred, whose private tuition furthered the mathematical careers of numerous young men in England, particularly such with Cambridge backgrounds, including Seth Ward and Christopher Wren, and whose *Clavis mathematicae* (1631) played a formative role in the early algebraic work of Wallis (Cf. Stedall, 2002, 63–64 and 68–73), allows us to witness the enthusiasm which the new method engendered in him, and no doubt also in many others, too. After having seen only the barest of accounts, he informs a certain Robert Keylway, in a letter written in late 1645, that he is able to divine that from Cavalieri's method "great enlargement of the bounds of the mathematical empire will ensue" (Oughtred to Keylway after October 26/[November 5], 1645, Rigaud, 1841, 65).[3] Indeed, such was the enlargement, real or perceived, that thirteen years later Blaise Pascal would claim that the doctrine of indivisibles could be rejected by no-one who aspired to status among contemporary geometricians.[4]

It is not necessary here to give an account of what Cavalieri's method actually involved, for notwithstanding the limitations of his own presentation this has been done already in the magnificent work of Alexandre Koyré (Cf. Koyré, 1973), Kirsti Andersen (Andersen, 1985; Andersen, 1986), Enrico Giusti (Giusti, 1980), Toni Malet (Malet, 1996; Malet, 1997), and François de Gandt (De Gandt, 1992a; De Gandt, 1992b), all of whom have also dealt with other contemporary approaches along similar lines. And in many ways it is irrelevant to the topic with which we are concerned. Despite the almost reverential references to Cavalieri as creator, hardly anyone actually employed the method in the way he had conceived it, mainly, but

---

[3] The complete passage of the letter is instructive: "I speak this the rather, and am induced to a better confidence of your performance, by reason of a geometric-analytical art or practice found out by one Cavalieri, an Italian, of which about three years since I received information by a letter from Paris, wherein was praelibated only a small taste thereof, yet so that I divine great enlargement of the bounds of the mathematical empire will ensue. I was then very desirous to see the author's own book while my spirits were more free and lightsome, but I could not get it in France." (Oughtred to Keylway, after October 26/November [5], 1645, Rigaud, 1841, 65) Evidently Oughtred refers to the same episode some ten years later, in his letter to Wallis of August 17/[27], 1655, relating how in a paper sent from France containing theorems demonstrated by Cavalieri's method he saw "a light breaking out for the discovery of wonders, to be revealed to mankind in this last age of the world." (Oughtred to Wallis on August 17/[27], 1655, Wallis, 2003, 160)

[4] "[...] qui sera le centre de gravité de la balance comme cela est visible par la doctrine des indivisibles, laquelle ne peut être rejetée par ceux qui prétendent avoir rang entre les géomètres." (Pascal, 1980, 134)

not exclusively, for the reasons already given. Suffice it to say that Cavalieri, who never explained precisely what he understood by the term "indivisible," used it to characterize the infinitely small elements he used in his method. He conceived the surface of a figure to be made up of an indefinite number of parallel lines and the volume of a solid to be composed of an indefinite number of parallel equidistant planes, these elements being designated as the indivisibles of the surface and of the volume respectively. The fundamental theorem which he then proceeded to employ was that two figures or two bodies could be said to be in the same ratio as "all their lines" or "all their planes."[5] He employed these concepts for their utility and almost without exception sought to avoid any kind of philosophical implications both regarding the nature of infinity and that of the construct most readily embodying it, the continuum.

Other mathematicians who either interpreted or defended Cavalieri's method generally cast this prudence aside and transformed his indivisibles into such which were conceived to compose the figure or body in which they were contained. Roberval, for example, who claimed to have had no other inspiration for his work than Archimedes (Cf. Roberval, 1736a, 366; Walker, 1932, 15–16), nevertheless defended the method of Cavalieri from critics such as Guldin by pointing out that the Italian mathematician did not consider a surface as composed of lines or a solid as composed of surfaces. But even if such a composition of the continuum were not intended, Cavalieri's method was on Roberval's opinion unable to escape this criticism, whereas his own approach which differed from it only to a small degree did.[6] The holder of the Ramus chair of mathematics at the Collège Royal achieved this compromise between utility and rigorosity by considering surfaces and solids to be built up of an infinite or indefinite number of surfaces and solids respectively, these infinite things being regarded "just as if they were indivisibles."[7] In other words, Roberval substituted indivisibles which were dimensionally homogeneous to the figures and bodies they

[5] See for example Cavalieri 1653, 113: "Figurae planae habent inter se eandem rationem, quam eorum omnes lineae juxta quamvis regulam assumptae; Et figurae solidae, quam eorum plana juxta quamvis regulam assumptae".

[6] "Est tamen inter clarissimi Cavallerii methodum & nostram, exigua quaedam differentia". (Roberval, 1736a, 368)

[7] "Nostra autem methodus, si non omnia, certe hoc cavet, ne heterogenea comparare videatur: nos enim infinita nostra seu indivisibilia consideramus. Lineam quidem tanquam si ex infinitis seu indefinitis numero lineis constet, superficiem ex infinitis seu indefinitis numero superficiebus, solidum ex solidis [...]". (Roberval, 1736a, 368–369)

were now understood to compose for the non-compositional heterogeneous indivisibles of Cavalieri.[8]

As we can gather from a letter which Charles Cavendish wrote to John Pell from Paris at the end of 1646, Roberval felt that he was not only defending, but also improving the method of Cavalieri.[9] An essential part of his approach to this, the reconciliation of the concept of the indivisible with traditional Aristotelian views on the continuum, clearly became part of the accepted understanding of the Cavalierian geometry of indivisibles in the mathematical community in France. Thus, Antoine Arnauld in his *Nouveaux Eléments de Géométrie* of 1667 has this to say of the "new method called the geometry of indivisibles":

> Quoique les Géomètres conviennent que la ligne n'est pas composée de points, ni la surface de lignes, ni le solide de surfaces, néanmoins on a trouvé depuis peu de temps un art de démontrer une infinité de choses, en considérant les surfaces comme si elles étoient composées de lignes, & les solides de surfaces.[10] (Arnauld, 1667, 306–307 = Arnauld, 1781, 327)

The far more influential interpretation of Cavalieri's method through Evangelista Torricelli likewise took indivisibles to be constituent of the figures or bodies they were supposed to make up, but in contrast to Roberval they were understood to be dimensionally heterogeneous (Cf. De Gandt, 1992b, 105). Paying little head to Cavalieri's precautions, Torricelli conceived a plane figure to be compositionally equal to a collection of lines and a solid to be compositionally equal to a collection of planes or surfaces. Thus, in order to discover one of the most important results contained in the treatise *De dimensione parabola solidique hyperbolici problematis duo*, which Torricelli published together with two other treatises in his *Opera*

---

8 See Roberval, 1736b, 207–209; Walker states: "Cavalieri compares figures through their geometric properties, while Roberval compares them through their numerical or algebraic properties, that is, he treats them by Cartesian analysis without the Cartesian symbolism". (Walker, 1932, 46)

9 "Mr. Robervall hath halfe promised to polish the geometrie by Indivisibles which Cavaliero hath begun, for he saies he invented & used that waie before Cavalieros booke was published; & that he can deliver that doctrine much easier & shorter; & shew the use of it in divers propositions which he hath invented by the help of it; but I doute it will be longe before he publish it; though I assure my self he is verie skillfull in it." (Cavendish to Pell on November 27/[December 7], 1646, Pell, 2005, 496)

10 "Although the geometers agree that the line is not composed of points, nor the surface of lines, nor the solid of surfaces, one has nevertheless recently found an art of demonstration for an infinity of things, by considering surfaces as if they were composed of lines, and solids as if they were composed of surfaces."

geometrica of 1644, the cubature of his 'acute hyperbolic solid', he treated this as a collection of concentric cylinders, whose surfaces could be added to produce the volume of the solid (Cf. Torricelli, 1919, 193–194; De Gandt, 1989, 159–161).

It was through Torricelli that Wallis first encountered Cavalieri's method of indivisibles. Soon after his appointment as professor of geometry, he came across the copy of the *Opera geometrica* contained in the mathematical library for the Savilian professors and over the next three years it inspired him to attempt similar quadratures and cubatures to those carried out by Torricelli. His decisive advance on the Italian mathematician was thereby to see that the necessary summation could be carried out arithmetically rather than geometrically.

3. Wallis' Employment of Indivisibles

Unlike most of his contemporaries, Wallis is explicit on his sources. In a letter dedicatory prefaced to his singular most important contribution to the development of modern analysis, the *Arithmetica infinitorum* of 1656, the Savilian professor of geometry explains how he came to develop the techniques which he employs in that work. The letter is addressed to William Oughtred, who he knew to share his interest in the new approach to quadratures:

> Exeunte Anno 1650 incidi in Torricellii scripta Mathematica, (quae ut per alia negotia licuit, anno sequente 1651, evolvi) ubi inter alia, Cavallerii *Geometriam Indivisibilium* exponit. Cavallerium ipsum nec ad manum habui, & apud Bibliopolas aliquoties frustra quaesivi. Ipsius autem methodus, prout apud Torricellium traditur, mihi quidem eo gratior erat quod nescio quid ejusmodi, ex quo primum fere Mathesin salutaverim, animo obversabatur.[11] (Wallis to Oughtred, July 19/[29], 1655, Wallis, 2003, 152)

Wallis' aim was to find a general method of quadrature and cubature. An important stage in this was the discovery of algebraic formulae for the parabola, the ellipse, and the hyperbola, enabling him to consider these curves

11 "Around 1650 I came across the mathematical writings of Torricelli (which, as other business allowed, I read in the following year, 1651), where among other things, he expounds the geometry of indivisibles of Cavalieri. Cavalieri himself I did not have to hand, and sought for it in vain at various booksellers. His method, as taught by Torricelli, moreover, was indeed all the more welcome to me because I do not know that anything of that kind was observed in the thinking of almost any mathematician I had previously met."

abstractly as figures in plano and thus to liberate them from what he called the "embranglings of the cone" (Wallis, 1685, 291) By overcoming reliance on geometrical representation, he sought to carry out summations arithmetically rather than geometrically, associating numerical values to the indivisibles of Cavalieri. In Wallis' view, his own method began where that of the Italian mathematician had ended.[12] Employing an often used misnomer, he says that just as Cavalieri had called his method the geometry of indivisibles, he might aptly term his own the arithmetic of infinites. However, in the process of transforming geometric problems to summations of arithmetic sequences, Wallis made liberal use of analogy and what he called induction. In so doing, he often neglected questions of rigor, although he would always claim that results achieved could if necessary be verified by the apagogic method of inscribed and circumscribed figures used by the Greeks.

The manner in which Wallis effected the transition from geometry to arithmetic is made plain in the proof, published in his tract *De sectionibus conicis* (Cf. Wallis, 1655a, prop. 3, 8–9 = Wallis 1695, 299), that the area of a triangle is the product of the base by half the altitude (Fig. 1).

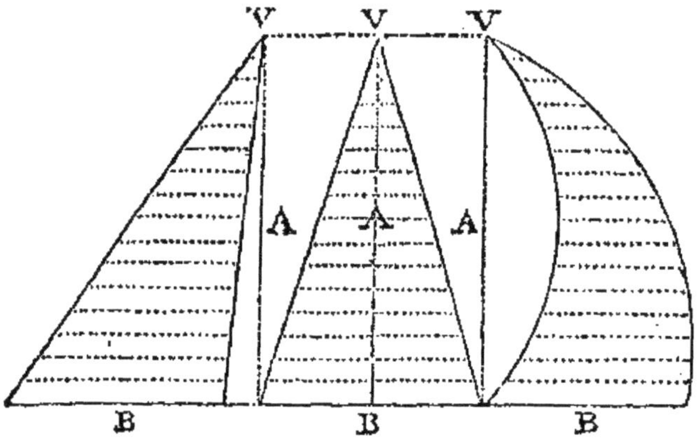

Figure 1.

He first assumes, as Torricelli had done, that a plane figure may be regarded as made up of an infinite number of parallelograms, the altitudes of which are equal, each being $1/\infty$ or an infinitely small aliquot part of the altitude of

[12] "Nempe inde ortum sumit haec nostra methodus ubi Cavallerii Methodus Indivisibilium definit. [...] ut enim ille suam, *Geometriam Indivisibilium*, ita Ego methodum nostram, *Arithmeticam Infinitorum*, nominandam duxi." (Wallis to Oughtred on July 19/[29], 1655, Wallis, 2003, 152) See also Wallis to Leibniz, July 30/[August 9], 1697, GM IV, 38.

the whole figure. (This is incidentally the first appearance of the characteristic loop symbol ∞ for infinity in mathematical literature.) A parallelogram whose altitude is infinitely small is, he writes, "scarcely anything but a line", except that this line is supposed extensible, or as to have such a small thickness, that by an infinite multiplication a certain altitude or width can be acquired.[13]

Wallis supposed the triangle to be divided into an infinite number of lines or infinitesimal parallelograms parallel to the base. The area of these, taken from the vertex to the base form an arithmetic progression beginning with zero. Moreover, there is a well-known rule, that the sum of all the terms in such a progression is the product of the last term by half the number of terms. Since, as Wallis tells us, "nulla enim discriminis causa erit"[14] (Wallis, 1655b, prop. 2, 2 = Wallis, 1695, 365), it can be applied in this context to the area in the triangle. If the altitude and the base of the triangle are taken as A and B respectively, the area of the last parallelogram in the progression will then be $1/\infty$A.B. The area of the whole triangle is therefore $1/\infty$A.B.$\infty/2$ or 1/2A.B (Fig. 2). Wallis then applied similar types of argument to numerous quadratures and cubatures involving cylinders, cones, and conic sections.

If in *De sectionibus conicis* Wallis' procedures were based largely on manipulations of his infinity symbol, in *Arithmetica infinitorum* he worked more fundamentally with methods similar to those of contemporaries such as Roberval and Simon Stevin and employing the limit concept. While achieving important results, including his celebrated formula for $4/\pi$ (the so-called Wallis product),[15] his employment of induction and interpolation subjected him to fierce criticism, particularly from Leibniz. But this is a story which is largely irrelevant to our present topic and will therefore only be dealt with very briefly.

As we have seen, Wallis, while speaking in a Cavalierian sense of composing plane figures from an infinite number of lines, prefers this composition to be understood as being from an infinite number of slender par-

---

13 "Suppono in limine (juxta Bonaventurae Cavalerii *Geometriam Indivisibilium*) Planum quodlibet quasi ex infinitis lineis parallelis conflari: Vel potius (quod ego mallem) ex infinitis Parallelogrammis aeque altis; quorum quidem singulorum altitudo sit totius altitudinis $1/\infty$, sive aliquota pars infinite parva; (esto enim ∞ nota numeri infiniti;) adeoque omnium simul altitudo aequalis altitudini figurae." (Wallis, 1655a, prop. 1, 4 = Wallis, 1695, 297)

14 "[...] there is no cause for discrimination between finite and infinite numbers".

15 Stedall, in the introduction to her translation of the *Arithmetica infinitorum*, sees the development of the method used to achieve this result as being "perhaps the one real stroke of genius" in Wallis' long mathematical career. Cf. Wallis, 2004, xviii.

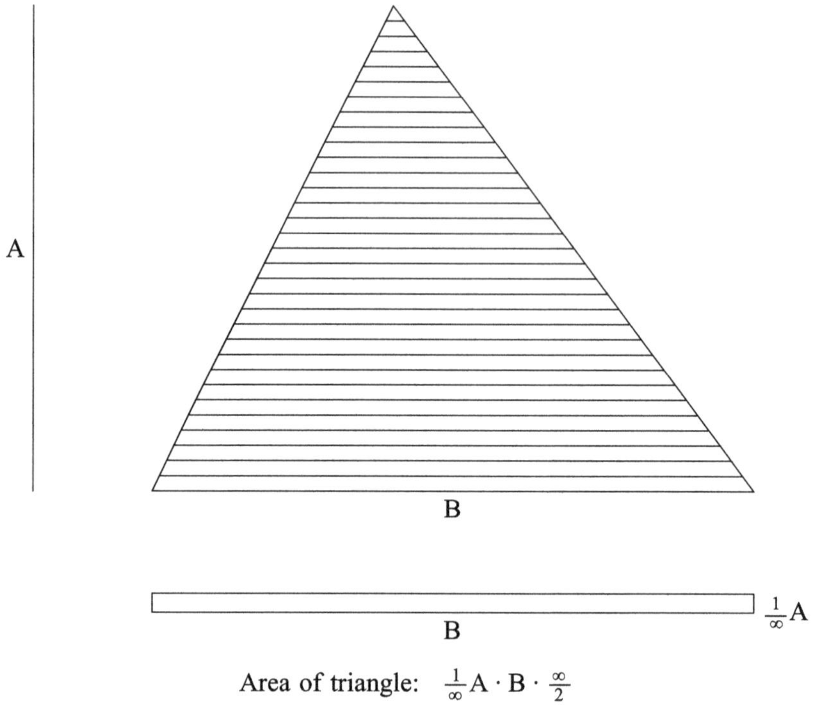

Figure 2.

allelograms each of whose altitude is an equal infinitely small part of the whole. In the dedicatory letter to Seth Ward and Lawrence Rooke which he prefaced to *De sectionibus conicis*, he suggests that this is an improvement to Cavalieri's method which however does not substantially change it.[16] While being infinitely small, his indivisibles are understood to be in a definite ratio to the altitude of the whole figure, so that when infinitely multiplied they make up the total altitude of the figure. Nor is this simply conceived as a useful mathematical technique (Cf. Malet, 1996, 68–69). In his major work on statics, the three volume *Mechanica* (1670–1), Wallis an-

[16] "Opus ipsum quod attinet; videbitis me, statim ab initio, Cavallerii *Methodum Indivisibilium*, quasi jam a Geometris passim receptam, tam huic quam tractatui sequenti (qui huic gemellus est) substernere; (ut multiplici figurarum inscriptioni & circumscriptioni, quibus in ἀπαγωγαῖς alias utendum saepius esset, supersedere liceat:) sed a nobis aliquatenus sive emendatam sive saltem immutatam: pro rectis numero infinitis, totidem substitutis parallelogrammis (altitudinis infinite-exiguae;) ut & pro planis, totidem vel prismatis vel cylindrulis; & similiter alibi." (Wallis, 1656a, sig. I2v = Wallis, 2003, 169)

nounces the proposition that every continuum can be understood in this way to be composed of an infinite number of indivisibles,[17] describing them as homogeneous particles in much the same way as Pascal had done, probably having himself lent this concept of homogeneity from André Tacquet (Cf. De Gandt, 1992b, 107).

A more explicit description of the nature of indivisibles according to this conception is to be found in his later work, the *Treatise of Algebra, both Historical and Practical* of 1685. Here he makes clear that the lines conceived qua indivisibles to compose a plain surface are themselves to be understood as infinitely narrow surfaces:

> According to this Method [sc. of indivisibles], a Line is considered, as consisting of an Innumerable Multitude of Points: A Surface, of Lines, (Streight or Crooked, as occasion requires:) A Solid, of Plains, or other Surfaces. [...] Now this is not so to be understood, as if those Lines (which have no breadth) could fill up a Surface; or those Plains or Surfaces, (which have no thickness) could compleat a Solid. But by such Lines are to be understood, small Surfaces, (of such a length, but very narrow,) whose breadth or height (be they never so many,) shall be but just so much as that all those together be equal to the height of the Figure, which they are supposed to compose.[18] (Wallis, 1685, 285–286)

Unfortunately, Wallis is not always consistent in his terminology and sometimes blurs important distinctions. Thus, he characterizes an infinitely small altitude on occasion also as being no altitude whatsoever, explaining that an infinitely small quantity is the same as a non-quantum or as we might say a non-quantifiable quantity – "nam quantitas infinite parva perinde est atque non-quanta" (Wallis, 1655a, prop. 1, 4 = Wallis, 1695, 297). While being a useful concept, his infinitesimal or infinitely small part, "pars infinitesima seu infinite parva" (Wallis, 1695, 367),[19] does not have the

[17] "Definitio. Continuum quodvis (secundum Cavallerii Geometriam Indivisibilium) intelligitur, ex Indivisibilibus numero infinitis constare. Ut, ex infinitis Punctis, Linea; Superficies, ex infinitis Lineis; & ex infinitis numero Superficiebus, Solidum: Item ex infinitis temporis Momentis, Tempus, &c. Hoc est; (ut nos idem explicamus in nostra Arithmetica Infinitorum, & Tract. de Con. Sect.) ex particulis Homogeneis, infinite exiguis, numero infinitis; Idque (ut plurimum) secundum unam saltem dimensionem aequalibus." (Wallis, 1670, part II, cap. 4, def., 110 = Wallis, 1695, 645)

[18] The distinction between what is and what is supposed to be infinite is crucial to Wallis' response to Hobbes' criticism of his concepts, as Malet has correctly pointed out, cf. Malet, 1996, 82–83.

[19] The term "pars infinitesima" is introduced only in this later reprint of the *Arithmetica infinitorum*. In the original 1656 edition, prop. 5, page 5, one finds only the expression "pars infinite parva".

same degree of sophistication as Leibniz's arbitrarily small but non-zero infinitesimal and correspondingly does not avoid the classical knots of the continuum problem.

4. Hobbes and Wallis

Wallis' *Arithmetica infinitorum* drew criticism from other leading mathematicians of his day, including Christiaan Huygens and Pierre de Fermat, but for reasons which do not pertain to our topic and which can thus here be safely ignored. Things are quite different in the case of the attacks launched by Thomas Hobbes in the course of his long drawn out dispute with Wallis which effectively began in the context of the Webster-Ward debate discussed at the beginning of this chapter and which has been the topic of excellent studies by Doug Jesseph (Cf. Jesseph, 1999) and Siegmund Probst (Cf. Probst, 1997). Not without justification, Hobbes attacked Wallis' employment of induction but also objected to his conception of indivisibles. As far as the latter were concerned, Hobbes found welcome opportunity to counter the numerous taunts which Wallis had made against him in respect of his mathematical endeavours in *De corpore* (1655), by pointing out what he saw as being serious inconsistencies in the Savilian professor's interpretation of the geometry of Cavalieri:

> To which I may add, that it destroys the method of *Indivisibles*, invented by Bonaventura; and upon which, not well understood, you have grounded all your scurvy book of *Arithmetica infinitorum*; where your Indivisibles have nothing to do, but as they are supposed to have Quantity, that is to say, to be *Divisibles.* [...] See here in what a confusion you are when you resist the truth. When you *consider* no determinate Altitude (that is, *no Quantity* of Altitude) then you say your Parallelogram shall be called a *Line*. But when the Altitude is determined (that is, when it is Quantity) then you will call it a Parallelogram. (Hobbes, 1656, 43 and 46 = EW VII, 300–301 and 309)

Wallis had, through his rather loose way of expression, invited such philosophical criticism. Of course, Hobbes saw no reason to excuse him his laxity of expression and when he picked up Wallis for having written in proposition 3 of *Arithmetica infinitorum* that a triangle consists "as it were" (quasi) of an infinite number of parallel lines in arithmetic progression, he did so by saying that "as it were" is no phrase of a geometrician (Hobbes, 1656, 46 = EW VII, 310).

Another weakness in Wallis' approach to quadratures in the eyes of Hobbes was the concept of continuity on which it rested. Not only did the implicit composition of continuous quantity from indivisibles conflate with the accepted doctrine of infinite divisibility, but also it bore little resemblance to what Cavalieri had actually written. The fact that Wallis had used Torricelli's interpretation as the starting point for his own work naturally made him open to such an attack, particularly as he always referred explicitly to Cavalieri as the originator of the approach. In one of the tracts published later in the course of the dispute, Hobbes accused Wallis on these grounds of having used fundamentally unsound principles, including that of composing the continuum out of indivisibles:

> Ad quam rem supponit duo Principia: alterum quidem (ut dicit) Cavallerii, nempe hoc, *Quod quantitas omnis continua constat ex numero infinito indivisibilium*, sive infinite exiguorum; quanquam ego Cavallerii libro lecto, nihil ibi in illam sententiam scriptum animadverti; neque Axioma, neque Definitionem, neque Propositionem. Nam falsum est. Quantitas enim continu, sua natura divisibilis est in semper divisibilia; nec potest esse aliquid infinite exiguum, nisi daretur diviso in Nihila.[20] (Hobbes, 1672, 7 = LW, V, 109)

But such objections, coming as they did from Hobbes, had little apparent impact on Wallis. In fact, in the course of the dispute he never really addressed the philosophical issues which the author of *De corpore* raised in respect of his understanding of indivisibles. Thus, in *Due Correction for Mr Hobbes* (1656) he seeks to explain what he understands be 'indivisible' by skirting the question of the infinite entirely:

> I do not mean precisely a line but a parallelogram whose breadth is very small, viz an aliquot part (divisor) of the whole figures altitude, denominated by the number of parallelograms (which is a determination geometrically precise). (Wallis, 1656, 47)

Being such an excellent controversialist as he was, Wallis could scarcely have done otherwise than reply to Hobbes' attacks on the central concepts he employed in his reformed version of Cavalieri's geometry of indivisibles. But fundamentally he felt that such philosophical criticisms of concepts were of little weight, so far as methods based on these concepts could

---

20 "To this end he assumes two principles. The first is one that, so he says, comes from Cavalieri, namely this: that any continuous quantity consists of an infinite number of indivisibles, or infinitely small parts. Although I, having read Cavalieri's book, remember nothing of this opinion in it, neither in the axioms, nor in the definitions, nor in the propositions. For it is false. A continuous quantity is by its nature always divisible into divisible parts: nor can there be anything infinitely small, unless there were given a division into nothing."

be shown to produce results which if necessary could be verified by the accurate but laborious classical method of exhaustions. In this way, his approach to criticism was very similar to that of the German mathematician and philosopher whose early career he decisively promoted in cooperation with his friend Henry Oldenburg, namely Gottfried Wilhelm Leibniz. For both Wallis and Leibniz utility and success of procedures in mathematics were decisive, not metaphysical qualms about the concepts which these procedures employed.

5. Leibniz and Wallis

As is well known, it was Wallis' *Arithmetica infinitorum* together with Cavalieri's so-called *Geometria indivisibilium* which the young Leibniz, through the concept of point he set forth in his *Theoria motus abstracti* of 1671, rather ambitiously claimed to have saved from the pernicious criticism leveled against them. "Punctum non esse, cujus pars nulla est, nec cujus pars consideratur; sed quod quolibet extenso assignabili minus est", he writes to Oldenburg on March 1/[11], 1670/1, and then roundly adds: "quod est fundamentum methodi Cavalerianae".[21] (A II, 1 (1926), 90; (2006), 147) About two months later in another letter to the secretary of the Royal Society he makes an even stronger claim: "Theoria motus Abstracti, invictas propemodum Compositionis continui difficultates expicat, Geometriam indivisibilium, et Arithmeticam infinitorum confirmat."[22] (April 29/ [May 9], 1671; A II, 1 (1926), 102; (2006), 166) Having little background in mathematics at that time, Leibniz had evidently gathered all he knew about Cavalieri and Wallis and the criticisms which had been directed against their respective methods from the polemical writings of Tho-

---

21 "There is no point whose part is nothing, nor whose part can be measured, but it is less than any assignable extended quantity. This is the foundation of Cavalieri's method". In the preface to the *Theoria motus abstracti* itself he claims to have placed both the Geometry of indivisibles and the Arithmetic of infinites, "the parents of so many excellent theorems", on a sound footing: "Geometriam Indivisibilium et Arithmeticam Infinitorum, tot egregiorum theorematum parentes, in solido locandas" (A VI, 2, 262). See also Leibniz to van Velthuysen, [April 25]/May 5, 1671, A II, 1 (1926), 97; (2006), 163–164); Leibniz to Carcavy, August [7]/17, 1671, A II, 1 (1926), 143; (2006), 236); Leibniz to Arnauld, beginning of November 1671, A II, 1 (1926), 172; (2006), 278).

22 "The Theoria motus abstracti explains the almost unconquerable difficulties of the composition of the continuum, confirming the geometry of indivisibles and the arithmetic of infinites."

mas Hobbes.[23] But Wallis was not one to be impressed by an inappropriate defense of his work. He scarcely expended more than a page in writing a rather reticent review of the *Theoria motus abstracti*, whereas the praise he heaped on the incomparably more coherent contemporaneous *Hypothesis physica nova* opened the doors of the Royal Society to Leibniz, who was elected member just two years later.[24] Not unimportantly, Wallis was able to refer to their fundamental agreement on various points of natural philosophy, including the question of the origin of resilience.

Leibniz's acquaintance with the genuine work of Cavalieri and Wallis did not take place until he got to Paris in 1672.[25] In the course of the momentous strides he made there leading to the production of the seminal tract *De quadratura arithmetica circuli* in 1675, Leibniz soon recognized the serious limitations both to Cavalieri's method and to Wallis' arithmetization of it (see DQA, 25, 69, 71). We can refer here, for example, to his letter to Jean Gallois of the end of 1675, in which he describes the utility of dividing a figure into an infinity of small or characteristic triangles rather than into an infinity of parallel ordinates or an infinity of tiny rectangles.[26]

Like other mathematicians, too, Leibniz saw in the employment of induction as well as in the use of interpolation essential weaknesses to Wallis'

---

[23] Leibniz read the works of Hobbes in Johann Christian von Boineburg's library with intense interest while he was in Mainz, as shown through his recently-discovered marginalia to the edition of *De corpore* (1655) and the *Opera philosophica* (1668) formerly contained in that library. The author should like to thank Ursula Goldenbaum for making her transcriptions of these marginalia available to him.

[24] See Beeley, 2004, 68–69. The reviews took the form of letters addressed to Oldenburg, dated April 7/[17] and June 2/[12],1671, and were published in the *Philosophical Transactions* No. 74 (August 14, 1671), 2227–2230, 2230–2231.

[25] Already by the end of 1672 Leibniz had read to some extent Wallis' *Arithmetica infinitorum*. See Leibniz for Jean Gallois, end of 1672 (A II, 1 (1926), 223; (2006), 343). In the first half of the following year he writes that there are many things concerning the arithmetic of infinites which probably had not been considered sufficiently until then, not even by Wallis: "Ad Arithmeticam infinitorum multa pertinent, hactenus fortasse, ac ne a Wallisio quidem satis considerata." (De arithmetica infinitorum perficienda, A VI, 3, 408)

[26] "La raison pourquoy ceux qui ont écrit de la Geometrie des Indivisibles, et de l'Arithmetique des infinis, n'ont pas fait la même remarque, est parce qu'on est accoustumé de ne resoudre les figures que par les ordonnées paralleles, et une infinité de petits rectangles, au lieu que j'ay trouvé un moyen general de resoudre utilement toute figure en une infinité de petits Triangles aboutissans à un point, par le moyen des ordonnées convergentes. [...] Ce theoreme a des grandes suites, et il suffit luy seul pour prouver par une seule demonstration Geometrique toutes les Quadratures de l'Arithmetique des infinis, que le celebre Mons. Wallis n'a trouvé que par induction." (Leibniz to Jean Gallois, end of 1675, A III, 1, 359). See also the first draft of the postscript to: Leibniz to Jacob Bernoulli, April 1703, GM III, 73.

approach to quadratures and cubatures in *Arithmetica infinitorum*.[27] If in both respects the Savilian professor rarely erred, it was because of his natural mathematical intuition: on the one hand his ability to recognize that an established pattern in a few cases could reasonably be assumed to continue indefinitely, and on the other hand his ability to interpolate between triangular, pyramidal and other figurate numbers (Cf. Wallis, 2004, xxiv-xxv). Leibniz, in his article *De la chainette*, which he published in the *Journal de Sçavans*, points both to Cavalieri's dependence on geometric figures and to Wallis' use of induction based on a certain sequence of numbers as central reasons for the superiority of his own analysis of infinites:

> C'est ce qu'il appelle l'*Analyse des infinis*, qui est entiérement différente de la Geometrie des indivisibles de Cavalieri, & de l'Arithmétique des infinis de Mr. Wallis. Car cette Geometrie de Cavalieri, qui est tres bornée d'ailleurs, est attachée aux figures, où elle cherche les sommes des ordonnées; & Mr. Wallis, pour faciliter cette recherche, nous donne par induction les sommes de certains rangs de nombres: au lieu que l'analyse nouvelle des infinis ne regarde ni les figures, ni les nombres, mais les grandeurs en general, comme fait la specieuse ordinaire.[28] (Leibniz, 1692, 148 = GM V, 259)

And similarly referring to Wallis' reliance on interpolation particularly in the second half of *Arithmetica infinitorum*, Leibniz emphasizes in one of the only recently published mathematical papers from the Paris period, *De progressionibus et de arithmetica infinitorum*, the general nature of his own approach: "Arithmetica infinitorum mea est pura, Wallisii figurata."[29] (A VII, 3, 102)

Not without reason Leibniz felt that the conception of the infinitely small employed by Wallis in his calculus was less than sophisticated, while the Savilian professor for his part sought to convince his younger German

---

27 See for example *Wallisii series interpolanda pro circulo. Fractionum resolutio dividendo per fractiones* (A VII, 1, 569–572); *De progressionibus et geometria arcana et methodo tangentium inversa* (A VII, 3, 55); Leibniz to La Roque, end of 1675 (A III, 1, 347–348); Leibniz to Gallois, end of 1675 (A III, 1, 359, 361); Leibniz to Tschirnhaus, end of June 1682 (A III, 3, 655). In the latter he refers to Ismaël Boulliau's proofs of results which Wallis had achieved by induction.

28 "It is this which he calls the *Analysis of infinites*, which is entirely different from the Geometry of indivisibles of Cavalieri and the Arithmetic of infinites of Mr Wallis. For that geometry of Cavalieri, which moreover is very restricted, is attached to figures where it seeks the sums of ordinates. And Mr Wallis, in order to facilitate this investigation, gives us by means of induction the sums of certain classes of numbers, whereas the new analysis of infinites considers neither figures nor numbers, but magnitudes in general, as does algebra."

29 "My arithmetic of infinites is pure, Wallis' is figurate".

friend of the fundamental identity of their approaches to tangents, quadratures, and cubatures. In correspondence exchanged in the late 1690s the two men traded their respective positions, after Wallis had taken up the topic in his letter to Leibniz of Juli 30/[August 9], 1697. Referring to the methods of tangents he had published in the March 1672 issue of *Philosophical Transactions* (Wallis, 1672, 4010–4016),[30] as well as in proposition 95 of his *Treatise of Algebra*, and which he had earlier used in his early mathematical tract *De sectionibus conicis*, he claims it to be evident that these methods rest on the same principles as those of Leibniz's differential calculus "sed diversa notationis formula"[31] (GM IV, 37). In particular, he seeks to equate the minute increment *a* which he employed in computing tangents with Leibniz's infinitesimals: "Nam meum *a* idem est atque tum *dx*, nisi quod meum *a* sit nihil, tuum *dx* infinite exiguum".[32] Since both of these quantities were incomparably small they could not on his view be other than identical. Thus what remains after one has disregarded those quantities which need to be ignored in order to shorten the calculation, is, he suggests to Leibniz, "tuum minutum triangulum, quod est apud te infinite-exiguum, apud me nullum est seu evanescens."[33] But this was precisely the point at issue: Wallis' increment *a* disappears from the calculation once it has effectively done the task of achieving the result sought under its supposition, whereas Leibniz's infinitesimals, irrespective of their metaphysical status, may continue to be computed.

In this respect Wallis adopted a strategy of equivocation which he also used when comparing Newton's and Leibniz's methods[34] and which ultimately provided at least part of the pretext and the literary basis for the grand priority dispute over the discovery of the calculus from the late 1690s onwards (Cf. Hall, 1980, 92–96). In his next letter to Leibniz he takes up the topic again, suggesting now that his conception of $a = 0$ has the advantage over Leibniz's *dx* of being more simple, since in contrast to multiples of differentials multiples of zero are always zero:

---

30 The article takes the form of a letter addressed to Oldenburg and dated February 15/[25], 1671/2.

31 "[…] though in different notational form".

32 "For my *a* is the same as your *dx*, except that my *a* is nothing and your *dx* is infinitely small".

33 "[…] your minute triangle which for you is infinitely small and for me nothing or disappearing."

34 See for example: "Et, ni fallor (sic saltem mihi nunciatum est), Newtoni Doctrina Fluxionum res eadem (vel quam simillima) quae vobis dicitur Calculus Differentialis: quod tamen neutri praejudicio esse debet." (Wallis to Leibniz April 6/[16], 1697, GM IV, 18)

> [...] mihi non opus sit tuis aliquot Postulatis de infinite-parvo in se ducto, aut in aliud infinite-parvum, in nihilum degenerante (quod nonnisi cum aliqua cautione adhibendum est), cum sit per se perspicuum (quod mihi sufficit), quod Nihili quodcunque multiplum est adhuc Nihil.[35] (Wallis to Leibniz July 22/[August 1], 1698, GM IV, 50)

For Leibniz such an interpretation was plainly inadequate to the tasks his calculus sets out to achieve. Rejecting Wallis' ultimate identification of infinitesimals with nothings ("nihili"), he points out that for his own mathematical practice it is necessary to have minute elements or momentary differentials considered as quantities, since they in turn have their differences and can also be represented by determinable proportional lines. Not only would Wallis' interpretation mean that all quantities divided by infinitesimals or all ratios of infinitesimals themselves reduce to infinity or zero, but would also exclude the possibility of higher order differentials. Moreover, to consider the indeterminable or characteristic triangle to be similar to a determinable triangle and yet devoid of quantity represents for him the introduction of an unnecessary obscurity. Then, as he points out in his reply to Wallis:

> Figuram sine magnitudine quis agnoscat? Nec video quomodo hinc auferri possit magnitudo, cum dato tali Triangulo intelligi queat aliud simile adhuc minus, si scilicet in linea alia simili omnia proportionaliter fieri intelligantur.[36] (Leibniz to Wallis, December 29, 1698/[January 8, 1699], GM IV, 54)

In the end the discussion between the two men turned on the questions of inassignable ratios and incomparable differences, with little room for reconciling different conceptions of the way mathematical analysis was to proceed. In part of a draft which he apparently omitted from the letter actually

---

35 "I do not need your particular postulate of some infinitely small, considered in itself or in relation to another infinitely small, degenerating into nothing (which concept is only to be used with a certain caution), as it is evident in itself (which suffices for me) that nothing multiplied as much as one pleases is just nothing."

36 "Who would accept a figure without quantity? Nor do I see how for this reason quantity could be taken away, since one such triangle can be considered yet smaller than another similar triangle, when namely in another similar line everything is understood to take place proportionally." – Leibniz had explained beforehand: "Putem praestare, ut Elementa vel differentialia momentanea considerentur velut quantitates more meo, quam ut pro nihilis habeantur. Nam et ipsae rursus suas habent differentias, et possunt etiam per lineas assignabiles proportionales repraesentari. Triangulum illud inassignabile, quod ego characteristicum vocare soleo, triangulo assignabili simile agnoscere tecum, et tamen pro nihilo habere, in quo retineatur species trianguli abstracta a magnitudine, ita ut sit datae figurae, nullius vero magnitudinis, nescio an intelligi possit, certe obscuritatem non necessariam inducere videtur."

sent, Leibniz brings in the idea of considering inassignable quantities as useful fictions which serve to shorten reasoning, and even allows that if necessary they be substituted by incomparably or sufficiently smaller quantities.[37] In view of his broader need for constructive dialogue with the Savilian professor he probably felt it important to keep as close to Wallis' position as possible. Thus in the final version he concedes to him that the form of a characteristic triangle in a curve can be correctly explained through the degree of the curve's declination, and simply points out that

> [...] pro calculo utile est fingere quantitates infinite parvas, seu ut Nicolaus Mercator vocabat, infinitesimas: quales, cum ratio eorum inter se utique assignabilis quaeritur, jam pro nihilis habere non licet.[38] (Leibniz to Wallis, March 30/[April 9], 1699, GM IV, 63)

Wallis on the other hand sought to argue that the employment of infinitely small differences as quantities cannot be justified mathematically, because such differences must be considered as evanescent and therefore ultimately as nothings. Characteristically, the Savilian professor argued that the classical concept of incomparable difference, employed since the time of Archimedes, was in itself quite sufficient:

> [...] quippe in omni genere Quantitatum, quae differunt dato minus, reputanda sunt Aequalia. Quo nititur Exhaustionum doctrina tota, Veteribus pariter et Recentioribus necessaria.[39] (Wallis to Leibniz, April 20/[30], 1699, GM IV, 66)

Wallis, finding his methods largely eclipsed by recent developments such as those brought about by Leibniz, sought to defend his approach to quadratures through its ancient origins. In his letter to Leibniz of April 6/[16], 1697 he describes Cavalieri's method as being nothing but a shortened version of the method of exhaustions. Furthermore, he sees his own understanding of the method of indivisibles, in which lines are understood as par-

---

37 "Verae interim an fictitiae sint quantitates inassignabiles, non disputo; sufficit servire ad compendium cogitandi, semperque mutato tantum stylo demonstrationem secum ferre; itaque notavi, si quis incomparabiliter vel quantum satis parva pro infinite parvis substituat, me non repugnare." (Leibniz to Wallis, March 30/[April 9], 1699, GM IV, 63) The background to this omission would appear to be that Leibniz by this time had come to have serious doubts about the reality of infinitesimals. See Jesseph, 1998, 27–28.

38 "[...] for the calculus it is useful to imagine infinitely small quantities, or, as Nicolaus Mercator called them, infinitesimals, such that when at least the assignable ratios between them is sought, they precisely may not be taken to be nothings."

39 "For in all kinds of quantity, those which differ to a degree smaller than any given quantity can be held to be equal. On this rests the whole doctrine of exhaustions, necessary for the ancient mathematicians just as it is for the more recent ones."

allelograms of infinitesimal width, as reconciling Cavalieri with the geometrical concept of continuity, thus enabling his method to be used more advantageously.[40] Similarly in his *Treatise of Algebra* he asserts that "the Method of Indivisibles, introduced by Cavallerius" is but a "shorter way of expressing that method of exhaustions", and that the "Arithmetick of Infinites" is a "further improvement on that method of Indivisibles."[41] (Wallis, 1685, 282, 285)

Leibniz for his part sought to correct Wallis on his views concerning historical continuity. In a wonderful play on words he remarks in a letter to Simon de La Loubère from October 5/15, 1691:

> "Car cette Methode sert principalement à traiter analytiquement les problemes physico-geometriques parce que mon Analyse est proprement l'Analyse des Infinis (infiniment differente de la Geometrie des indivisibles de Cavalieri et de l'Arithmetique des infinis de Wallis) et la nature va tous jours par une infinité de changemens."[42] (A I, 7, 400)

---

40 "Quando autem ego alicubi insinuaverim Cavallerii *Geometriam Indivisibilium* non aliam esse quam Veterum Methodum *Exhaustionum* compendiosius traditam, nolim quis id a me dictum putet in ejus derogationem, sed in ejus confirmationem. Cum enim objecerint aliqui, non id esse Geometriae consonum, ut (verbi gratia) ex Lineis Rectis (nullius latitudinis) compleri censeatur Superficies Plana: per Rectas hasce (commoda interpretatione) intelligenda dixerim Parallelogramma, quorum latitudo sit infinitesima pars Altitudinis totius figurae, qualibus, numero infinitis, compleri posse spatium illud, satis Geometrice dici possit; saltem, ex talibus fieri figuram vel inscriptam vel circumscriptam, quae inter se differant (adeoque et ab exposita figura) dato minus. [...] Qua benigna interpretatione non laesum iri putem Cavallerii methodum, sed adjutum, ut quae compendiosius tradat, aliorum prolixiores Exhaustiones." (Wallis to Leibniz, April 6/[16], 1697, GM IV, 19)

41 See also Wallis to Leibniz, April 6/[16], 1697, GM IV, 19; Wallis to Leibniz, July 30/[August 9], 1697, GM IV, 37. In his letter to Oldenburg of February 11/[21], 1674/5, Wallis asserts in respect of Boulliau's attempt to give a more rigorous proof of his *Arithmetica infinitorum* that he employed his method of induction in such a way that the demonstrations could easily be put into a rigorous form. Moreover, if he had wished to introduce demonstrations according to the form of the ancients the business would have been very long drawn out and as such foreign to his purpose. See Oldenburg, 1977, 188–189.

42 "For this method serves in general to treat the physico-geometrical problems analytically because my Analysis is truly the analysis of the infinitesimals (infinitely different from geometry of indivisibles of Cavalieri and from arithmetic of infinites of Wallis) and nature always goes through an infinity of changes." – Already during his stay in Paris Leibniz felt confident to claim that with his method, then in its inception, everything could be demonstrated which previously had been demonstrated by the geometry of indivisibles and more besides. See *De differentiis progressionis harmonicae* (A VII, 3, 126). See also *De geometria recondita et analysi indivisibilium atque infinitorum* (Acta eruditorum, June 1686, 292–300, 298 = GM V, 231–232); *De la chainette* (Leibniz 1692, 148 = GM V, 259); *Solutio illustris*

He reiterates this opinion in his long review of the first two volumes of Wallis' monumental *Opera mathematica*, which he published in the June 1696 issue of *Acta eruditorum*, where he goes on to compare the difference between his own and Wallis' arithmetic of infinites to that between algebra and arithmetic (Leibniz, 1696, 252).[43] The following year, in a letter to Wallis, he sets out reasons why the geometry of indivisibles cannot strictly be reduced to the ancient method of exhaustions, noting that the one operates with finite quantities, the other with quantities incomparably smaller than the whole (GM IV, 24–25).[44] His aim thereby was not so much to question Wallis' conception of his place in a tradition stretching from classical antiquity to the present, but rather to emphasize just how much his own efforts represented a considerable leap beyond what Cavalieri and Wallis had achieved. While so much of Leibniz's philosophy falls within the scope of his law of continuity this certainly did not apply to scientific endeavors, least of all his own.

6. Conclusion

In mid seventeenth-century Europe no-one in the avant-garde of mathematics could afford to ignore the possibilities presented by the geometry of indivisibles for devising new approaches to quadratures. Much of the history will inevitably remain obscure, lines of dependency uncertain, but through the painstaking record of sources provided by John Wallis, we

---

problematis a Galilaeo primum propositi de figura chordae aut catenae e duobus extremis pendentis, pro specimine novae analyseos circa infinitum (*Giornale de' Letterati*, 1692, 128–131, 128–129 = GM V, 263): "Ediderat is [sc. Leibnitius] Analysin quandam novam circa infinitum a Cavaleriana Geometria indivisibilium, et Wallisiana Arithmetica infinitorum plane diversam."

43 "Ex his patet, Arithmeticam infinitorum sensu Wallisii longe diversum significare ab Analysi infinitorum, seu calculo differentiali, qui ita se habet ad illam, ut Analysis speciosa ad Arithmeticam." (Leibniz, 1696, 249–259, 252)

44 "Dixi aliquando in Lipsiensibus Eruditorum Actis, mihi omnes Methodos Tetragonisticas ad duo summa genera reducendas videri: vel enim colliguntur in unum quantitates infinitae numero, quantitate incomparabiliter minores toto; vel semper manetur in quantitatibus toti comparabilibus, quarum tamen numerus infinitus est quando totum exhauriunt. Utriusque Methodi specimina jam dedit Archimedes, sed nostrum seculum utramque longius produxit. Itaque, strictius loquendo, Methodos Exhaustionum a Methodo Indivisibilium distingui potest: tametsi commune omnibus sit principium demonstrandi, ut error ostendatur infinite parvus, seu minor quovis dato, Euclidis jam exemplo." (Leibniz to Wallis, May 28/[June 7], 1697, GM IV, 24–5)

know that his arithmetical reform of Cavalieri's method, which was to be a decisive step in the growth of modern analysis, took its starting point in the version of that method handed down by Torricelli. At the University of Oxford political and educational reform were reflected primarily through advances in mathematics carried out by Wallis, who together with Seth Ward contributed decisively to making that university one of the great centers of science in early modern Europe. The geometry of indivisibles played a central role in Wallis' mathematical career and there is a sense of irony in the fact that the young Leibniz at the beginning of his career should have sought somewhat naively to save Cavalieri's and Wallis' approaches by means of the innovative concept of point which he had then developed. Leibniz eventually moved far beyond the geometry of indivisibles in his own work on analysis leading up to and beyond the discovery of his infinitesimal calculus. The two men, whose biographies had been interwoven since 1671, eventually addressed the history of their own work. Their perspectives, non-adjacent moments in a line of development stretching back to classical antiquity, for the one continuous, for the other less so, inevitably soon themselves began to recede into infinity.[45]

[45] The author should like to thank Doug Jesseph for very useful discussions on themes associated with this chapter and Christoph Scriba for his critical comments on an earlier version. Jörg Dieckhoff kindly assisted in preparing the illustrations.

Ursula Goldenbaum

Indivisibilia Vera – How Leibniz Came to Love Mathematics[1]

> *"I have a use for all that too; but I cannot do the same thing with it."*[2]

Leibniz had the powerful gift of razor-sharp logical thinking and the ability to immediately grasp an entire argument into its most remote consequences. He was not, however, a born mathematician and he came very late to mathematics.

> It was neither in adolescence – like Pascal and Huygens – nor, like Torricelli and Newton, in his student days at university, nor even – as John Wallis before him – in his first graduate years that he entered the mathematical area, but rather in full intellectual maturity, his doctorate gained and with a developed awareness of his abilities and creative potentialities. (Hofmann, 1974, 1)

Because he started so late, Leibniz never acquired the familiarity with common mathematical techniques that his professional colleagues had.[3] When he finally turned to mathematics he had already been celebrated as a doctor of law and had obtained the position of a lawyer at the court of the second most important ruler in the German Empire, the Archbishop of Mainz. But only a few years later, after intense studies of the most recent mathematics of France and England during his stay in Paris, Leibniz invented the calculus and became one of the leading mathematicians, if not the leading one, in Europe. At that time, he was 29. The intellectual miracle of this late but thrilling transformation from mathematical ignoramus to mathematical in-

---

1 I should like to thank Stephen P. Farrelly for his improvement of the English in this paper.

2 Lessing said this to Jacobi in their discussion about reason and faith, Spinoza and Leibniz, in the summer of 1780 (Lessing, 2005, 251).

3 "With the more primitive things – such as a typical proof in elementary geometry or a lengthy transformation in algebra – he never even in later years found it easy to cope, and errors in calculations are no rarity in his writings." (Hofmann, 1974, 9)

novator within four or five years cries for an explanation. Although Leibniz was a genius, "having mathematics in his blood even if he is still ignorant of its detail" (Hofmann, 1974, 2), the question arises: why did his genius not show any signs of the "*grande passion*" (Hofmann, 1974, 9) any earlier? Why – all of a sudden – did he turn to mathematics at the end of 1669 or in early 1670, but then in such an intense way?

Nowadays, scholars have come to accept that Leibniz's interest in mathematics arose somewhat before his departure to Paris in March 1672.[4] The growth of this insight can even be seen in the contrast between the German and English editions of Hofmann's book *Leibniz in Paris*, the first published in 1949, "completed in manuscript in 1946" (Hofmann, 1974, IX), and the second in 1974. Between the two editions of his book, as Hofmann himself emphasizes, he had finished his editorial work on the first volume of Leibniz's Mathematical Writings in the Akademieausgabe.[5] However, this first volume still starts with Leibniz's first mathematical manuscript from Paris. Only recently have scholars acknowledged that

[4] Hofmann, when canvassing the various moments in which Leibniz gained access to mathematical knowledge, does not distinguish between the lectures at the university that Leibniz very likely listened to and Leibniz's own arduous efforts to obtain mathematical knowledge since the very end of 1669. It is the latter which I want to call his mathematical turn of late 1669.

[5] Cf. Hofmann, 1974, IX. The main difference is the enormously extended amount of footnotes giving e.g. access to all the writings of Hobbes that Leibniz must have studied in addition to *De corpore* according to his own writings. According to Hofmann's footnotes, around 1670–1, Leibniz was acquainted with "the separate Latin parts of Hobbes' *Elementa philosophiae* (1655, 1658, 1642) as well as the complete edition including the *Leviathan* (1651), the *Examinatio* (1660) and *De principiis* (1666)." (Hofmann, 1974, 7, fn. 31) This extended list itself ironically undermines Hofmann's intention of minimizing Hobbes' impact on the young Leibniz. The "complete edition", i.e. the *Opera philosophica*, appeared in Amsterdam in 1668. Leibniz's references to his books in Leipzig, mentioned by Hofmann as well, do not help much to date his acquaintance with Hobbes. All we know is that he wrote them before the fall of 1666. Hoffmann emphasizes Leibniz's alleged critique of Hobbes' arbitrary definitions. In fact, Leibniz takes this view himself; he attributes it to Galileo and agrees (in the *Accessio*, A II, 1 (2006), 350). He only rejects Hobbes' conclusion that the truth of all sentences would likewise be arbitrary. Hofmann's introduction to the volume of the *Akademieausgabe*, dedicated mainly to Leibniz's alleged plagiarism of Newton, spends one sentence on Hobbes (A III, 1, LI). Although I will criticize Hofmann for his prejudiced minimization of Leibniz's debts to Hobbes, I do not question Hofmann's enormous insight into Leibniz's mathematical development. But his prejudice against Hobbes hinders open-mined research just because of his great and otherwise justified authority. Almost no work on the young Leibniz takes Hobbes' enormous influence into account.

some of Leibniz's mathematical manuscripts stem from his time in Mainz before 1672.[6]

This rather belated acknowledgment of Leibniz's mathematical efforts prior to Paris was to some extent due to the fact that these efforts were closely connected to Leibniz's studies of Thomas Hobbes. Because of Hobbes' bad mathematical reputation, scholars usually have rejected out of hand the idea that Leibniz could have acquired anything of merit from a study of Hobbes' mathematics.[7] In the meantime, however, scholars generally acknowledge that, in particular, it was Hobbes' specific conception of the conatus which Leibniz enthusiastically, though in a critical way, embraced after 1669.[8] Nevertheless, given the fact that Leibniz studied Hobbes' *De cive* around 1663 (taking up Hobbes' foundation of law),[9] *De corpore* around 1666 (taking up ideas of logic),[10] and again in 1668 (this time taking up Hobbes' principles of mechanical philosophy) – how could another study of Hobbes' *De corpore* in 1670 cause Leibniz to study mathematics?

In fact, Leibniz's choice of Hobbes' *De corpore* as his text book for mathematics, especially with regard to the method of indivisibles, has often been deplored by historians of mathematics because this book was "written by a man lacking proper mathematical expertise." (Hofmann, 1974, 7) As many scholars saw it, Leibniz had to overcome the confused mathematical understanding he allegedly inherited from Hobbes before he could start his new career as a mathematician in Paris.

Several years ago, however, Douglas Jesseph argued that Leibniz's mathematical studies of Hobbes had not been as ephemeral as usually supposed. Rather, they shaped some prominent concepts and formulations in Leib-

---

6 See Siegmund Probst's paper in this volume.

7 Couturat argued vehemently against any meaningful impact of Hobbes on Leibniz's logic as claimed by Tönnies; he dedicated a special *Appendix* of his book to this refutation (Couturat, 1901, 457–472). Couturat's view is still widely accepted. Hofmann (Hofmann, 1974, 7) refers to Couturat's *Appendix* in a footnote added in 1974 as the only source for the relation of Leibniz to Hobbes. Loemker although supporting Tönnies' claims in general (Loemker, 1956, 105) does not accept Hobbes' influence on Leibniz's logic, nor does Bernstein following Loemker (Bernstein, 1980, 37), all in spite of the statements of Leibniz himself.

8 Tönnies, 1887, 570–71; Hannequin, 1908, 74–107; Bernstein, 1980, 25–37; Beeley, 1996, 229–31; Jesseph, 1998, 7–16; Ross, 2007, 24–26.

9 For Hobbes' influence on Leibniz's philosophy of law see Goldenbaum, 2002b, 209–231.

10 For the logical impact of Hobbes on Leibniz (against Couturat) see Dascal, 1987, 31–45 and 61–79.

niz's mature mathematics[11] (Jesseph, 1998, 11–14). In the same year, I found Leibniz's marginalia in Boineburg's copies of Hobbes' *De corpore* (Hobbes, 1655) and his *Opera philosophica* (Hobbes, 1668). The evidence I discovered further[12] justifies the claim that Hobbes worked a deep and important influence on Leibniz's development as a mathematician. The latter edition includes not only the main works of Hobbes in Latin but also his controversial mathematical writings against John Wallis up to 1668. In this essay I would like to outline answers to the following questions. First, why did Leibniz begin a study of mathematics in the end of 1669 or in early 1670? Second, how did Hobbes' conception of the conatus, related to the method of indivisibles, become so fascinating for Leibniz in the end of 1669 or early 1670 even though he had ignored it in his earlier studies of Hobbes? Third, how did Leibniz benefit in mathematics from the mathematical trouble-maker Hobbes?

1. Why did Leibniz turn to Mathematics Just at the End of 1669?

There is no question that it is Leibniz's *Theoria motus abstracti* (TMA) which shows, for the first time in his career, a knowledge of the most recent developments in mathematics, even if he still confuses incompatible concepts. Moreover, this treatise shows Leibniz to be "completely under the spell of the concept of indivisibles" (Hofmann, 1974, 8). This awakening of the "grande passion" was closely connected with Leibniz's first studies of mechanical theory, as shown by Hannequin.[13] Today it is the generally ac-

[11] Jesseph emphasizes: "It would doubtless be going too far to claim that the whole of Leibniz's calculus is simply the application of Hobbes' ideas." (Jesseph, 1998, 15)

[12] There is some more evidence about Leibniz's enthusiasm for Hobbes in his time in Mainz. In Boineburg's copy of Seth Ward's *Thomae Hobbii philosophiam exercitatio epistolica* (Oxoniae: Richard Davis 1656), now in the Boineburg collection of the *Universitätsbibliothek Erfurt* (Call number: UB Erfurt, Dep. Erf. 03-Pu. 8° 1434), I found a table sorting the partisans of Hobbes and his opponents. Leibniz is listed as L. among the partisans. This book has been purchased by Boineburg in 1669 according to Boineburg's entry in the book beside his signature. See the picture of the table in the appendix with the transcription of the marginalia of Leibniz in Hobbes' *Opera philosophica* and *De corpore*. I should like to thank the *Universitätsbibliothek Erfurt* for their permission to publish these marginalia, and Frau Dr. Kathrin Paasch and Herrn Thomas Bouillon for their great support during my stays at Erfurt exploring the library of Baron von Boineburg.

[13] "Mais le souci de trouver et de suivre dans leurs dernières conséquences les lois du mouvement, au lieu de se contenter d'une croyance vague au principe que tout s'y réduit dans la

cepted view among scholars, that Leibniz's turn to a more serious study of mathematics was at first prompted by his acquaintance with Huygens' (and Wren's) critique of Descartes' rules of motion in August, 1669. He got hold of their papers in the *Philosophical Transactions* by Mauritius, a lawyer acquainted with Baron von Boineburg, when they all stayed at the spa in Bad Schwalbach (cf. A VI, 2, XXXI). But, given the fact that Leibniz had not yet carefully read either Descartes's laws of motion or anything else about the laws of motion, why was he so seriously disturbed by this controversy among experts?

Leibniz disagreed neither with Huygens' formulation of the law, nor with his descriptions of experiments. Rather, Leibniz was concerned with the implicit contradiction between Huygen's law and the first principle of mechanical philosophy as he saw it. According to Huygens' presentation, a body in motion, upon collision with another body at rest, would lose all its motion, and transfer that motion to the body at rest. This second body would start to move with the same speed as the previously moving body (now at rest). According to Leibniz, this would mean that rest itself could cause something. Thus Leibniz's protest is directed against this violation of mechanical philosophy: "Quies nullius rei causa est, seu corpus quiescens alii corpori nec motum tribuit, nec quietem, nec directionem, nec velocitatem." (A VI, 2, 161) Thus it was clearly a metaphysical interest which made the young lawyer sit down still in Bad Schwalbach in August 1669 and write a reply within days, as he described it himself to Oldenburg in September, 1670 (A II, 1 (2006), 101). Not lacking in self-confidence, he handed it to Mauritius and asked him to transfer it to his friend Martin Vogel in Hamburg, who in turn was in correspondence with the secretary of the Royal Society, Henry Oldenburg.[14]

Leibniz's readiness to write a critique of Huygens within days clearly shows that he did not at all need a serious study of geometry or mechanics in order to argue with the most advanced mechanical theorist of the time. He simply did so on the basis of his commitment to a mechanical philo-

nature, l'amène vers la fin de 1669 à la résolution d'établir une sorte de mécanique rationnelle ou de géométrie du mouvement." (Hannequin, 1908, 22)

[14] Whether to Leibniz's benefit or detriment, Vogel refused to do so, seeing the author as incompetent and largely ambitious. It seems to me more than probable that Vogel's remark in the end of his letter to Oldenburg from February 1670/1 is related to Leibniz and his draft on motion (Oldenburg VII, 455). He did not even answer Leibniz's following letters. See A, II, 1, N. 38 and 79. (The letter N. 41 was not forwarded by Conring.)

sophy to which he had turned while still a student in Leipzig.[15] Although he abandoned atomism, at least in its strict form,[16] soon after he came to Mainz (*Confessio naturae contra atheistas*, 1668), he clearly retained mechanical philosophy in general, even if he now tried to reconcile it with Aristotle's principles (as in his letter to Thomasius from April 1669 (A II, 1 (2006), N. 6)). Although his letters to his former teacher are sometimes interpreted as an expression of Leibniz's Aristotelianism, Hannequin has clearly shown how much Aristotle's metaphysics had to be twisted and violated by Leibniz in order to fit the mechanism of the moderns.[17]

In fact, Leibniz strongly urged his former teacher to accept a mechanical philosophy softened by reconciliation with Aristotle. Why did he press Thomasius (as well as the other German Aristotelian, Hermann Conring) so strongly to "convert"? According to Leibniz, mechanism could not be rejected because of its wonderful capacity to defend Christian religion; he sees it even as a "munus Dei [...] senectae mundi datum velut unicam tabulam, qua se viri pii ac prudentes in incumbentis nunc Atheismi naufragio servaturi sunt."[18] (A II, 1 (2006), 37) What seems paradoxical at the first glance, since mechanism was usually seen as dangerous to Christian faith, is a sure thing for Leibniz: mechanical philosophy offered the strict passivity of the body as a great advantage for a defense of Christian religion.

The letter to Thomasius from April, 1669 also displays how Leibniz's approach to geometry proceeded from a completely metaphysical perspective. By stressing the *mechanical* construction of geometrical figures, producing lines by the motion of points, surfaces by the motion of lines and bodies by the motion of surfaces,[19] Leibniz argues that geometry fulfills Aristotle's criterion of a true science: it explains everything by its cause

---

15 Leibniz's own report that he had embraced the moderns and turned down the Aristotelian school philosophy at the age of 15 (i.e. by the end of June 1662) has been doubted by Kabitz (Kabitz, 1909, 49–53) and increasingly so since then. However, I do not see any serious reason to hinder us from believing Leibniz. He went to the university at Leipzig at Easter of 1661, before he turned 15 and after having studied Suarez, Zabarella, and other rather heavily metaphysical literature.

16 See for the specifics of Leibniz's engagement with atomism the illuminating paper of Richard Arthur (Arthur, 2003, 183–227).

17 "On voit, dés la première lecture, par tout ce qui précède, que toute cette tentative n'est qu'une perpétuelle violence faite à la philosophie d'Aristote pour le mettre d'accord avec les modernes, bien loin qu'il soit l'inspirateur de l'ingénieuse doctrine développée par Leibnitz." (Hannequin, 1908, 49)

18 "[...] armor from God [...] given to the aging world as the only life-boat by which pious and prudent men can still save themselves from shipwrecking in the overtaking Atheism."

19 Compare Hobbes' definition of philosophy in *De corpore* I, 1, 2.

(A II, 1 (2006), 31).[20] Moreover, he attributes a central function to geometry or pure mathematics (dealing with the forms or figures of things) in the order of sciences because it has to mediate between theology or metaphysics on the one hand and physics on the other. Theology or metaphysics deals then with the mind as the "rerum efficiens" (ibid.). He adds: "Mens enim ut bonam gratamque sibi rerum figuram et statum obtineat, materiae motum praebet."[21]

On the basis of *this* metaphysical position, Leibniz is ready to criticize Huygens (A VI, 2, N. 38_1). He argues precisely against the idea that a body at rest could cause something. The draft starts with an explanation of Leibniz's methodological approach, distinguishing between a theoretical demonstrative foundation of the laws of motion that disregards the phenomena which we observe by our senses and description of the phenomena on the basis of observation and experiment. When experience and reasoning conflict, however, we have to follow reason alone (#1–9). In the next three articles (#10–12) Leibniz gives the explication of the principle of inertia. In articles 13–14 he explains the relativity of softness and hardness, claiming that the differences are given only to our senses but caused simply by the motion of the superficial parts of a body against our body. From article 15 on, Leibniz treats collision, stating in article 19 that a body which impacts another body at rest becomes one with it, both becoming "continuous" bodies. The body composed by the collision will continue to move with the same speed and direction. Then Leibniz prepares the ground for the moving mind: according to #22 no other reason can be given for acceleration than curvilinear motion, and #23 adds that curvilinear motion presupposes incorporeal entities.

It is article 25 where Leibniz turns to the problem of cohesion. This is the thorny metaphysical problem he started to struggle with in the *Confessio* in 1668 and which will continue to trouble him at the time of his first letter to Hobbes in July, 1670 (A II, 1 (2006), 92). In August, 1669 however, Leibniz claims: "Duo corpora eatenus tantum concurrunt, quatenus continuato

---

20 "But if we consider the matter more accurately, it will be seen that it does demonstrate from causes. For it demonstrates figures from motion; a line arises from the motion of a point, a surface from a motion of a line, a body from the motion of a surface. The rectangle is generated by the motion of one straight line along another, the circle by the motion of a straight line around an unmoved point, etc. Thus the constructions of figures are motions, and therefore the properties of figures, being demonstrated from their constructions come from motion, and hence, a priori, from a cause."

21 "It is the mind, which provides motion to matter in order to obtain a good figure and state of things agreeable to itself."

impetu se penetrarent."[22] (A VI, 2, 163) At the first glance, the solution that he provides does not differ greatly from his later explanation of cohesion in the TMA, – except that in the latter it is no longer the simple bodies that press or penetrate each other, but rather some indivisible parts of the bodies which have the endeavor/*conatus* to penetrate each other, thus becoming one body as long as the endeavor of their penetrating parts will last. In order to justify this conclusion, Leibniz would indeed have to appropriate Hobbes' conception of *conatus* and therefore learn more about the mathematics of indivisibles. But I cannot see any particular inner theoretical reason forcing him to do so at this stage. He seemed to be quite satisfied in the summer of 1669.

The other explanation for why Leibniz turned to mathematics in spring 1670, according to which Leibniz's Hobbes studies at that time made him embrace Hobbes' conception of conatus, which in turn enabled him to write the TMA, is not sufficiently explanatory either. As a matter of fact, Leibniz's understanding of mechanical philosophy, as in the *Confessio*, the letters to Thomasius, and in his first drafts against Huygens are *already* deeply influenced by Hobbes.[23] The central position of geometry in the system of sciences, its justification as a true science because it explains from causes, i.e. moving points, lines, surfaces, and bodies, the methodological distinction between a theoretical mechanics working with definitions and demonstrations and physics depending on sense experience thus never being demonstrative, the relativity of softness and hardness of bodies, the infinite divisibility of bodies causing the problem of cohesion, the mechanical explanation of cohesion by mutual pressure of bodies, and, last but not least, – the absolute passivity of bodies in themselves: all that can be found in Hobbes' *De corpore*, which Leibniz had been studying at least since 1666. Thus the great impact of *De corpore* on Leibniz had happened much earlier and was already virulent in his very first critique of Huygens. (It goes without saying that Leibniz adapted Hobbes' materialistic philosophy to his own metaphysical or theological goals.)

After all, the question is still how Leibniz overcame his well-known aversion to studying what could not be understood by simple reading, namely

---

22 "Two bodies run only as long together as they penetrate each other by a continuous impetus."

23 It was the sociologist Ferdinand Tönnies who first saw the great and general influence of Hobbes on Leibniz, against the prejudice of most scholars of his time against Hobbes. See Tönnies, 1887, 561–573 (=Tönnies, 1975, 151–167). More recently, Catherine Wilson convincingly argued for a much more general influence of Hobbes on the young Leibniz (Wilson, 1997, 339–351). See also Goldenbaum, 2002a, 204–10.

mathematics and mechanical theory in a strict sense. Here is what I think: Leibniz got his hands on the *Opera philosophica* of Thomas Hobbes in October, 1669, probably at the book fair in Frankfurt (Müller/Krönert, 1969, 17), when Boineburg purchased the book. As we know Leibniz, he would start reading immediately, particularly those parts he had not read before, such as the *Leviathan*, *De homine*, and the mathematical writings against Wallis. Of course, this new study of Hobbes was now framed by his new and sharpened interest in the rules of collision, the laws of motion, and by the question of cohesion. Nevertheless, the driving interest was still his metaphysical project of a philosophy of mind, related to the *Conspectus Catholicus*.

The study of the *Opera philosophica* allowed Leibniz to discover two important new things about Hobbes, both sufficient to spur him to take the effort to study Hobbes's conception of *conatus* and therefore mathematics more seriously. On the basis of Leibniz's marginalia in Hobbes' *Opera*, I dare to claim that Leibniz became aware of Hobbes' conception of *conatus* only by his reading of *De homine* in the fall of 1669.[24] Boineburg's copy of the *Opera* contains marginalia from Leibniz's hand in almost all chapters of *De homine* (unlike the *Leviathan* which was also new for Leibniz). Given Leibniz's work on a philosophy of mind and his intention to make the mind the moving principle of the body, in close connection with his project of the *Conspectus Catholicus*, Hobbes' mechanical conception of *conatus* as a striving through a *point*, causing sense perception, had to catch Leibniz's interest.

The second most marginalia can be found in the mathematical writings (including *De corpore*). Above all, Leibniz simply learned about Hobbes' failure in mathematics. We can find the expression of his astonishment, excitement and even triumph about this discovery in the conclusion of the TMA where he greatly laments Hobbes' incredible mistake: "Hobbes in dubium revocat inventum Pythagorae hecatomba dignum, 47^{lmi} Euclidis, fundamentum Geometriae: negat radicem quadrati [...] coincidere numero partium lateris, fundamentum non Algebrae tantum, sed et Geodaesiae".[25] (A, VI, 2, 275) However, this lamenting, apparently expressing surprise about the amazing failure of the great Hobbes, had also appeared

---

24 I see this claim confirmed by Probst's paper in this volume, see especially fn. 10 there.

25 "Hobbes raises doubts about Pythagoras' invention having been worth a sacrifice [of an ox], about the 47th Theorem of Euclid, the fundament of geometry. He denies that the square root coincides with the number of the side's parts, not only the fundament of Algebra but also of Geodesy."

more than a year previously in one of Leibniz's printed writings, with the same excitement. It can be found, at the very end of Leibniz's *Dissertatio Praeliminaris* to his edition of Marius Nizolius' book *De veris principiis et vera ratione philosophandi libri IV* (A VI, 2, N. 54). And indeed, if we look at the respective passages in Hobbes' mathematical writings in Boineburg's copies of the *Opera philosophica* (Hobbes, 1668) and *De corpore* (Hobbes, 1655), we find comments and underlined passages from Leibniz's hand exactly in "De magnitudine circuli" within the *Problemata physica,* in *De principiis et ratiocinatione, Examinatio et emendatio mathematicae hodiernae*, and in chapters 20–22 of *De corpore*, in both available editions! Given Leibniz's early and great admiration for Hobbes's logical skills, despite both Hobbes' materialism and his bad faith, Leibniz must have been curious to *understand* this unbelievable, catastrophic failure. Moreover, understanding this failure could even help him to learn how to defeat and to refute this smart but dangerous thinker: he grasped immediately that Hobbes' mathematical failing was closely connected to his materialism, i.e. his denial of the minds, the "indivisibilia vera" (A VI, 2, 275).

Given that Leibniz's *Nizolius* came out at the Easter book fair in Frankfurt in 1670, between April 16th and the 22nd (Müller/Krönert, 1969, 19) it is obvious that Leibniz had learned the news about Hobbes before the spring of 1670, as the editors of the Akademieausgabe suggest and as is generally accepted. This is confirmed as well by Leibniz's references to Euclid's *Elements* several times after January 1670.[26] He certainly could not get through the technical mathematical parts of Hobbes without finally getting into the more technical parts of geometry.[27] We find him at that time even paralleling the *Elements* of Euclid with Hobbes' *Elements* on motion and his own still unwritten *Elements on the mind*, all three praised by him for their strict demonstrations.[28]

---

26 See Leibniz to Conring in January 1670 (A, II, 1 (2006), 49), to Velthuysen in April 1670 (A II, 1 (2006), 63), to Chapelain within the first half of 1670 (A II, 1 (2006), 87), and also in his letter to Oldenburg in September 1670 (A II, 1 (2006), 104)

27 "Sed demonstrationes ipsae tumultuario sermone exponi nec possunt, nec si possent, debent. Merentur enim non lectionem cursoriam, sed patientiam attentionis: [...] quemadmodum Geometris Euclidis demonstrationes non percurrendae sed examinandae et in prima usque Elementa resolvandae sunt, donec in clara et a nemine negabilia incidatur." (A, II, 1 (2006), 182)

28 E.g. to Johann Friedrich from May 1671 (A II, 1, N. 58). Also, he certainly studied the relevant parts of Descartes' *Principia*, as is evident from his discussion of the laws of motion in his letter to Oldenburg from September 18, 1670 (A II, 1 (2006), 102).

2. How did Leibniz Grasp Hobbes' *conatus* Conception at the End of 1669?

Leibniz mentions Hobbes for the first time in a letter to Thomasius from Jena in September 1663 (A II, 1 (2006), N.1). In that letter, he not only asks for more advice about Hobbes' political philosophy but already discusses the main ideas of *De cive*.[29] It is then in his *Ars combinatoria* from 1666 that we can see the results of Leibniz's enthusiastic reading of Hobbes' *De corpore* and that we can recognize his obvious familiarity with Hobbes' logic. At a minimum, Leibniz picked up the idea of thinking as reckoning (A VI, 1, 194).[30] His general outline of mechanical philosophy in his letters to Thomasius from 1668 and 1669, as well as his first critique of Huygens in August 1669, displays his deep debt to Hobbes' view of mechanical philosophy in spite of his rejection of Hobbes' materialism and atheism.

Thus when Leibniz comes to read Hobbes' *Opera philosophica* in the fall 1669, he is already familiar with this author. Given the fact, that Leibniz had studied Hobbes at least three times before the fall of 1669 with respectively different interests in law, logic and mechanical philosophy, he seems to have seen his opponent as a model, even if a negative one, for his own work on a new philosophical system as thoroughgoing and consistent as that of the admired Hobbes. But of course, whereas Hobbes did not accept anything in the world except bodies, explaining thinking as reckoning with words, seeing even God as a body, it was Leibniz's goal to explain the whole natural world of bodies as originated by minds (or God). If there is any continuity in Leibniz's philosophical development it is his fervor to install the mind as an active and immortal thing whereas the body had to play a passive role and was subject to corruption.

But how could Leibniz embrace the Monster of Malmesbury again and again, praising Hobbes even to the rather horrified Aristotelians Conring and Thomasius if he so arduously desired to refute materialism and atheism? How could he learn from Hobbes when he was trying to defend Christian religion? What looks paradoxical at first glance becomes quite clear if we take a closer look. The first time Leibniz took the side of Hobbes

---

29 There are sufficient reasons to assume that the student of law received the requested instruction from his teacher after his return to Leipzig. It was certainly at this time that Leibniz adopted, against Grotius, Hobbes' view of the striving of human individuals for their own sake as the necessary starting point for law. See Goldenbaum, 2002b, 215–16. See also Goldenbaum, 2008, forthcoming.

30 This is emphasized most recently by Ross, 2007, 21–22.

in terms of the foundation of law was against Hugo Grotius, who had claimed that the natural law would be valid even if God did not exist. Grotius thus gave natural law an objective status independent of God. Hobbes, on the opposing side, argued largely in agreement with the stricter Protestant position that, unfortunately, human beings would not follow natural law but rather their own self preservation (although he did not go as far as to claim the corruption of the humans by the original sin). When Leibniz then studied Hobbes' logic it helped him to develop combinatorics as an instrument which could parallel his atomism. He saw the *ars combinatoria* as a possible tool of God when creating the manifold of the world from a few principles and elements. This underlying assumption of an isomorphic structure of concepts and creatures was very similar to the view of the protestant Bisterfeld with whose ideas Leibniz was quite familiar at that time.[31] As mentioned above, even mechanical philosophy was seen by Leibniz as a spare-anchor against the waves of atheism, naturalism and Socinianism, although, of course, he had to revise Hobbes' approach. Thus, in all these readings of Hobbes, it was precisely Leibniz's ardor to affirm his Protestant and more generally his Christian view which pushed him to adopt and to adapt Hobbesian ideas.

In the fall of 1669, Leibniz knew for sure that studying Hobbes was worth the effort regardless of his faithless and materialistic approach.[32] But this time he certainly got more than he had expected. As mentioned above, his reading of Hobbes' *Opera philosophica* after October 1669 offered him on the one hand the conception of *conatus* as a foundation for sense perception, which thus led him to a revision of his mechanical philosophy on behalf of his philosophy of mind; on the other hand he grasped the surprising news about Hobbes' failing quadratures of the circle and his incredible doubts of the Pythagorean theorem, which revealed to Leibniz the mathematical Achilles heel of the admired philosophical opponent. Both discoveries caused Leibniz to study mathematics.

Of course, Hobbes' theory of sensation and emotion was a mere mechanical and materialistic theory and therefore clearly not to Leibniz's lik-

---

31 For the close connection of Alsted's and Bisterfeld's encyclopedic ideas with millenarianism see Hotson, 2000.

32 His first letter to Hobbes on July 13/23, 1670 (A II, 1 (2006), 93) asks eagerly for the favor to learn of his newest writings, and Boineburg repeats this request. We learn of this from Oldenburg's answer to Boineburg, which claimed that Hobbes would no longer publish after the edition of the *Opera philosophica*, being "more than eighty years old", seeking "quiet and repose", refusing "to be drawn by the lively sallies of younger antagonists" (Oldenburg VII, 108).

ing. In this theory, however, although sense perception is reduced to pressure (related to cohesion) and resistance to the conatus of different bodies which touch each other, these bodies do so at a *point*. Hobbes defines sensing as the phantasm caused by the reaction of the momentary lasting *conatus* in the sensory organ to the external conatus, stemming from the external object toward the internal (*De corpore* IV, 25, 2). Such a phantasm is produced in an *instans*, according to Hobbes, i.e. again at a *point* of time. That Leibniz indeed became aware of the exceptional status of the "points" in Hobbes' theory of perception becomes clear in his rather ironical comment in the margins of the 25th chapter of *De corpore*. There he points out that Hobbes had in fact denied the actual existence of points earlier in the book. Although we can see by this remark as well as by other critical comments in the marginalia that Leibniz is definitely critical of Hobbes' materialistic intention and rejects it, in *De homine* and then in chapter 25 of *De corpore* he could definitely acquire the tools he needed for his own philosophy of mind. In particular, Hobbes offered him here the *conatus*, the point and the *instans* as the places where thinking, starting with sense perception, would occur. In addition, I take the mostly positive character of Leibniz's comments (in comparison with the thoroughgoing critical ones in the mathematical writings) as a clear expression of his definite excitement about his findings. Of course, Leibniz mined Hobbes' materialist explanation of sense perception by means of *conatus* for his own purposes, turning the ideas upside down and, designing *conatus* in his own way in order to start his mechanical explanation of the material world from the activity of ideal minds in his TMA. However, Leibniz's revision of his mechanical philosophy, as it can be seen in the TMA, was not so much caused by Leibniz's study of mechanical theory in Huygens or Hobbes, but rather by his study of Hobbes' conception of sense perception based on the conception of *conatus*. Therefore he again took up *De corpore*, focusing this time on Hobbes' mechanical conception of *conatus* as it was closely related to his presentation of the method of indivisibles.

The most striking argument for the truth of this claim is in my eyes Leibniz's famous definition of the body as a momentary mind in the TMA (A VI, 2, 266). It is borrowed from Hobbes as well but immediately modified to fit Leibniz's idealistic intention (*De corpore*, I, 25, 5). For example, the materialist Hobbes had a hard time to explain why *conatus*, if always producing phantasmata or phenomena whenever bodies mutually exerted pressure on one another, obviously did not produce such phenomena in all bodies. Hesitating to attribute sense perception to all bodies, a consequence drawn by Spinoza, Hobbes is quite defensive in his answer. He finds it rather unlikely to assume sense perception in, for example, stones although he admits that he could not prove the lack of sense perception in inanimate bodies. Then he explains the lack of sense perception in less complex bodies by their lack of memory. Simple bodies would not be able to keep more than one sense impression at a time, and thus would be unable to compare them and to *sense* the change of them which alone caused conscious sense perception. Therefore sense perceptions of such simple bodies could only last *momentarily*.

This was immediately recognized and grasped by Leibniz, who was looking for an explanation of the mind in distinction to the body and for a mental moving power of the natural world besides bodies. He did not have to change much in order to arrive at his distinction between bodies and minds. That Leibniz indeed found the inspiration for his famous definition of bodies as momentary minds in Hobbes is even confirmed by his own remark in the *Conspectus Catholicus* from the very same time. Although we should not expect a mention of Thomas Hobbes in this theological project we read there: "Omnis sensio reactio durans, v. Hobbes, sed haec in corporibus nulla."[33] (A VI, 1, 495) This critical appropriation of Hobbes shows again how well the Christian idealist philosopher Leibniz could embrace the philosophical ideas of the unbeliever and materialist. But of course, in taking up his ideas, Leibniz adapted them to his own purposes.

That Leibniz was thrilled and convinced he had succeeded in transforming the foundations of Hobbes' philosophy into his own idealistic metaphysics, which in turn would serve as the long-intended bridge between modern mechanical theory and revealed religion, is evident from all the triumphant letters to important scholars like Velthuysen and Oldenburg, to the Catholic theologian Arnauld as well as to the Duke of Braunschweig-Lüneburg Johann Friedrich after the publication of the two parts of his *Hypothesis* in 1671.[34] But as he never tires of emphasizing, he does not triumph about his *Hypothesis physica nova* for its own sake but rather for its capacity to prove the possible accordance of the Christian mysteries with modern science. In addition, it would be able to rescue the active power of the mind, which was thus capable of spontaneity and freedom, traits very much needed on behalf of Christianity as well.

[33] "Each sensuous reaction is lasting, see Hobbes, but there is none lasting in bodies."

[34] Cf. the numbers 56a, 57, 58 and 87 in A II, 1 (2006).

3. What Could Leibniz Learn from Hobbes for his Mathematical Career?

Having argued for the metaphysical, i.e. theological motivation as the driving engine of Leibniz's several critical appropriations of Hobbes between 1663 and 1668 as well as of his turn to Hobbes' conception of *conatus* in late 1669 or early 1670, I want now to investigate the possible mathematical outcome of this last turn to Hobbes, particularly by looking at Leibniz's first Parisian paper of late 1672, the *Accessio ad Arithmeticam Infinitorum* (A II, 1 (2006), N. 109).

Before turning to the *Accessio* I want to point to the simple fact that Leibniz had to study Hobbes' mathematical arguments and constructions in order to understand Hobbes' failure in squaring the circle and solving other problems. He had to study Hobbes' mathematical writings in order to understand how the admired thinker could doubt the Pythagorean theorem. Thus it does not come as a surprise that the most marginalia in *De corpore* (besides those on sense perception in part III and IV) occurs precisely in chapters 20 (squaring the circle), 21 (on circular motion) and in 22 (on other varieties of motion), in *both* editions available to Leibniz. He clearly read these parts with a pen in his hand, examining Hobbes' geometrical demonstrations and constructions step by step. This fact can be seen by his little critical comments in the margins. Simply by doing this work, and also by going back to Euclid and his commentators for help, Leibniz could certainly make a great step forward in his technical skills. The marginalia certainly confirm Leibniz's turn to mathematics through an intense process of studying and penetrating Hobbes' unsuccessful mathematical work, especially his squaring of the circle.

There are even more marginalia in Hobbes' mathematical writings – in the *Examinatio et emendatio mathematicae hodiernae*, in the *Problemata Physica*, and in *De Principiis et Ratiocinnatione Geometrarum*. All of these writings were directed against John Wallis, questioning his understanding of the method of indivisibles but also discussing basic mathematical concepts such as quantity, number, limit, whole and part, demonstration and induction, and so on. Because Hobbes followed the classical *methodus polemica* in his controversial mathematical writings – although mocking Wallis and Seth Ward – he regularly provided the reader with his opponent's argument before replying to it. Thus, by studying Hobbes' *Opera philosophica*, Leibniz gained access to the whole fascinating mathematical discussion concerning geometrical rigor and the foundation of the method of indivisibles. Moreover, by reading Hobbes, he is directed immediately to the most disputed and essential questions of modern mathematics, particularly to the metaphysical status of indivisibles. In addition, he could not only learn about Hobbes' and Wallis' controversial views but also about those of other mathematicians and physicists mentioned in the works, contemporaries as well as ancients, people such as Archimedes, Cavalieri, Roberval, Galileo, Vieta. From this discussion of the philosophy of mathematics he profited enormously; Leibniz was aware of this and grateful even after recognizing the end of Hobbes' mathematical career. He still referred to Hobbes as an authority in his *Accessio ad arithmeticam infinitorum* and wrote his great letter to Hobbes in July 1670, even though he knew for sure of Hobbes' mathematical disaster. This appraisal goes far beyond anything he ever wrote to an admired scholar and certainly cannot be attributed merely to politeness.

The *Accessio*, written for Gallois in late 1672, is Leibniz's first known mathematical paper from his time in Paris and is generally seen as documenting his entry into real mathematics. Hofmann is rather irritated that Leibniz still speaks in this work of Hobbes as a great mathematician, naming him together with Gregoire S. Vincent, Pascal, Cavalieri and Galileo (Hofmann, 1974, 20). However, the presence of Hobbes in this paper is overwhelming, and, significantly, in precisely those points seen by Hofmann as the beginning of a promising mathematical career. This presence begins to appear with the explanatory subtitle of the *Accessio* which clearly recalls Leibniz's readings of Hobbes: "ubi et ostenditur Numerum maximum seu numerum omnium numerorum impossibilem esse sive nullum; item quae pro axiomatis habentur, demonstrabilia esse evincitur exemplis."[35] (A II, 1 (2006), 342) Both topics, the proof of the impossibility of the greatest number and that of Euclid's axiom that the whole is greater than the part are widely discussed by Hobbes (*De corpore*, I, 7). Before turning to Leibniz's discussion of the two topics, I want to point to the opening paragraph of the paper which emphasizes a strict distinction between rational and sensuous knowledge:

> Constat Scientiam Minimi et Maximi, seu Indivisibilis et Infiniti, inter maxima documenta esse, quibus Mens humana sibi vendicat incorporalitatem. Quis enim sensu duce persuaderet sibi, nullam *dari* posse lineam tantae brevitatis, quin in ea sint non tantum infinita puncta, sed et infinitae lineae (ac proinde partes a se in-

[35] "Where it will be shown that the largest number or the number of all numbers is impossible or zero; it will also be proved by examples that what is taken for axioms can be demonstrated."

vicem separatae actu infinitae) rationem habentes finitam ad *datam*; nisi demonstrationes cogerent.[36] (A, II (2006), 342)

This is exactly Hobbes' methodological approach in *De corpore* and Leibniz's in the TMA, although Leibniz is not yet clear about his approach to the indivisibles in the latter.

Turning to the two announced topics, the greatest number and Euclid's axiom that the whole is greater than the part, Leibniz clearly was aware of his agreement with Hobbes' argument. Hobbes had criticized precisely the idea of an infinite number (although never mentioning Galileo's name). It is particularly in chapter 7 of *De corpore* where Hobbes spends articles 11–13 on the discussion and refutation of the idea of an infinite number. He starts with a definition of the whole and of its parts: "Quod autem pro omnibus ex quibus constat, sic ponitur, vocatur *totum*, et illa singula, quando ex totius divisione rursus seorsim considerantur, partes ejus sunt. Itaque *totum et omnes partes*, simul sumptae, idem omnino sunt".[37] (OL I, 86) Then he goes on and concludes: "His intellectis manifestum est, totum nihil recte appellari, quod non intelligatur ex partibus componi, et in partes dividi posse; ideoque si quid negaverimus dividi posse, et habere partes, negamus idem esse totum."[38] (OL I, 86)

From there he denies the existence of an infinite number by referring to the fact that any mentioned number would always be finite. Therefore any talk of an infinite number could only mean something indefinite but no particular number:

> Numerus autem infinitus dicitur, qui quis sit non sit dictus; nam si dictus sit binarius, ternarius, millenarius, &c. semper finitus est; sed cum nihil sit dictum praeterquam *numerus est infinitus*, intelligendum est idem dictum esse ac si diceretur nomen hoc *numerus* esse nomen indefinitum.[39] (OL I, 87)

---

[36] "Who could ever convince himself, led by the senses, that no line can be *given* so short that it did not include infinitely many points, even infinitely many lines (and consequently actually infinitely many parts) being separated from each other) having a finite ratio to the given line – if it were not by constraining demonstrations."

[37] "And that which is so put for all the severals of which it consists, is called the whole; and those severals, when by the division of the whole they come again to be considered singly, are parts thereof; and therefore the *whole* and *all the parts taken together* are the same thing." (EW I, 97)

[38] "This being well understood, it is manifest, that nothing can rightly be called a whole, that is not conceived to be compounded of parts, and that it may be divided into parts; so that if we deny that a thing has parts, we deny the same to be a whole." (EW I, 98)

[39] "When we say number is infinite, we mean only that no number is expressed; for when we speak of the numbers two, three, a thousand, &c. they are always finite. But when no more

Another example of Hobbes' critique of an infinite number can be found in the first part of *De corpore* where he argues against Zeno who had claimed that a line capable of being divided into infinitely many parts would be itself infinite (Cf. OL I, 56–7; EW I, 63). Moreover, Hobbes claims and proves that nothing unlimited can be a whole, an argument often used by Leibniz even in his mature period (cf. Breger 1990a, 59). He then also proves all this for the smallest number, "non datur minimum" (OL I, 89; EW I, 100). Whereas Leibniz already held the latter position in the TMA, it is only in the *Accessio* that he argues against the largest number. Last but not least, Leibniz also agreed with Hobbes in reserving access to infinity of whatever kind exclusively to God alone (OL I, 335f.; EW I, 411f.). It is in fact more than probable that Leibniz studied these passages with great care. He obviously refers to Hobbes' nominalistic argument about the collective whole (without mentioning his name) while criticizing the nominalist Nizolius in his introductory *Dissertatio* (A II, 2, 430–31).

Leibniz gives his proof of the impossibility of an infinite number by reference to Euclid's axiom that the whole is greater than the part. After discussing the well known paradox of several infinite numbers (such as those of natural numbers, square numbers, cubic numbers etc.) being smaller or greater and thus in fact no longer the greatest number, he argues that the axiom that a whole is greater than a part would no longer be valid if an infinite number existed (A II, 1 (2006), 349). Then he points explicitly to Hobbes' demonstration of it: "At vero cum Hobbius, quod unum ego ab eo inprimis recte praeclareque factum arbitror, demonstraverit atque in numerum theorematum hoc axioma reduxerit totum esse majus parte".[40] (A II, 1 (2006), 350).

When discussing the *Accessio*, Hofmann, in spite of Leibniz's reference to Hobbes, does not mention Hobbes' proof at all. Rather, he tries anxiously to reduce or to avoid the influence of Hobbes as well as to exaggerate Leibniz's critique of Hobbes (following Couturat). Ironically, the footnotes added to the English edition of 1974 often speak a different language. Hofmann sees Leibniz starting with the principle that the whole is greater than the part, emphasizing its heuristic significance for Leibniz's approach to infinite series (Hofmann, 1974, 20). Then he mentions Hobbes in

is said but this, number is infinite, it is to be understood as if it were said, this number is an indefinite name." (EW I, 99)

40 "But Hobbes demonstrated this axiom according to which the whole is greater than the part and reduced it into the number of theorems, and I might judge that he did so in a correct and illuminating way."

passing, although in a misleading way: "Leibniz had taken a closer look at Euclid's axiom," "[p]ersuaded by reservations expressed by Hobbes" (Hofmann, 1974, 12). I read this passage to claim that Hobbes had had reservations against Euclid's axiom. But as we know, Leibniz learned about the significance of this axiom and its disputed validity in the case of the angles of contact as well as in the realm of infinity through his reading of Hobbes' *Opera philosophica.* This is evident from his mention of Hobbes' proof and the connected problems in his introductory *Dissertatio* to Nizolius, written before April 1670 (A II, 2, 432). But according to Hofmann, Leibniz assumed the demonstrability of Euclid's axiom because of the discussion about the quantity of the angle of contact, thus ignoring the mathematical impact of Hobbes.

Concerning the discussion of the angles of contact, I should like to add that Leibniz did not see Hobbes' "quite ingenious solution"[41] to the mathematical problem of the quantity of angles (after his death adopted even by Wallis (Jesseph, 1999, 172)). His interest in angles was again rather metaphysical. He saw angles as sections of a point bringing about the "partes indistantes" of the unextended but indivisible *point*: "doctrina de Angulis non est alia quam doctrina de quantitatibus puncti"[42] (A II, 1, 103), as he explains to Oldenburg in September 1670.

Likewise, it seems as if Hofmann is not at all aware that Leibniz adopted his general understanding of a demonstration as a mere chain of definitions from Hobbes, even though Leibniz himself mentions this quite often between 1666 and 1670 (A II, 1 (2006), 153; A VI, 1, N. 12). But it is precisely because of this understanding that Hobbes demands a demonstration even of axioms. Leibniz agrees with Hobbes' position (A II, 1 (2006), 281; A VI, 2, 480). When Hofmann complains that Leibniz's demonstration of Euclid's axiom was not acknowledged by Bernoulli and other mathematicians, "his contemporaries failed to see" its "effectiveness" (Hofmann, 1974, 14), I should like to add that Leibniz was certainly aware, even in 1696, that Hobbes would have appreciated it.[43]

Hofmann goes on to show how fruitful this proof was for Leibniz, in developing, at first, his main theorem on the summation of consecutive terms of a series of differences. Then Hofmann continues:

---

[41] Jesseph, 1999, 169; see also 168–172.

[42] "The doctrine of the angles is nothing else than the doctrine of the quantities of the point".

[43] Hofmann (Hofmann, 1974, 14) comments especially on Bernoulli's reaction in 1696 (cf. GM III, 329–30).

> Considerations of this sort led him to the conviction that we should be able to derive the sum of any series whose terms are formed by some rule, even where one has to deal with infinitely many terms – assuming only that the expected sum approaches a finite line. (Ibid., 14)

This credit might then be given to Hobbes, as well, from whom Leibniz had adopted precisely such logical "considerations of this sort". This is certainly clear in the case of Euclid's axiom whose heuristic role for Leibniz's development of the calculus is well known and emphasized by Hofmann,[44] Knobloch,[45] Breger,[46] Bassler,[47] and others. Leibniz's deep understanding of its crucial role in mathematics, especially for the mathematics of infinitesimals, can certainly be attributed to his studies of Hobbes's *Opera philosophica*.

But there is another principle which likewise provided great heuristic value for Leibniz in developing the calculus, i.e. the principle of continuity. I want to argue that he learned about this principle as well from Hobbes' "considerations of this sort". Of course, this principle was only explicitly formulated by Leibniz, but I am talking here about the use of this principle *avant la lettre*. Leibniz himself certainly undertook considerations in the spirit of this principle before he had formulated it. In a similar manner, Hobbes uses this kind of consideration of continuity, especially when he has to argue against intuition (he would rather say "imagination"), focusing on reasoning alone. I want to provide only a few examples. When he discusses the quantity of an angle between crooked lines in *De corpore* (II, 14, 9) he demands that

> anguli quantitas in minima a centro sive a concursu distantia aestimanda est; nam minima distantia (quia linea curva intelligi nulla potest, qua recta non sit minor) tanquam recta linea consideranda est.[48] (OL I, 161–62)

When explaining his conception of *conatus* Hobbes writes:

---

[44] See Hofmann, 1974, 13–15.

[45] Knobloch, although rightly stressing Leibniz's great debt to Galileo (and Nicholas of Cusa) (Knobloch, 1999, 91), also emphasizes the important *difference* regarding the infinite number as well as Leibniz's rejection of the infinite number because of Euclid's axiom (Knobloch, 1999, 94).

[46] See Breger, 1990a, 59.

[47] See Bassler 1999, 162.

[48] "[...] the quantity of the angle is to be taken in the least distance from the centre, or from their concurrence; for the least distance is to be considered as a strait line, seeing no crooked line can be imagined so little, but that there may be a less strait line." (EW I, 186)

> Quanquam autem hujusmodi conatus, perpetuo propagatus, non semper ita appareat sensibus tanquam esset motus aliquis; apparet tamen ut actio, sive mutationis alicujus efficiens causa. Nam si statuatur, exempli causa, ante oculos objectum aliquod valde exiguum, ut una arenula, quae quidem ad certam quandam distantiam sit visibilis; manifestum est eam removeri longius tanto posse ut sensum fugiat, nec tamen desinere agere in videndi organum, ut jam ostensum est, ex eo quod conatus omnis procedit in infinitum.[49] (OL I, 278–79)

As my last example I want to point to Hobbes' view of the relativity of hardness and softness of bodies. It is well known how important this will become for Leibniz and how this importance increases in his mature philosophy and physics. Hobbes, defining force as being "*impetum multiplicatum sive in se, sive in magnitudinem moventis, qua movens plus vel minus agit in corpus quod resistit*"[50] (OL I, 179) is then going to demonstrate

> *quod punctum quiescens, cui aliud punctum quantulocunque impetu usque ad contactum admovetur, ab eo impetu movebitur.* Nam si ab eo impetu a loco suo nihil *omnino removeatur, neque ab eo impetu duplicato removebitur, quia duplum nihil, nihil est*[51]. (OL I, 179)

Hobbes' use of such considerations in the spirit of the principle of continuity are of particular interest in regard to Leibniz's mathematical development because he uses them specifically for his understanding of indivisibles such as the point, *conatus* and the instant, all of which are inaccessible to imagination. The enormous significance of the principle of continuity for the development of the calculus is emphasized by Leibniz himself.[52]

---

49 "Now although endeavour thus perpetually propagated do not always appear to the senses as motion, yet it appears as action, or as the efficient cause of some mutation. For if there be placed before our eyes some very little object, as for example, a small grain of sand, which at a certain distance is visible; it is manifest that it may be removed to such a distance as not to be any longer seen, though by its action it still work upon the organs of us, as it is manifest from that which was last proved, that all endeavour proceeds infinitely." (EW I, 342)

50 "[...] the impetus or quickness of motion multiplied either in it itself, or into the magnitude of the movent, by means whereof the said movent works more or less upon the body that resists it" (EW I, 212)

51 "[...] that if a point moved come to touch another point which is at rest, how little soever the impetus or quickness of its motion be, it shall move that other point. For if by that impetus it do not at all move it out of its place, neither shall it move it with double the same impetus. For nothing doubled is still nothing". (EW I, 212)

52 Leibniz emphasizes the significance of the metaphysical principle of continuity for the understanding of the calculus (justifying it) as enabling us to go beyond the imagination of geometry: "prenant l'egalité pour un cas particulier de l'inegalité et de repos pour un cas particulier du movement, et le parallelisme pour un cas de la convergence etc. supposant non pas que la difference des grandeurs qui deviennent egales est déjà rien, mais qu'elle est

Finally, I want to draw some attention to Leibniz's enigmatic underlining of one single word in the very end of the 3rd article of the 1st chapter of *De corpore* in Boineburg's copy (*De corpore*, I, 1, 3) – of "considerare". Hobbes defines this quite common word as follows:

> Rem autem quamcumque addimus vel adimimus, id est, in rationes referimus, eam dicimur *considerare*, Graece λογίζεσθαι, sicut ipsum computare sive ratiocinari συλλογίζεσθαι nominant.[53] (OL I, 5)

This sentence might simply seem to be tacked on by Hobbes at the end of the discussion, standing as a single paragraph in the Latin (and earlier) version; he had explained his well known opinion that thinking is computing at full length before, in this and the foregoing article. What this single sentence adds besides the Greek words for the Latin "ratiocinari" and "computare" is nothing but the definition of "considerare" as reasoning which refers to ratios! Introducing this rather common word as a technical term makes perfect sense regarding Hobbes' copious application of it in his approach to "indivisibles", i.e. to the point, the conatus and the *instans*. According to this definition "considerare" means referring to a ratio or proportion of things! I take Leibniz's underlining of this single word as an expression of his consciousness of the crucial meaning of this term for Hobbes in dealing with the "indivisibles" point, *conatus*, and *instans*, all of which, according to Hobbes, have a quantity even though they are not *considered*, due to the extreme ratio between *incomparables*.[54] I feel supported in my claim of such an early awareness of the particularity of Hobbes' use of "considering" by Douglas Jesseph's paper in this volume who shows that Leibniz picked up exactly this expression and continued to use it as a mature mathematician when pressed to justify his calculus and to clarify the status of infinitesimals.

Leibniz indeed immediately adopted this Hobbesian formulation of the unconsidered quantity of the point, as can be seen in his definition of the point in the TMA (A II, 2, 265). Hobbes uses his formulation in order to deal with indivisibles, abstracting from the quantity and divisibility of a

dans l'acte d'evanouir, et de même du mouvement, qu'il n'est pas encore rien absolument, mais qu'il est sur le point de l'estre." (GM IV, 105)

53 "Now such things as we add or subtract, that is, which we put into an account, we are said to consider, in Greek λογίζεσθαι, in which language also συλλογίζεσθαι signifies to compute, reason, or reckon." (EW I, 5)

54 For an interesting discussion of Leibniz's use of the term "incomparable" see Ishiguru, 1990, 86–90.

body.[55] Leibniz keeps the divisibility of points (as does Hobbes), denies their extension, pace Hobbes, but then still attributes a quantity to them, although one which is not to be *considered*. He wants to place *conatus* as well as the activities of the mind, within unextended points (therefore containing parts), as thoughts which come together in a point. They are thus able to be compared in a moment and to produce a sensation. The confusion of these ideas in the *Fundamenta praedemonstrabilia* is in no way due to Hobbes but rather to Leibniz's metaphysical project which got in his mathematical way. What Leibniz had to abandon when he finally dedicated himself to mathematics in Paris was his metaphysical approach to indivisibles, trying to implement minds into mechanics. This is exactly what he did.[56]

Hobbes was certainly a stubborn loser in mathematics but he was nevertheless a very thoughtful philosopher of mathematics.[57] Nobody in the history of philosophy – and even less in the history of mathematics – was more aware of this than Leibniz, who wrote the following after having realized the mathematical failure of the Monster of Malmesbury:

> Desinam igitur, cum illud testatus fuero, et profiteri me passim apud amicos, et Deo dante etiam publice semper professurum, scriptorem me, qui Te et exactius et clarius et elegantius philosophatus sit, ne ipso quidem divini ingenii Cartesio demto, nosse ullum.[58] (A II, I (2006), 94)

If any harm came to Leibniz by studying a few mistaken quadratures of the circle, these studies also eventually made him study mathematics. Likewise, this harm was certainly compensated by Hobbes' philosophical introduction to the methodological significance of Euclid's axiom and to the value of considering continuities when it came to problems beyond imagination. Moreover, Leibniz became aware of the thorny problems with the infinite before he even studied the recent methods of indivisibles in Paris. Al-

---

55 Jesseph, 1999, 184.

56 Breger shows how Leibniz abandons indivisibles around 1672/3 (Breger, 1990a, 59). I should like to thank Daniel Burckhardt for providing me a copy of his Magisterarbeit (Technische Universität Berlin) on Leibniz's DQA which was very helpful for understanding the process of Leibniz's invention of the calculus. The advisor of this thesis was Eberhard Knobloch.

57 "Hobbes rightly pointed out the obscurity of infinitesimal mathematics, and although he did not have a fully developed alternative, his objections were not the ravings of a madman." (Jesseph, 1999, 188)

58 "And I shall always profess, both among friends and, God willing, also publicly (since I am myself a writer), that I know no one who philosophized more exactly, clearly, and elegantly than you, not even excepting that man of divine genius, Descartes himself."

though he was soon to enter the contemporary mathematical discussion then greatly dedicated to infinitely small quantities he seems to have been reluctant to adopt such contradictory ideas as "infinite quantities",[59] i.e. wholes without limits from the very beginning – as did Hobbes.

Appendix: Leibniz's Marginalia in Boineburg's Copies of Hobbes' *Opera Philosophica* (1668) and *De Corpore* (1655)

I found these marginalia in 1998 while exploring the books of the former Boineburg library to which Leibniz had full access during his years in Mainz. He even produced a catalogue of this library in the winter 1670/71 (Müller/Krönert, 1969, 21). Boineburg's son Philipp von Boineburg, while being a Statthalter of Erfurt on behalf of the Archbishop of Mainz, gave the library of his father to the then-founded University of Erfurt. When the university was closed by Napoleon in 1806, it was no longer used and more and more forgotten. (The Prussian State Library took some pieces out to Berlin.) After World War II it came to the city of Erfurt, as part of their special collections. It is now in the possession of the library of the newly founded Universität Erfurt. The Call number of Hobbes' *Opera philosophica* (1668) is UB Erfurt, Dep. Erf. 03-Pu 8°1430, the Call number of Hobbes' *De corpore* (1655) UB Erfurt, Dep. Erf. 03-Pu 8° 1432.

It was not by chance that I traveled to Erfurt. When I went first to the almost forgotten Boineburg books I was looking for Spinoza's *Tractatus theologico-politicus*. The suspicion that Leibniz had studied Spinoza's TTP seriously at the time of its publication came to my mind when I was work-

[59] That Leibniz as a mature mathematician and philosopher did not take infinitesimals to be real entities, but rather as finite quantities, was clarified as early as 1972 by Hidé Ishiguru. She pointed to Leibniz's statement in the *Theodicy*: "every number is finite and assignable, every line is also finite and assignable. Infinites and infinitely small only signify magnitudes which one can take as big or as small as one wishes, in order to show that the error is smaller than one that has been assigned" (§70). Obviously, the *Theodicy* is not a mathematical text book but certainly it is the only book Leibniz published during his life. Given the strategic character of this publication and the great care of all its formulations, it can be taken as Leibniz's Credo. See Ishiguru, 1990, 83.

THOMÆ HOBBES
Malmesburiensis
OPERA
PHILOSOPHICA,
Quæ Latinè scripsit,
OMNIA.
Antè quidem per partes, nunc autem, post cognitas omnium
Objectiones, conjunctim & accuratiùs Edita.

AMSTELODAMI,
Apud IOANNEM BLAEU,
MDCLXVIII.

ELEMENTORUM

Philosophiæ

SECTIO PRIMA

DE

CORPORE

Authore

THOMA HOBBES

Malmesburiensi.

LONDINI

Excusum sumptibus Andreæ Crook sub signo Draconis viridis in Cœmeterio B. Pauli, 1655.

ing on Leibniz's *Commentatiuncula de judice controversiarum*. This text seemed to me a detailed discussion of some particular arguments in Spinoza's TTP. But given the traditional resistance of Leibniz scholars against any Spinoza influence I was thinking of a more robust argument to make my point, as the presence of this book in Boineburg's library would be. I simply expected everybody to agree that Leibniz had not hesitated to read this book if at all available to him. The book was indeed there. Moreover, when opening the book I was overwhelmed by marginalia on the title page and the opposite one, continuing throughout the entire book. They stemmed from two hands. One of them was certainly that of Boineburg, according to his signature in the books. The other hand was very likely that of Leibniz. After consulting Heinrich Schepers from the Leibniz Akademieausgabe, we know now for sure. I published these marginalia in the TTP in 1996 (Goldenbaum, 1999).

Since then, I dreamed of going back to this library in order to do a more extended and systematic research in the Boineburg collection. I could manage to stay there again for two weeks in 1998 and then more often. I looked through almost all books which were mentioned by Leibniz during his years in Mainz and which are still present in this library, now in Erfurt. The surprising outcome of my research was the lack of marginalia in almost all books considered. What was frustrating at first glance was actually thrilling. There are some other books with a few underlined words – but only the books of Spinoza and Hobbes are so full of marginalia. This gives great evidence to the intense presence of these thinkers in Leibniz's mind in his years in Mainz. This is especially true for Hobbes, given the many books of his Leibniz studied.

I will give here the transcription of the marginalia found in Hobbes' *Opera philosophica* (1668) and in *De corpore* (1655) in order to support the thesis of my paper in this volume. I will publish other marginalia in my book on *Leibniz in Mainz* I am currently working on. What is interesting about Leibniz's marginalia in these two volumes is the fact that I did not find any in *De cive* and only a few in the *Leviathan*. The most of them can be found in *De homine*, in almost all chapters. In *De corpore*, they are concentrated in those chapters which deal with mathematics (ch. 20–22) and with sense perception (ch. 25–29). While he had read and critically appropriated Hobbes' political philosophy, logic and mechanical philosophy before, in 1663, 1666, and 1668, he started in the end of 1669 to study Hobbes' theory of sense perception, discovered the crucial role of *conatus*, and turned then again to *De corpore* and the mathematical writings in the *Opera philosophica* in order to study the geometrical-mechanical theory of *conatus* and the mathematics of indivisibles. The outcome of *this* Hobbes study was Leibniz's *Hypothesis physica nova*, especially the part of the *Theoria motus abstracti*.

I will not only give Leibniz' notes and comments in the margins or between the printed lines but also the underlined words. Being aware of the difficulty to determine the authorship of underlining, I feel however confident in almost all cases to recognize the "author." Boineburg writes mostly with a silver pen and always with the confident swing of the owner of the books. Leibniz's lines are quite different. He uses almost always a common pen, writing with ink, and he underlines careful and modestly. In addition, it is obvious that Boineburg did neither study Hobbes' theory of *conatus* nor mathematics. Also, I will keep the line breaks of Leibniz's comments in the margins and explain, where he wrote comments between the printed lines. The underlined words are also given in quotation marks if not entire paragraphs (which are rather easily to be found in current editions). I point even to one single double dogs ear which was clearly produced by intention.

I should like to thank the *Universitätsbibliothek Erfurt* for their permission to publish these pictures and marginalia, and Frau Dr. Kathrin Paasch and Herrn Thomas Bouillon for their great support during my stays at Erfurt exploring the library of Baron von Boineburg.

Thomae Hobbes, Malmesburiensis Opera philosophica, quae latinè scripsit, Omnia. Antè quidem per partes, nunc autem, post cognitas omnium Objectiones, conjunctim & accuratiùs Edita. Amsterdam: Blaeu 1668.

Signatur/Call Number: UB Erfurt, Dep. Erf. 03-Pu. 8° 1430

The title page has the signature of the first owner: "J.Chr.v. Boineburg"

De corpore

Cap. 1, article 3, p. 3
Underlined in line 19:
"considerare"

p. 6 [Boineburg's copy has a figure here which is not the case in all copies. Leibniz refers to it below.]

ſive Logica. Pars 1. 3

unam quadrati. Patet ergo quo modo animus concepta componit. Rurſus ſiquis hominem aſtantem videat, concipit totam ejus ideam, ſi verò recedentem ſequatur oculis tantùm, amittet ideam earum rerum quæ ſigna erant rationis, attamen idea animati oculis inhærebit, itaque ex tota idea *hominis*, id eſt, *corporis animati rationalis* ſubducitur idea *Rationalis*, reſidua eſt *Corporis Animati*; Deinde paulò pòſt in diſtantiâ majore amittetur idea *animati*, reſidebit idea *corporis* tantùm, tandemque cum præ diſtantia ampliùs conſpici non poteſt, tota idea ab oculis evaneſcit. Atque his exemplis qualis res ſit animi ratiocinatio interna ſine vocibus, ſatis exemplificatum eſſe puto.

Non ergo putandum eſt computationi, id eſt, ratiocinationi in numeris tantùm locum eſſe, tanquam homo à cæteris animantibus (quod cenſuiſſe narratur *Pythagoras*) ſola numerandi facultate diſtinctus eſſet, nam & magnitudo magnitudini, corpus corpori, motus motui, tempus tempori, gradus qualitatis gradui, actio actioni, conceptus conceptui, proportio proportioni, oratio orationi, nomen nomini (in quibus omne Philoſophiæ genus continetur) adjici adimique poteſt.

Rem autem quamcumque addimus vel adimimus, id eſt, in rationes referimus, eam dicimur *conſiderare*, Græcè λογίζεσθαι, ſicut ipſum computare ſive ratiocinari συλλογίζεσθαι nominant.

4. Effectus autem & Phænomena ſunt, corporum Facultates ſive potentiæ quibus alia ab aliis diſtinguimus, id eſt, unum alteri æquale aut inæquale, ſimile vel diſſimile eſſe concipimus; ut in ſuperiore exemplo, Poſtquam ad corpus aliquod ſatis appropinquatum eſt, ad percipiendum motum & inceſſum ejus, diſtinguimus ipſum ab arbore, à columna & ab aliis certis corporibus defixis, unde inceſſus ille, *proprietas* ejus eſt, quippe animalium proprius, quo à cœteris corporibus diſtinguitur.

5. Quomodo autem effectûs cognitio, ex cognita generatione acquiri poteſt, circuli exemplo facile intelligetur. Propoſitâ enim figurâ planâ, ad figuram circuli proximè accedente, ſenſu quidem circulus necne ſit cognoſci nullo modo poteſt; at ex cognita figuræ propoſitæ generatione, facillimè; facta enim ſit figura ea, ex cujuspiam corporis circumductione, cujus alter terminus maneat immotus; ſic ratiocinabimur, corpus circumductum eâdem ſemper longitudine applicat ſeſe primò ad unum radium, deinde ad alium, & ad tertium, quartum, & ſucceſſivè ad omnes; itaque ab eodem puncto, attingit circumferentiam eadem longitudo undequaque, id eſt omnes radii ſunt æquales. Cognoſcitur itaque à tali generatione procedere figuram à cujus puncto uno medio ad omnia extrema æqualibus radiis attingitur.

Similiter à cognita Figura perveniemus ratiocinando ad generationem

Bbb 2 ali-

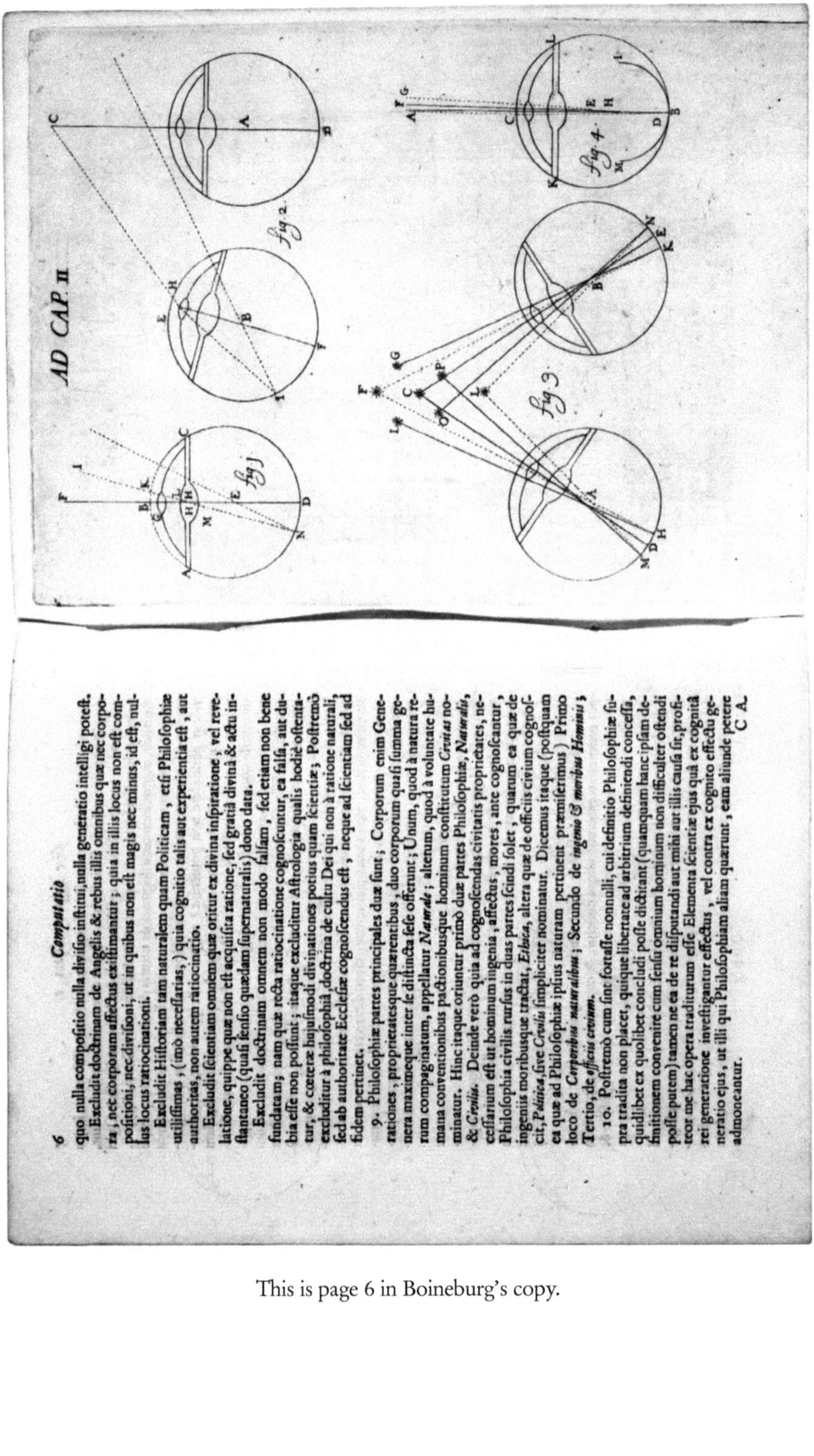

6 *Computatio*

quo nulla compoſitio nulla diviſio inſtitui, nulla generatio intelligi poteſt.

Excludit doctrinam de Angelis & rebus illis omnibus quæ nec corpora, nec corporum affectus exiſtimantur; quia in illis locus non eſt compoſitioni, nec diviſioni, ut in quibus non eſt magis nec minus, id eſt, nullus locus ratiocinationi.

Excludit Hiſtoriam tam naturalem quam Politicam, etſi Philoſophiæ utiliſſimas, (imò neceſſarias,) quia cognitio talis aut experientia eſt, aut authoritas, non autem ratiocinatio.

Excludit ſcientiam omnem quæ oritur ex divina inſpiratione, vel revelatione, quippe quæ non eſt acquiſita ratione, ſed gratiâ divinâ & actu inſtantaneo (quaſi ſenſio quædam ſupernaturalis) dono data.

Excludit doctrinam omnem non modo falſam, ſed etiam non bene fundatam; nam quæ recta ratiocinatione cognoſcuntur, ea falſa, aut dubia eſſe non poſſunt; itaque excluditur Aſtrologia qualis hodiè oſtentatur, & cœteræ hujuſmodi divinationes potius quam ſcientiæ; Poſtremò excluditur à philoſophiâ, doctrina de cultu Dei qui non à ratione naturali, ſed ab authoritate Eccleſiæ cognoſcendus eſt, neque ad ſcientiam ſed ad fidem pertinet.

9. Philoſophiæ partes principales duæ ſunt; Corporum enim Generationes, proprietateſque quærentibus, duo corporum quaſi ſumma genera maximeque inter ſe diſtincta ſeſe offerunt; Unum, quod à natura rerum compaginatum, appellatur *Naturale*; alterum, quod à voluntate humana conventionibus pactionibuſque hominum conſtitutum *Civitas* nominatur. Hinc itaque oriuntur primò duæ partes Philoſophiæ, *Naturalis*, & *Civilis*. Deinde verò quia ad cognoſcendas civitatis proprietates, neceſſarium eſt ut hominum ingenia, affectus, mores, ante cognoſcantur, Philoſophia civilis rurſus in duas partes ſcindi ſolet, quarum ea quæ de ingeniis moribuſque tractat, *Ethica*, altera quæ de officiis civium cognoſcit, *Politica*, ſive *Civilis* ſimpliciter nominatur. Dicemus itaque (poſtquam ea quæ ad Philoſophiæ ipſius naturam pertinent præmiſerimus) Primo loco de *Corporibus naturalibus*; Secundo de *ingenio & moribus Hominis*; Tertio, de *officiis civium*.

10. Poſtremò cum ſint fortaſſe nonnulli, cui definitio Philoſophiæ ſupra tradita non placet, quique libertate ad arbitrium definiendi conceſſa, quidlibet ex quolibet concludi poſſe dictitant (quamquam hanc ipſam definitionem convenire cum ſenſu omnium hominum non difficulter oſtendi poſſe putem) tamen ne ea de re diſputandi aut mihi aut illis cauſa ſit, profiteor me hac opera traditurum eſſe Elementa ſcientiæ ejus quâ ex cognitâ rei generatione inveſtigantur effectus, vel contra ex cognito effectu generatio ejus, ut illi qui Philoſophiam aliam quærunt, eam aliunde petere admoneantur.

C A.

This is page 6 in Boineburg's copy.

Cap. 2, article 10, p. 11
Underlined in line 20:
"intentio prima"

Underlined in line 21:
"posterior & secunda cura fuerit"

Cap. 3, article 20, p. 24
Underlined in line 12:
"Praemissas esse causas Conclusionis"

Cap. 4, article 7, p. 27
Underlined in line 17:
"legitimus fiet syllogismus,"
Leibniz comments in the margins: "Hinc Sturmiana."

Cap. XX, Prop. 1: De rationibus motuum, p. 147
Leibniz notes:
"vid. figur. ante pag. 159"
This refers to the figures 1 and 2 between pp. 158 and 159 [which are not bound at the same place in other copies of the Opera philosophica.

p. 150, Prop. 2
Leibniz comments in the margins:
"Hoc ostendum
est, quod omnes
sinus compositi ter-
minentur in re-
ctam XF."

Cap. XXI, article 3, p. 160
Leibniz comments in the margins:
"Paralogismus:
non succedunt,
nisi quae sunt
in circulo in quo
gyratur corpus
motu circulari
simplice."

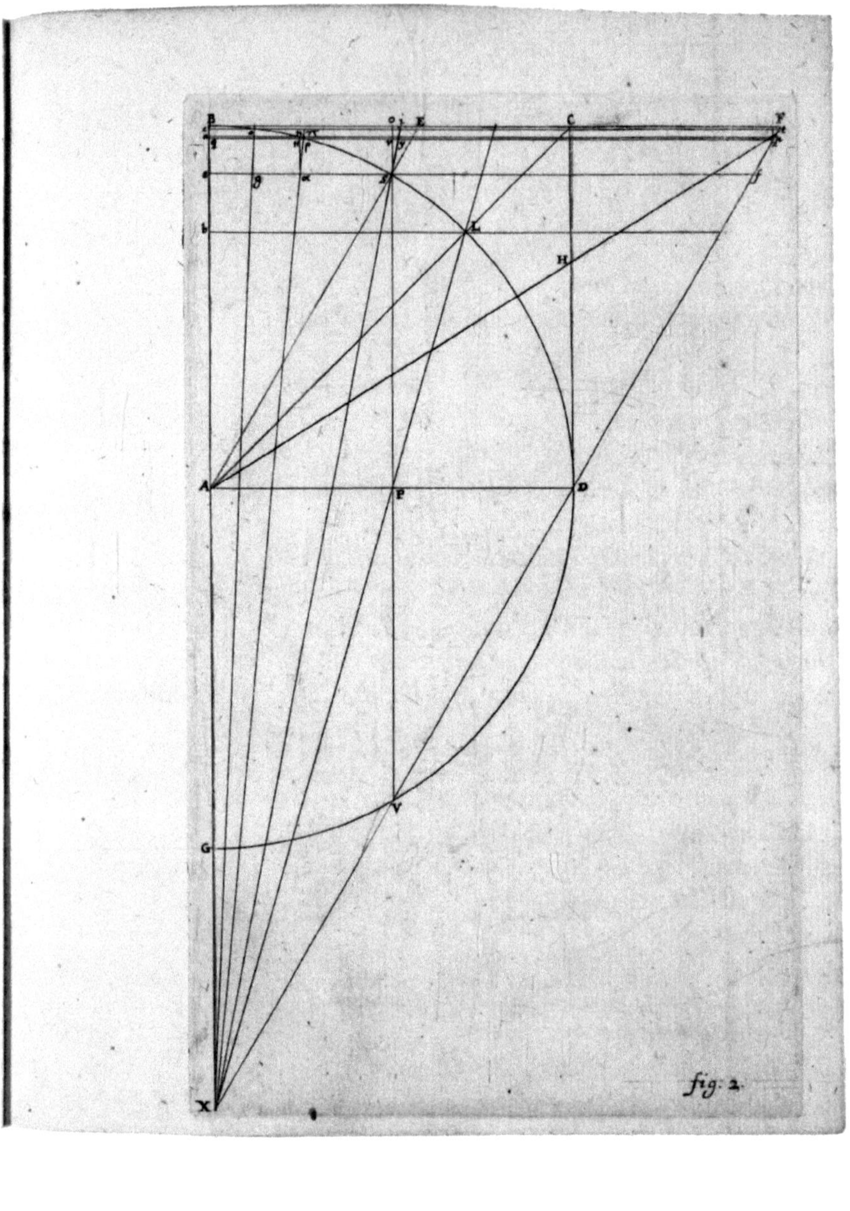

fig. 2

This is the figure before page 159 in Boineburg's copy.

Cap. 25, article 10 (in the end), p. 201
Underlined in line 10–11:
"si impressio fuisset levior, atque inde major fit Idea."

Cap. 26, article 9, p. 216
Underlined in line 1:
"rotabitur recta q r,"
Leibniz inserted in the printed line, after the underlined words:
"unde hoc?"

Cap. 29, article 18, p. 250
Leibniz comments in the margins of the three last lines of the chapter:
"at ipse suprà ne-
gavit punctum
esse, nisi
non expositum"
In addition, Leibniz drew a vertical line in the margin of these lines.

De homine

Cap. 2, article 1, p. 9,
At lines 15–17, Leibniz comments in the margins on Hobbes' reference to a figure:
"Fig. 1 vid. pag. ~~6~~ 6. de corpore."
See picture above.

Cap. 3, article 3, p. 17
Leibniz comments Hobbes' Q.E.D. within the free space of the printed line 9:
"eleganter"

Cap. 3, article 8, p. 20
Underlined in lines 36 and 37:
"ut in unam compactae multo apparerent majores quam aut luna, aut ipse sol,"

p. 21
Underlined in line 5:
"minus sit quam oculi pupilla"
Leibniz writes his comment in the free space of the last line of the 3rd chapter:
"elegantius ab apertura pupillae, cui toto somni tempore, in tenebris assuevimus."

Cap. 5, article 5, p. 32
Underlined in lines 34–37:
"(Quomodo autem à puncto dato recta linea ita duci possit ut ad aliud punctum reflectatur datum.) Problema solidum est, & fieri potest ope Hyperbole, sed ipsa Hyperbola non fit nisi per puncta, id est,Mechanicé."
In addition, Leibniz drew a double vertical line from line 36 to 37 (from "solidum est" to "Mechanice".

Cap. 7, article 2, p. 41
Leibniz drew a double vertical line from line 28 to 32:
"Qualis autem illa linea sit, difficile est determinare. Si GK, HL, IM essent omnes aequales perpendiculari EA, tunc quidem linea AM esset conchois vulgaris: nunc verò non est conchois, sed tamen quia, etsi in infinitum procederet atque ad lineam EI semper accederet, nunquam tamen illam attingeret, videtur ea inter species innumeras linearum conchoeideùn rectè numerari posse."

Cap. 8, article 9, p. 51
Underlined in the end of the chapter:
In line 24: "Hyperbolica"
In line 26: "Ellipticis"
In line 27: "ad comburendum"

Cap. 9, article 1, p. 52
Double vertical line in the margin from line 25 to 29:
"(Nam ut punctum valdè parvum discerni, id est, distincte videri possit, impossibile est, nisi omnes radii ab uno puncto ad unum quoque punctum refringi possent, id quod nulla figura earum quas hactenus Geometrae consideraverunt efficere potest.)"

Cap. 10, article 2, p. 59
Leibniz underlined in the 2nd paragraph those words in line 14 to 16 which are here given in quotation marks:
Nam "convenisse quondam in consilium Homines ut verba verborumque contextus quid significarent, Decreto statuerent, incredibile est."

Cap. 11, article 5, p. 64
Underlined in line 10:
"Pulchra"

Cap. 14, article 6, p. 79
Underlined in line 14:
"misericordem"
In addition, Leibniz commented in the margins:
"imò DEUS non po-
test intelligi misericors, si vera
est definitio misericordiae cap. 12.
num. 10."

Problemata physica

Cap. II, p. 13
Underlined in line 1:
"& ad litora Palaestinae valde"
Leibniz writes in the margins:
"Monconisius" [i.e. Balthasar Monconys, Journal des voyages, Lyon, Bd. 3, 1666].

Leibniz comments in the margins, at line 30–32:
"alicujus momenti
esset haec respon-
sio si aequalis esset
vis rejiciendi"

Cap. VII, p. 36
Underlined in line 15:
"absolutè vo[e]locitas eadem"

Propositiones XVI. De magnitudine circuli [Part of Problemata physica, extra page numbering]

p. 40,c
Underlined in line 5:
"per praecedentem"
Underlined in line 6:
"atque etiam"
Underlined in line 8:
"sive"

Leibniz writes in the free space of line 21 the following two lines:
"Non sequitur, sed hoc, similem esse in omnibus partium componentium rationem."

In addtion he comments to the same paragraph (line 16–22):
"Est paralogismus
non est recta haec
ad arcum ut
arcus ad arcum,
etsi ratio partium
recta inter se
~~ut~~ sit ut ratio
partium arcus
inter se."

Examinatio et emendatio mathematicae hodiernae, qualis explicatur in libris Johannis Wallisii … 6 dial.

Dialogue 1, p. 25
Underlined in line 32 and 33:
"idem Magnitudine corpus, locum modo majorem, modo minorem occupare possit."

p. 33
Underlined in line 1 and 2:
"vocem illam *numerum* non esse Numerum?"

De Principiis et Ratiocinatione Geometrarum, ubi ostenditur incertitudinem falsitatemque non minorem inesse scriptis eorum, quam scriptis Physicorum & Ethicorum, contra fastum Professorum Geometriae

Cap. XVIII, p. 37
In the margins Leibniz's notes (responding to Hobbes' mention of "Librum quem inscripsit Mesolabium"):
"Slusius." (Zeile 6)

Cap. XX: De dimensione circuli, p. 41–42
This sheet has a double dogs ear at the bottom corner.

p. 42
Leibniz inserts the sign "#" in line 22, between "recta b a." and "Eadem methodo". The sign is repeated in the margins where he comments:
"# ~~Restat~~ ostenda-
omnes sinus/
tur √ esse semper
minores ~~arcu~~
recta BS. et
omnes tangentes
semper ma-
jores. Seu arcum
BD neque esse majo-
rem neque minorem
quam recta BS."
The words "omnes sinus" *are meant to be inserted at the sign* "√".

Leibniz underlines two times in line 26:
"Minor autem esse non potest, cum locus nul-"
He inserts above these underlined words:
"hoc nondum demonstratum est."
And he comments to this in the margins:
"Leotaudus
hanc □ turam
omnium, quas
novit, praeci-
sissimam
ait, etsi falsam"
In addition he inserts between line 27 and 28, in the middle:
"sed horum computationi ipsemet non fia/ent [?]"

CAP. XXIII, p. 48
A vertical line is drawn in the margins from line 21 to 22 framing the words which are included in quotation marks:
proportio-
"nalis inter AB sive CD & ejusdem duas quintas. Secetur enim AD (quae aequalis est Radio) in quinque partes aequa-"
les

p. 49
Two times underlined in line 1:
"& edidit Josephus Scaliger"

p. 50
Underlined in line 2:
"sed dubitans nil pronuntio".
In addition a double vertical line from line 1 to line 2.

Elementorum Philosophiae Sectio Prima de Corpore, Authore Thoma Hobbes Malmesburiensi. Londini: Andrea Crook sub signo Draconis … 1655

Signatur/Call Number: UB Erfurt, Dep. Erf. 03-Pu. 8° 1432
The back of the title page has the signature of the first owner: "J.Chr.v. Boineburg"

Cap. 1, article 7, p.4–5
Vertical line from the beginning of article 7 on p. 4 until p. 5, including the words: "Harum ergo omnium utilitatum causa est Philosophia." (Silver pen, rather drawn by Boineburg)
There is however another shorter vertical line on p. 4, from line 40–42, emphasizing the words:
"corpora quam eorum motus; Movendi gravissima pondera; Aedificandi; Navigandi; instrumenta ad omnem usum"

p. 5
Underlined, rather by Boineburg:
In line 19: "Bellorum & Pacis causae ignorantur"
In line 22: "moralis philosophia."
In line 23 and 24: "a nemine clarâ & rectâ methodo hactenus tradita sit?"
In line 28: "officia sua"

p.6
Vertical line by ink at:
"personarum circumstantiis non minus saepe ad sceleratorum consiliorum confirmationem"

until
"Recti Regula aliqua & mensura certa constituta sit,"

Vertical line by silver pen:
"(quam hactenus nemo constituit) inutile est."
until
"quanta sit utilitas."

Article 8, p. 7
The words within quotation marks are underlined by ink:
de "cultu Dei qui non à ratione naturali, sed ab authoritate Ecclesiae cognoscendus" est

Art. 9, p.7
Underlined with silver pen
"appellatur Naturale" *until* "civitas nominatur Philosophiae, Naturalis, & Civilis."
quarum "ea quae de ingeniis moribusque tractat, *Ethica*, altera quae de officiis civium cognoscit, *Politica*, sive *Civilis* simpliciter nominatur."
"primo loco de corporibus naturalibus; secundo de ingenio & de moribus Hominis; Tertio, de officiis civium."

Art. 10, p. 7
Vertical line at:
"me hac opera Traditurum esse Elementa scientiae ejus quâ ex cognitâ rei generatione investigantur effectus, vel contra ex cognitu effectu generatio ejus, ut illi qui Philosophiam aliam quaerunt, eam aliunde petere admoneantur."

Cap. 14, article 19, p. 118
In line 30, Leibniz inserted a "+" *between* "duae lineae quaelibet," *and* "aut parallelae sunt"
He adds in the margins:
"in eodem plano"

Cap. 21, article 3, p. 184
Underlined in line 38:
"minima"
And in the margins of line 38:
"ἄτομον"

Cap. 22, article 6, p. 193
Underlined in line 14 of the article:
"(Cap. 16. Art. 8.)"
Leibniz comments in the margins:
"falsum ibi"

Article 8, p. 195
Underlined in line 7:
"(ut artic. 6. ostensum est"
Leibniz comments in the margins:
"erratum ibi,
movebitur potius
per arcum ----[*stroken*]
a b versus g con-
tra a et ita
servatur expe-
rientia de plau-
stello."

p. 196
Underlined in line 6:
"Velocitas"
Underlined in line 8:
"EA ad DA"
Leibniz comments in the margins:
"hac non conce-
do. Aliud enim
velocitas aliud
vis ictus. velo-
citas est in
tota linea, vis
est impetus in momento im-
pactus, qui solus efficiens[?]"

p. 304 (last page), last paragraph
Underlined with a silver pen:
"Transeo nunc ad Phaenomena corporis Humani. Ubi de [hidden by binding]rica, item de Ingeniorum, Affectuum, Morumque; human [hidden by binding] (Deo vitam tantis per largiente) causas ostendemus."
Below the last sentence are written a few words by Leibniz:

This is a table I found on the last page in Boineburg's copy of Seth Ward's *Thomae Hobbii philosophiam exercitatio epistolica* (Oxoniae: Richard Davis 1656), now in the Boineburg collection of the *Universitätsbibliothek Erfurt* (Call number: UB Erfurt, Dep. Erf. 03-Pu. 8° 1434). After buying the book Boineburg wrote his name on the title page in order to mark his ownership and added the date of its purchase – 1669. I should like to thank the *Universitätsbibliothek Erfurt* for their permission to publish these marginalia, and Frau Dr. Kathrin Paasch and Herrn Thomas Bouillon for their great support during my stays at Erfurt exploring the library of Baron von Boineburg.

"Est forte Civis eius."
In the same line slightly higher:
"Utinam iam prodissent: prodirentur"
Then indented below:
"Corpus politicum"
"Leviathan"
"Principia Justi ac decoris[?]"

SIEGMUND PROBST

Indivisibles and Infinitesimals in Early Mathematical Texts of Leibniz[1]

The main purpose of this article is to present new material concerning Leibniz's use of indivisibles and infinitesimals in his early mathematical texts. Most of these texts are contained in hitherto unpublished manuscripts and are soon to be printed in volume VII, 4 of the Academy Edition.[2] They present examples which illustrate how Leibniz operated with concepts such as indivisibles and infinitesimals in that period of his development.[3]

It does not need to be stressed that the employment of the term "indivisible" by Leibniz and his contemporaries does not in itself imply that they understood this term in a Cavalierian sense. Already among Cavalieri's disciples we find the term "indivisible" being used in such a way as to mean infinitely small parts of the same dimension as the whole.[4] This later became standard practice among most contemporary mathematicians.

---

1 I should like to thank Ursula Goldenbaum, Walter S. Contro and Eberhard Knobloch for permission to use unpublished material they are preparing for print, Herbert Breger and Staffan Rodhe (Uppsala) for critical remarks on an earlier version, and Philip Beeley for his assistance in producing the present English translation from the original German draft.

2 All manuscripts to be published in the forthcoming volume A VII, 4 of the Akademie-Ausgabe, edited by Walter S. Contro and Eberhard Knobloch, will be indicated in the text by their numbers in A VII, 4, and by Cc 2 numbers. A VII, 4 will contain about 800 pages in print of Leibniz's papers on infinitesimal mathematics, nearly all from the year 1673, and will offer the opportunity for a more detailed analysis than that presented here. Some of these texts have been studied by Gerhardt, 1891, Child, 1920, Mahnke, 1926, and by Pasini, 1986 and 1993; see also Eberhard Knobloch's paper in this volume.

3 The terminology of Leibniz varies: in most cases he uses "smaller than any assignable" ("minor assignabili" or "inassignabilis") which occurs nearly one hundred times (for a detailed analysis of Leibniz's use of "inassignabilis" see the paper of Eberhard Knobloch in this volume); after this comes "infinitely small" (about eighty times). In contrast, "indivisible" and "infinitesimal" ("infinitesima") are relatively rare, occurring about twenty-five and ten times respectively.

4 See Giusti, 1980, 47–49.

In 1671 Leibniz was convinced that he had given a foundation for the theory of indivisibles in the *Theoria motus abstracti* as is evident from the preface to that work and several letters from the time before his sojourn at Paris.[5] The importance of this achievement for Leibniz can be understood in view of the high esteem he had for the theory of indivisibles in mathematics which he regarded as a source of inventions and demonstrations.[6]

Since up to now no genuine mathematical manuscripts of Leibniz concerning infinitesimals from this time have been known, it has not been possible to ascertain to what extent his theoretical evaluation of the theory of indivisibles was based on mathematical practice.[7]

Only very recently has a manuscript long believed to belong to the Parisian years been able to be dated to around this period (1670/71) and as a consequence it can now be considered as the earliest known mathematical text by Leibniz on indivisibles (Cc 2, N. 817; A VII, 4 N. 4). The jottings concerned are written on paper which is identical with that which was originally used for work on the *Corpus juris reconcinnatum* (A VI, 2, XXIf.). The contents themselves also support the earlier dating of the text. Leibniz starts with a short remark on his geometrical instrument.[8] Another topic is the apparatus for grinding lenses, especially hyperbolic lenses, which we know played an important role in his correspondence from autumn 1670 onwards.[9] In addition, the mathematical passage also provides clear pointers to the text having been written earlier than has previously been thought. This short text reveals Leibniz's confidence at the time of being able to solve all problems concerning curves by the theory of indivisibles. He declares explicitly that the investigation of the hyperbola and all other curves can be carried out easily with the help of indivisibles.

Leibniz starts with the example of the hyperbola and considers thereby a right triangle [ABC] in which one of the legs is the base and the other the altitude. By rotating the triangle around the altitude [AB] a (finite right) cone is generated.[10] The altitude is divided into an arbitrary number of parts that represent the indivisibles: "altitudinem divide in partes quotcun-

---

5 See A VI, 2, 262; for further references see note 21 in Philip Beeley's paper in this volume.

6 "Geometria indivisibilium, id est, fons inventionum ac demonstrationum" (A II, 1, (1926), 172; (2006), 278).

7 For Leibniz's early studies in mathematics see Hofmann, 1974, 1–11.

8 This "geometrical" instrument is mentioned alongside the "arithmetical" calculating machine in Leibniz's letters to Duke Johann Friedrich of October 1671 (A II, 1, (1926), 160f.; (2006), 262), and to Antoine Arnauld of November 1671 (A II, 1, (1926), 180; (2006), 286).

9 Cf. A II, 1, N. 34, 38, 43, 46, 47, 57, 69, 80, 86, 89, 91, 99.

10 Cf. A VI, 2 N. 38_5, § 21 [bis], 184, and Euclid, *Elements*, book XI, def. 18.

que, hae sint indivisibilium seu punctorum loco". The cone is then deemed to consist of as many circles parallel to the base as there are points in the altitude.

In order to construct the hyperbola, Leibniz takes an arbitrary point [D] on the surface of the cone between base and vertex and draws the perpendicular [DE] to the base. The perpendicular line [DE] is the altitude of the future hyperbola; the chord [FG] which – passing through the perpendicular foot – intersects the diameter [CH] of the base circle at right angles, is the base of the hyperbola.

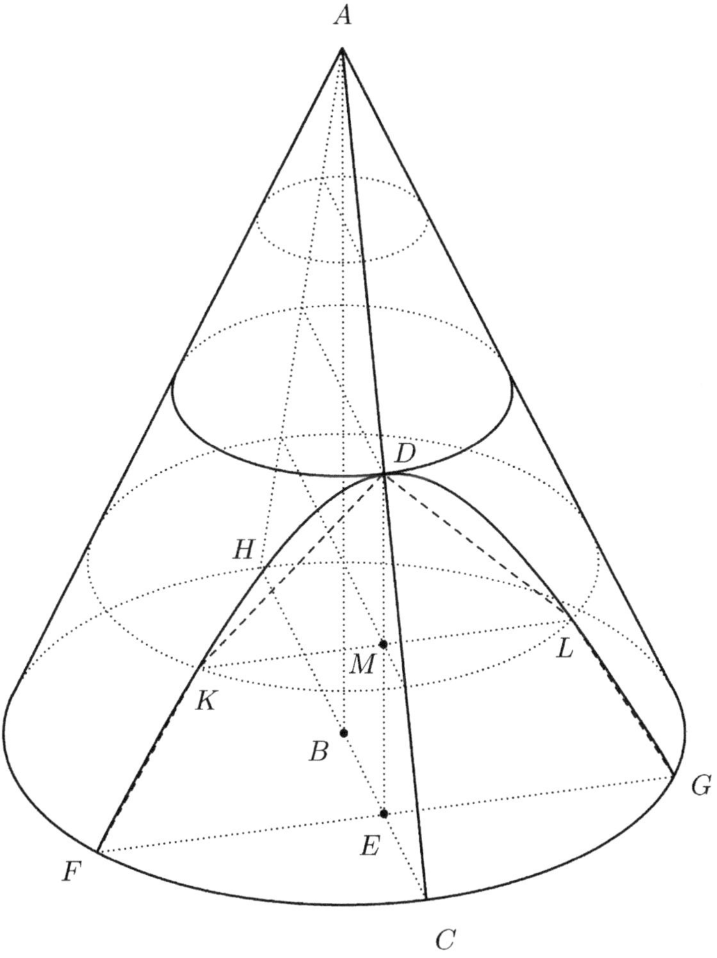

Leibniz does not provide any figure. This figure is added by me, on the basis of Leibniz's explanation. – S.P.

Similar chords [e.g. KL] can be drawn in all the parallel circles making up the cone (below [D]), and when the endpoints of these chords are connected they constitute the hyperbolic line. In the case of the number of these circles being finite, the hyperbola is constructed not in a geometrically exact manner, but rather "mechanically" i.e. pointwise and represented by a broken straight line: "mechanice describetur per puncta seu rectam fractam" [e.g. FKDLG].[11] Although not explicitly stated by Leibniz, it is clear that the inscribed polygon approaches the hyperbolic segment when the subdivision is refined.

Leibniz immediately states that the line elements of the hyperbola are to the line elements of the altitude ("minimum hyperbolicum ad minimum rectae") as the straight lines connecting the endpoints of the chords are to the parts of the altitude connecting the centers of the chords [i.e. as DL to DM, LG to ME]. But as this ratio is not constant but varying and cannot be determined generally ("in universum neque numeris neque lineis exhiberi potest"), he concludes that an exact quadrature of the hyperbola is impossible. In the final sentence of the passage, he announces his intention of carrying out an investigation in the near future on the question of whether or not the quadrature of the parabola is possible.

In the final passages of the text Leibniz expresses confidence that this investigation will help also in deciding the question of the usefulness of hyperbolic lenses in dioptrics, which had been doubted by Hobbes,[12] and he

---

11 In 1998, Ursula Goldenbaum discovered that the copy of Hobbes' *Opera philosophica*, 1668, formerly in possession of Johann Christian von Boineburg, contains marginal notes by Leibniz (now at University Library Erfurt, call number Pu 1430). Cf. these marginalia in this volume, published by Goldenbaum. In Boineburg's copy of Hobbes' *Opera* (1668), in *De homine*, Cap. 5, § 5, 32 (= OL, II, 46), the following sentence is underlined by Leibniz: "(Quomodo autem à puncto dato recta linea ita duci possit ut ad aliud punctum reflectatur datum.) Problema solidum est, & fieri potest ope Hyperbole, sed ipsa Hyperbola non fit nisi per puncta, id est, Mechanicè." (By friendly permission of Ursula Goldenbaum.) – It is possible that this peculiar construction of the hyperbola is inspired by a proposition of John Wallis, freely quoted by Hobbes in his discussion of Wallis' use of indivisibles and also printed in the *Opera philosophica* of 1668: "Planum coni sectionem efficiens, si unum ex parallelis in cono circulis secet secundum rectam ipsius diametro perpendicularem, etiam reliquos illi parallelos circulos secabit secundum rectas, quae ipsorum Diametris parallelis sunt perpendiculares" (*Examinatio et emendatio mathematicae hodiernae*, Hobbes, 1668, 111 = OL IV, 175). The original source is Wallis, 1655, prop. 7, 17 = Wallis, 1695, 304.

12 In the corresponding paragraph of *De homine* in the Boineburg Copy used by Leibniz (Hobbes, 1668, Cap. 8, § 9, 51 = LW, II, 78) some words are underlined. Interestingly, Leibniz does not touch on this in his letter to Hobbes of July 1670, but he does in his letter to Spinoza of October 1671 (A II, 1, N. 25 and N. 80).

declares also that several kinds of lenses can be produced using a single adjustable instrument. Later on, however, Leibniz dismissed the whole text as youthful nonsense, appending the expression "nugae pueriles".

It does not need to be pointed out that the mathematical deliberations which Leibniz conducts here were not sufficient in order to carry out a quadrature or for that matter to prove the impossibility of a quadrature. From the wording of the text it is not even possible to reconstruct to any reasonable degree of certainty the method which Leibniz wanted to employ. Even if the reduction of the cone into parallel circular surfaces ("Constabit conus ex tot circulis basi parallelis, quot sunt puncta altitudinis") reminds us initially of the method of Cavalieri, the relation which he establishes between the line elements of the hyperbola and the line elements of the altitude imply rather an infinitesimal consideration.

For such an attempt at a quadrature it would be sufficient to sum the areas of the parallelogramms consisting of the chords and the (equal) elements of height. The length of the elements of the arc of the hyperbola would not be necessary for this. A meaningful application could consist in considering the limits of the infinitesimals to be trapezoids. The parallelograms would then be supplemented on both sides by right-angled triangles and the area under the curve would thus be approached to a higher degree of accuracy – in the finite case – than by means of parallelograms alone. (Strictly speaking the elements of the arc would also not be required for this, since these triangles are already determined by the element of height and by the differences of the chords.) Incidentally, Leibniz carries out such a move from infinitesimal parallelograms to trapezoids in *De functionibus* (Cc 2, N. 575; A VII, 4, N. 40).

A possible source of the two kinds of infinitesimal reduction could have been Thomas Hobbes' critique of John Wallis' approach to quadratures, in which he points out that, for example, triangles are not composed of parallelograms but rather of trapezoids: "Neque enim trianguli constant ex parallelogrammis, sed ex Trapeziis."[13] But how is the final sentence of the mathematical passage to be understood? Was Leibniz still unaware that the quadrature of the parabola had already been carried out by Archimedes or was it for him a question of solving the problem by a new method?

In fact, it is unlikely that Leibniz had studied Archimedes' writings by that time. Therefore other sources must be considered. It is possible, for instance, that he knew the result from Bonaventura Cavalieri's *Geometria in-*

---

[13] *Examinatio et emendatio mathematicae hodiernae*, Hobbes, 1668, 110 = OL, IV, 174.

divisibilibus continuorum, although later on he recalled that he had only consulted the book a few times.[14] There is at least one other possibility: Hobbes mentions the quadrature of the parabola by Archimedes in *De Corpore*.[15]

Leibniz's interest in arc elements could be interpreted first of all as part of an attempt to provide a solution to the hitherto unsolved problem of the rectification of the hyperbola through their summation. The rectification of the parabola by means of its reduction to the quadrature of the parabola was quite a new result at the time and was probably still unknown to Leibniz.[16] Opposed to this is however the wording of the text. Leibniz employs explicitly the term "squaring" (quadrari) on both occasions.

It is noteworthy that in this manuscript Leibniz combines two approaches: a finite approach leading to an approximation, and an infinite approach aimed at obtaining a geometrically exact determination.

In fact, the latter argument combines in a rudimentary way aspects of three different methods: Cavalieri's method of indivisibles, the method of exhaustion, and the method of infinitesimals. Moreover, it displays a certain similarity to the procedure adopted by Pascal which Leibniz would hold in high esteem later on.[17]

Although Leibniz modified his view concerning indivisibles in 1672,[18] his attitude towards the mathematical use of indivisibles seems to have remained unchanged, as is indicated by his letter to Jean Gallois of December 1672 where he ranks the method of indivisibles among the things that vindicate the incorporeality of the human mind by referring to the works of Archimedes, Cavalieri, Galilei, Wallis, and J. Gregory.[19] Obviously he does not separate strictly the method of exhaustion of the ancients from the

---

14 Cavalieri, 1653, book IV, theorema 1, 1–3. Leibniz also remarks that he had been delighted by the method he had found in Cavalieri. See *Historia et origo calculi differentialis*, GM V, 398; further references in Hofmann, 1974, 5 n. 26.

15 Hobbes, 1668, 155 = OL I, 254. There are marginal notes by Leibniz in the copy of Boineburg, on pp. 147, 150, 160, and a reference to the table between pp. 158 and 159.

16 Solutions of the rectification by Neil, van Heuraet, and Fermat had been published in 1659 and 1660, as well as by James Gregory in 1668; see Hofmann, 1974, 101–117; Leibniz possibly knew about the failed attempt of the rectification of the parabola in chapter 18 of Thomas Hobbes' *De corpore*.

17 See the paper of Herbert Breger in this volume.

18 See A VI, 3, N. 5, and Leibniz, 2001, 8–19.

19 "Constat Scientiam Minimi et Maximi, seu Indivisibilis et Infiniti, inter maxima documenta esse, quibus Mens humana sibi vendicat incorporalitatem" (A II, 1, (1926), 222f.; (2006), 342).

method of indivisibles (and infinitesimals).[20] But in mathematical texts of the same period he repeatedly addresses the limitations and shortcomings of the method of indivisibles even when he defends it by stating that the arithmetic of infinites and the geometry of indivisibles do not lead to error any more often than surd roots, imaginary dimensions, and negative numbers do.[21] By this time, however, he had a new preference, namely the method of differences with which he expected to produce all the results hitherto achieved by the geometry of indivisibles and a few more besides. Prominent among these new possibilities was the rectification of curves which in his opinion was impossible to achieve using the method of indivisibles.[22] This claim (which is true in respect of the method of Cavalieri) shows that Leibniz was not yet acquainted with the rectification of several curves by recent methods which were in fact infinitesimal.[23]

In autumn 1672 he had been successful in using the method of differences for summing the series of reciprocal figurate numbers (Probst, 2006a, 164–173), but soon it turned out that he was not yet able to achieve similar results with series whose terms were not discrete numbers but continuous magnitudes like the ordinates of a curve.[24]

By April 1673 a new method had attracted his interest when Christiaan Huygens published his *Horologium oscillatorium*. Leibniz received a personal copy "ex dono authoris" as he recorded on the titlepage.[25] The results obtained by evolutes of curves impressed him deeply and he immediately studied the work as well as van Heuraet's *Epistola de transmutatione* (1659). Although a first attempt to produce new results of his own in this field failed (Cc 2, N. 609; A VII, 4, N. 7), he remained optimistic and noted that the method of exhaustion and the method of indivisibles were equally sub-

20 This feature, criticized by J. E. Hofmann (Hofmann, 1974, 7), possibly has its roots in Hobbes who attributes the use of indivisibles to Archimedes (Hobbes, 1668, 156 = OL I, 254).

21 "Sed arithmetica infinitorum et geometria indivisibilium, non magis fallunt quam radices surdae et dimensiones imaginariae et numeri nihilo minores" (A VII, 3, N. 6, 69).

22 "Hac methodo ea omnia possunt demonstrari, quae hactenus per geometriam indivisibilium, et nonnulla ampliora. Non enim possunt exhiberi curvae rectis aequales per geometriam indivisibilium, at hac methodo exhiberi possunt tales infinitae" (A VII, 3, N. 8, 126).

23 See for example Pascal, 1659, or Wallis, 1659.

24 See for example LH XXXV, II, 1, Fol. 299–300; Cc 2, N. 547; A VII, 4, N. 16_3, where he investigates the quadrature of the logarithmic curve. See also Pasini, 1993, 56f.

25 Hannover, Gottfried-Wilhelm-Leibniz-Bibliothek, Leibn. Marg. 70; publication of the marginal notes forthcoming in A VII, 4, N. 2.

ject to errors.[26] Other attempts, this time to solve the quadrature of the circle, based on a proposition which Leibniz found in Honoré Fabri's *Synopsis geometrica* (1669), contained an error of reasoning, as Leibniz subsequently detected.[27] Later in spring of the same year, Leibniz extended his studies to Blaise Pascal's *Lettres de Dettonville* (1659), and probably as a result he now returned to the explicit use of the term "indivisibles" (Cc 2, N. 544; A VII, 4, N. 10),[28] and introduced the concept of an indivisible unit:

> Nota quemadmodum in aequationibus Geometriae quando comparantur lineae cum superficiebus, vel superficies cum solidis, vel lineae cum solidis, necesse est dari unitatem (unde in numeris aequationes inter dimensiones diversorum graduum libere admittuntur), ita in Geometria indivisibilium, cum dicitur summam linearum aequari cuidam superficiei, vel summam superficierum cuidam solido, necesse est dari unitatem, dari scilicet lineam quandam cui applicatae intelligantur, seu in cuius partium infinitarum aequalium, unam, quae *unitatem* exhibet, ducantur, ut infinitae inde fiant superficies, etsi qualibet data minores.[29]

Clearly, Leibniz follows here in the footsteps of Pascal, as Mahnke and Pasini have indicated,[30] but there is an important difference: Pascal generally defends his method of indivisibles on account of its being in accordance with pure geometry. In order to preserve the dimension e.g. of a required area, he multiplies the ordinates of a curve into infinitely small parts of the axis and adds the resulting rectangles to get the area between the axis and

---

26 "NB. Ideo calculus per polygona aeque obnoxius erroribus, <id>eo calculus per indivisibilia" (A VII, 3, N. 16, 199). Further remarks concerning the problems of the method of indivisibles occur for example in A VII, 3, 227: "Ergo valde cavendum ne indivisibilibus abutamur" and in Cc 2, N. 547; A VII, 4 N. 16_2.

27 See A VII, 1, 63–66; A VII, 3, 225–227; Cc 2, N 500; A VII, 4, N. 8. An edition of Leibniz's marginal notes in his exemplar of the *Synopsis Geometrica*, Lyon, 1669 (Hannover, Gottfried-Wilhelm-Leibniz-Bibliothek, Leibn. Marg. 7, 1), is forthcoming in A VII, 4, N. 1.

28 Cc 2, N. 544, is partly printed in Gerhardt, 1891; engl. translation in Child, 223–227.

29 "Note: in the same way as it is necessary in equations in geometry, when lines are compared with surfaces or surfaces with solids or lines with solids, that a unity is given (whence in numbers equations between dimensions of different degrees are freely admitted), so it is necessary in the geometry of indivisibles, when it is said that that the sum of lines is equal to some surface or the sum of surfaces to some solid, that a unity be given, that some line is given, of course, as whose applicates they are understood, or that they are multiplied into one of the infinitely many equal parts of that line each of which denotes the *unity*, so that infinitely many surfaces are generated, though they are smaller than any given surface." (Quoted with Italian translation in Pasini, 1993, 53.)

30 See Pascal, 1659, (first pagination) 10–12 = Pascal, 1904–1914, VIII, 351–355; cf. Mahnke, 1926, 32, and Pasini 1993, 54.

the curve. The same can be done in higher dimensions by multiplying the ordinates into infinitely small squares or cubes and even n-dimensional cubes with $n>3$. Leibniz agrees with Pascal but tries to proceed a step further by explicitly introducing the n-dimensional cubes as n-dimensional units and thus transforms the procedure into an arithmetical approach.

Shortly afterwards Leibniz tried to apply the method to squaring the circle, but he met with as many difficulties as he had in his earlier attempts. Nevertheless, he hoped to overcome these difficulties by a systematic theoretical approach:

> Unde apparet quam necessaria sit ista profundior contemplatio indivisibilium atque infiniti, sine qua occurrentibus in infiniti atque indivisibilium doctrina difficultatibus occurri non potest. Nota: *Indivisibilia* definienda sunt infinite parva, seu quorum ratio ad quantitatem sensibilem [...] infinita est.[31] (Cc 2, N. 546; A VII, 4, N. 16_1)

Leibniz seems to have coined the term "infinitesimal" in late spring 1673 and he uses it most more frequently in the summer of that year.[32] Apparently the term is a fruit of his study of Nicholas Mercator's *Logarithmotechnia* (1668), as he was to recall more than thirty years later.[33] However there is an interesting difference in this respect between Mercator and Leibniz: the former does not use the term "infinitesimal", but instead "pars infinitissima" and he does so both for numbers and for lines (Mercator, 1668, 30–34). Mercator's expression signifies a minimal quantity and is therefore terminologically still close to Cavalieri's indivisibles, although he in fact employs infinitesimal quantities. By switching to the term "infinitesima", which effectively paraphrases Wallis' symbolic expression $\frac{1}{\infty}$ (Wallis, 1655, prop. 1, 4 = Wallis, 1695, 297), Leibniz restores agreement between terminology and usage.

The study of Pascal's works finally led Leibniz to the discovery of the characteristic triangle and this concept proved to be of eminent importance for his future results, including his method of transmutation and the

---

31 "From whence it appears how necessary that more profound contemplation of the indivisibles and the infinite is; for without this it is impossible to cope with the difficulties that occur in the doctrine of the infinite and the indivisibles. Note: *Indivisibles* are to be defined as infinitely small, or whose ratio to a sensible quantity [...] is infinite."(Partly printed in Pasini, 1986, App. N. 5, fol. 9–14, quotation fol. 12f.)

32 See for example Cc 2, N. 546, 547, 695, 697, 696, 612, 638, 575, 614; A VII, 4, N. 16, 22, 26, 27, 34, 38, 40, 44 in chronological order.

33 See Leibniz to Wallis, March 30/[April 9], 1699 (GM IV, 63), quoted in note 38 of Philip Beeley's paper in this volume.

arithmetical quadrature of the circle achieved in autumn 1673 (Probst, 2006b).

At the beginning of this development, Leibniz pursued a twofold approach, creating two lists of propositions based on the properties of similar right triangles. The first of these lists he derived from finite triangles whose sides resulted from various constructions in a circle (*Catalogus propositionum, quibus ductus curvilineorum ex circulo natorum, comparantur*, Cc 2, N. 697; A VII, 4, N. 26), while the second list derived from the comparison of an infinitely small right triangle whose hypotenuse is an infinitely small part of the arc of the circle with similar finite triangles (*Trigonometria inassignabilium*, Cc 2, N. 696; A VII, 4, N. 27). Leibniz explicitly referred to the second list when he described his method of transmutation in summer 1673:

> Tota res nititur triangulo quodam orthogonio laterum infinite parvorum, quod a me appellari solet *characteristicum*, cui alia communia, laterum assignabilium, similia, ex proprietate figurae constituantur. Ea porro triangula similia characteristico comparata, exhibent propositiones multas, pro tractabilitate figurae, quibus diversi generis curvae inter se comparantur. *Pauca* sunt, quae ex hoc *triangulo characteristico* non deducantur.[34] (*Fines geometriae*, Cc 2, N. 552; A VII, 4, N. 36)

Although Leibniz does not use the term "indivisibles" in the two lists, but rather speaks of "infinitesimals", "infinitely small parts" and "parts smaller than any given part", shortly afterwards, in *Triangulum characteristicum, speciatim de trochoidibus et cycloide*, he identifies his method with the method of indivisibles:

> *Analysis indivisibilium* (quatenus ab arithmetica infinitorum separatur) in eo consistit maxime, ut data qualibet linea curva, aut superficie curva, eam ad spatium quoddam unum plurave reducamus, a quorum quadratura eius mensura pendeat. Quod per varias methodos hic praescriptas facile fiet: Porro ut spatium datum quadremus, examinanda primum ratio progressionis, an sit summae capax ex arithmetica infinitorum. Si hanc methodum respicit, ad analysin indivisibilium veniendum est, id est constituendum triangulum characteristicum

34 "The whole thing is based on some right triangle with infinitely small sides, usually called by me *characteristic*, in relation to which other common triangles with given [finite] sides, that are similar to it, are constituted from the qualities of the figure. Furthermore, these similar triangles compared with the characteristic triangle produce many propositions which are dependent on the tractability of the figure; with help of these propositions curves of different kinds can be compared to each other. There are *few* that cannot be deduced from this *characteristic triangle.*"

figurae, eique quotcunque fieri potest triangula similia, quod fieri potest tum ductibus rectarum in figura, tum calculo.[35] (Cc 2, N. 549; A VII, 4, N. 29)

In the famous *De functionibus*, dating from August 1673, Leibniz avoids the term "indivisible" entirely (although he uses something similar, the "figura syntomos", see below), and "infinitesimal" appears only twice. In one case Leibniz has changed the text afterwards. The first version reads:

> Intelligatur figura ex infinitis parallelogrammis aeque altis constare, et curva ex infinitis numero rectis infinite parvis, quorum parallelogrammorum unum intelligatur esse EFGH. Eritque recta EF. vel GH. infinite parva, eademque erit infinitesima rectae AE. abscissae.[36] (Cc 2, N. 575; A VII, 4, N. 40)

By contrast, the second version is formulated thus:

> Intelligatur abscissa AE dividi in partes aequales infinitas, quales sunt EF. FG. easque proinde infinite parvas, constat figuram intelligi posse compositam ex infinitis trapeziis quales sunt EFHD et FGKH.[37]

In this revised version, the omission of the term "infinitesimal" is however probably less significant than the change from parallelograms to trapezoids. While the parallelograms only represent the area of the figure, the trapezoids, whose upper sides coincide with the infinitely small parts of the curve and of the respective tangents, represent the area and (with their upper sides) the arc of the curve. And this is of importance in the case concerned, as part of what Leibniz sets out to do here is to solve the inverse tangent problem.

As Mahnke already pointed out in his account, in the course of the manuscript Leibniz introduces infinitely small lines of higher degree by in-

---

35 "The *analysis of indivisibles* (insofar as it is separated from the arithmetic of infinites) consists mainly in reducing some given curved line or curved surface to some single space, or several spaces, from whose quadrature its own measuring depends. This can easily be done by several methods described here. Moreover, to square a given space, first of all the law of its progression has to be examined, in order to see whether its sum is capable of being determined by the arithmetic of infinites. If it defies this method one has to proceed to the analysis of indivisibles, i.e. the characteristic triangle of the figure has to be constituted, and as many triangles similar to the characteristic triangle as possible. And this can be done by constructing straight lines in the figure or by calculating."

36 "The figure is to be conceived so as to consist of infinitely many parallelograms of equal altitude, the curve of infinitely many infinitely small straight lines. Let EFGH be one of these parallelograms. The straight line EF or GH will be infinitely small and it will be an infinitesimal part of the abscissa AE."

37 "The abscissa AE is to be conceived so as to be divided into infinitely many equal parts like EF, FG, and these are therefore infinitely small. It is clear that the figure can be conceived as being composed of infinitely many trapezoids like EFHD and FGKH."

vestigating the second differences of ordinates (Mahnke, 1926, 46f). A first attempt is already to be found in Leibniz's investigation of the cycloid from spring 1673 (Cc 2, N. 609–611; A VII, 4, N. 7). Without doubt this represents a significant extension of the concept of infinitely small parts.

The use of the term "indivisible" nearly ceases in the second half of 1673. But there are exceptions, namely *De invenienda curva cuius data est elementorum progressio* (Cc 2, N. 607; A VII, 4, N. 51), which is dated the end of 1673 by Leibniz himself. It appears that only from November 1675 onwards does Leibniz again use the term "indivisible" a few times.[38] In *De invenienda curva*, Leibniz works with something similar to Cavalieri's indivisibles, namely pairs of figures that produce equal sections and which he therefore calls "syntomo[i], seu aequisecabil[es]".[39] He extends the application of these from comparing areas under curves to comparing areas and arc lengths since by this time he knows that the rectification of a curve with abscissa x and ordinate y can be reduced to the quadrature of the curve with abscissa x and ordinate $\sqrt{1 + \left(\frac{dy}{dx}\right)^2}$ (in modern notation). Finally, he points out that the use of the method of indivisibles in the problem of finding a curve from its given arc elements can now be replaced by the method he developed in *De functionibus*, i.e. the solution of the inverse tangent problem by a series expansion using differences of higher order. As is well known, it took Leibniz two years to put the beginnings of this new method into practice.

38 Nearly all the occurrences of "indivisible" in texts concerning infinitesimal mathematics between 1674 and 1676 can be found in the manuscripts dating from November 1675 in which Leibniz develops his calculus. See for example Child, 87, 96, 104, 108. There is also only a handful occurrences of "infinitesimal" and of "smaller than any given", whereas "infinitely small" occurs about eighty times.

39 See also *De functionibus* (Cc 2, N. 575; A VII, 4, N. 40) and Leibniz's later definition (A III, 1, 142); further occurrences are to be found in A VII, 3, 314, 480f., and in LH XXXV, XIII, 3, Fol. 243 (A VII, 4, N. 46).

Samuel Levey

Archimedes, Infinitesimals and the Law of Continuity: On Leibniz's Fictionalism

Actual infinitesimals play key roles in Leibniz's developing thought about mathematics and physics between 1669 and 1674.[1] But by April of 1676, with his early masterwork on the calculus, *De Quadratura Arithmetica*,[2] nearly complete, Leibniz has abandoned any ontology of actual infinitesimals and adopted the *syncategorematic* view of both the infinite and the infinitely small as a philosophy of mathematics and, correspondingly, he has arrived at the official view of infinitesimals as *fictions* in his calculus. This picture of Leibniz on infinitesimals owes largely to the pioneering work of Hidé Ishiguro,[3] Eberhard Knobloch[4] and Richard Arthur.[5] The interpretation is worth stating in some detail, both for propaganda purposes and for the clarity it lends to some questions that should be raised concerning Leibniz's fictionalism. The present essay will consider three. Why does Leibniz abandon actual infinitesimals in mid 1676? What does the new view of infinitesimals as fictions come to? Does Leibniz have an integrated fictionalism at work across his philosophy of mathematics? In each of the answers to be offered below, Leibniz will emerge at key points to be something of an Archimedean. But we begin by considering the syncategorematic infinite.

---

1 See Richard Arthur, 2008c.

2 Translations of Leibniz follow those of Child, L, DLC and NE, as noted in the List of Abbreviations of this volume, though I have sometimes modified translations without comment. Responsibility for uncited translations is my own, though in many cases I have relied on translations supplied to me by Richard Arthur, which I gratefully acknowledge.

3 Ishiguro, 1990, Chapter 5.

4 Knobloch, 1994, and Knobloch, 2002.

5 Arthur, 2008a; Arthur, 2008c.

1. The Syncategorematic Infinite and Infinitesimal

In describing his view of the infinite, Leibniz recalls the distinction between categorematic and syncategorematic terms:[6]

> A proprement parler il est vray qu'il y a une infinité de choses, c'est à dire qu'il y en a tousjours plus qu'on n'en peut assigner. Mais il n'y a point de nombre infini ny de ligne ou autre quantité infinie, si on les prend pour des veritables Touts, comme il est aisé de demonstrer. Les écoles ont voulu ou dû dire cela, en admettant un infini syncategorematique, comme elles parlent, et non pas l'infini categorematique.[7] (A VI, 6, 157)

On the traditional account, a categorematic term is one that *predicates*, that is, has reference or a semantic content of its own. By contrast, a term is syncategorematic when it predicates only in conjunction with other terms: it has no referent or semantic content of its own, but rather contributes to the meaning of sentence only by virtue of its links with other terms in the expressions to which it belongs. (*Syncategorematic* literally means 'jointly predicating'; its Latinate equivalent is *consignificantia*.) The distinction is not perfectly sharp independently of a given semantic theory, but it is easy to illustrate by examples. 'Apple', 'wise' and 'gold' are categorematic terms; 'if', 'some' and 'any' are syncategorematic.

A familiar contemporary example of syncategorematic analysis *par excellence* is Russell's technique for contextual definition of definite descriptions as quantifier phrases. Recall the present king of France:

(1) The present king of France is bald.

Russell's analysis of the meaning of (1) construes it as 'One and only one thing is a present king of France and it is bald'. Or in symbols:

(1*) $(\exists x)(\forall y)((y \text{ is a present king of France} \leftrightarrow x = y)\ \&\ x \text{ is bald})$.

---

[6] The term 'syncategorematic' descends from a distinction drawn by Priscian (6th century C.E.), in *Institutiones grammaticae* II, 15, between *categorematic* and *syncategorematic* expressions, though its employment in the diagnosis of fallacies was made famous by the 13th century *Syncategoremata* of (the mysterious) Peter of Spain; William Heytesbury, the 14th century logician-mathematician and fellow of Merton College, was perhaps the first explicitly to defend an analysis of the infinite as syncategorematic. See *sophisma* xviii of his *Sophismata*, in Heytesbury, 1994.

[7] "It is perfectly correct to say that there is an infinity of things, i.e. that there are always more of them than one can specify. But it is easy to demonstrate that there is no infinite number, nor any infinite line or other infinite quantity, if these are taken to be genuine wholes. The Scholastics were taking that view, or should have been doing so, when they allowed a 'syncategorematic' infinite, as they called it, but not a 'categorematic' one." (NE II.xvii.1, 157)

The definite article 'the' is syncategorematic: it does not refer to the *the*, nor to a property of the-ness. Rather, its contribution to the semantic value of an expression containing it is a matter of the system of logical relations it imposes among the semantic values of other terms in that expression. Russell claims, more strongly, that the definite description as a whole lacks a meaning of its own[8] – and so it is, in the Scholastic term, syncategorematic. The phrase 'the present king of France' does not predicate or have any meaning apart from its occurring within the context of a sentence; only conjointly with a predicate, such as 'is bald' in (1), does it predicate.

In parallel fashion, a syncategorematic analysis of the infinite and the infinitely small denies that the terms 'infinite' and 'infinitesimal', and so on, carry semantic values of their own and instead represents their semantic contributions in terms of the meanings of larger expressions in which they are embedded. To say, for instance,

(2) There are infinitely many Fs,

is not to assert, for instance, that there is some (infinitary) number that counts the Fs and is itself greater than any finite number. Rather, on the syncategorematic analysis, the expression 'infinitely many' in (2) is understood to introduce a wide-scope universal quantifier ranging over finite numbers and, thereby, limiting the range of the existential quantifier 'there are' to finite values as well. Thus on analysis (2) proves to be a claim that refers only to finite numbers:

(2*) For any (finite) number n, there are more than n Fs.

('More than n Fs' cashes out as there being a one-one map from the natural numbers up to n into the Fs, but not vice versa.) In interpreting (2) as (2*), the order of the quantifiers is crucial. The wide scope of the universal quantifier ensures that any specific claim about the multitude of Fs is always fixed to a pre-assigned, or given, finite number. Given a number n, there can be no one-one map of the naturals up to n onto the multitude of Fs, and this result holds for any (finite) value of n. By contrast, to reverse the order – i.e. to say that there is a number of Fs such that it is greater than all finite numbers – would involve referential commitment to infinite quantities, a "categorematic" infinite.

The syncategorematic analysis of the infinitely small is likewise fashioned around the order of quantifiers so that only finite quantities figure as values for the variables. Thus,

---

[8] Cf. Russell, 1905; and Russell, 1919, 72ff.

(3) The difference $|a - b|$ is infinitesimal,

does not assert that there is an infinitarily small positive value which measures the difference between a and b. Instead it reports,

(3*) For any finite positive value ε, the difference $|a - b|$ is less than ε.

Elaborating this sort of analysis carefully allows one to articulate the now-usual epsilon-delta style definitions for limits of series, continuity, etc., without any reference to fixed infinite or infinitely small quantities. Indeed the so-called 'rigorous reformulation' of the calculus that emerged from the nineteenth century can be viewed as a wide-scale syncategorematic analysis of its seventeenth-century formulations that replaced expressions for infinities and infinitesimals with systems of logical relations among finite terms. This is not to trivialize the effort, which required great subtlety of insight and involved genuine clarification of the mathematics itself. Yet for all that, it was also a project of systematic interpretation of the key terms, and one motivated by a concern to sidestep conceptual commitment to infinite and infinitely small quantities – i.e. to escape the perplexities of a categorematic interpretation.

But the seventeenth century was not devoid of efforts at clarifying the mathematics behind infinitary expressions in finite terms. Leibniz himself provides some wonderfully clear examples in his own works, as in this passage from April of 1676 when he writes:

> Quandocunque dicitur seriei cuiusdam infinitae numerorum dari summam, nihil aliud dici arbitror, quam seriei finitae cuiuslibet eadem regula summam dari, et semper decrescere errorem, crescente serie, ut fiat tam parvus quam velimus.[9] (*Numeri infiniti*, A VI, 3, 503)

As has been noted by commentators, this closely anticipates Cauchy's definition of the sum of an infinite series as the limit of its partial sums. It is worth observing in this case how the syncategorematic analysis may be developed from a statement involving apparently infinitary terms – an analysis that allows a systematic replacement of those terms by variable expressions that refer only to finite quantities. Take the sequence $a_1, a_2, a_3, \ldots$ *ad infinitum*, and its related series $a_1+a_2+a_3+ \ldots$ *ad inf.* What, then, is the sum of our series? Consider the following as a provisional definition:

9 "Whenever it is said that a certain infinite series of numbers has a sum, I am of the opinion that all that is being said is that any finite series with the same rule has a sum, and that the error always diminishes as the series increases, so that it becomes as small as we would like." (DLC, 99)

The sum of the infinite series is L if, and only if, the difference between L and the sum of the terms up to a_n becomes infinitely small as $n \to \infty$.

This provisional definition appears to refer to infinitely large and infinitely small values. The finitary, syncategorematic formulation is distilled in a few steps. To parse the expression of the infinitely small we set a finite variable 'ε' and say that the difference $|L - (a_1 + \ldots + a_n)|$ always eventually becomes less than ε as $n \to \infty$. The expression '$n \to \infty$' is then parsed as a variable expression whose value is dependent upon that of the variable 'ε' thus: for any ε, there is a sufficiently large n such that $|L - (a_1 + \ldots a_n)| < \varepsilon$. Last, the stepping-stone indefinite expression 'sufficiently large' is also reduced to a relational expression between finite variables: there is an N such that $n \geq N$. In modest shorthand the definition becomes:

L is the limit of the series a_n if, and only if, for any ε, $|L - (a_1 + \ldots a_n)| < \varepsilon$, for $n \geq N$.

Although this equation is not likely to be misinterpreted in the practice of mathematics, there remains an ambiguity of the scope of the final quantifier phrase 'for $n \geq N$', and in fact that phrase actually subsumes a pair of quantifiers. With fuller disambiguation, the right side of the equation would read:

for any $\varepsilon > 0$, there exists an N such that, for any $n \geq N$, $|L - (a_1 + \ldots a_n)| < \varepsilon$.

No mathematician would write that out in practice. In life mathematical equations drop their quantifiers, letting the variables be interpreted as the theory demands. Potentially ambiguous formulae are read correctly by virtue of a grasp of the relevant theory, gaining in economy of expression what is lost in explicitness. When the underlying theory is not yet perfectly understood, however, mathematical formulae can give rise to a host of interpretations corresponding to different scope readings of the unstated quantifiers. The idioms of quantificational logic, when carried far enough, eventually force one to make explicit the relations among the variables. Clarified in this way, the rigorously finitary, syncategorematic readings of 'infinitely small', '$n \to \infty$', etc., become evident. Neither infinitely large numbers nor infinitely small differences are supposed by the formulae.

This matters, since Leibniz's usual practice in finding sums of infinite series involves the "fiction" that the series itself is a whole with a terminal element and that this terminal element itself is both the infinitieth term in the series and infinitely small.[10] With the definition in terms of finite quantities on hand

10 For discussion, see Hofmann, 1974, 14ff.; Mancosu, 1996, 153ff.; Levey, 1998, 72f.

to be substituted for the fictions, however, we can dispatch with the unwanted ontology of infinitary quantities, large and small, while retaining the fictional, infinitary expressions for their convenience.

The systematic application of the syncategorematic view of infinitesimal terms in Leibniz's mathematics allows us to interpret most if not all of that mathematics consistently with a rejection of any infinitarily small quantities – and to do so in a way that is 'rigorous' and honors his own philosophical remarks about the infinite and infinitely small. As I shall indicate below, the elements of this view are in place already in mid 1676 and Leibniz does not later abandon them. Thus after early 1676 infinitesimals are only fictions in Leibniz's philosophy of mathematics.[11]

2. The End of the Actual Infinitesimal

The end of the actual infinitesimal in Leibniz's writings comes in the Spring of 1676. In *De arcanis sublimium vel De summa rerum*, written in February of that year, Leibniz still imagines that liquid matter might be "dissolved" into a powder of infinitesimal points (A VI, 3, 474). And with his infinitesimal calculus now well along in construction, Leibniz contemplates whether its infinitesimals might indeed be realities in nature and not simply artifacts of the mathematical formalism. He writes: "Cum videamus Hypothesin infinitorum et infinite parvorum praeclare consentire ac succedere in Geometria, hoc etiam auget probabilitatem esse revera."[12] (A VI, 3, 475) Yet this appears to be the actual infinitesimal's last moment of glory. Something happens in mid-March to change Leibniz's mind, apparently for good. What it is that happens, exactly – that is, just what brings Leibniz to change his mind – remains something of a mystery. The change is not trumpeted. But there are some signs. In a note, *De infinite parvis*, dated to 26 March 1676, Leibniz remarks:

> Videndum exacte an demonstrari possit in quadraturis, quod differentia non tamen sit infinite parva, sed quod omnino nulla, quod ostendetur, si constet eousque inflecti semper posse polygonum, ut differentia assumta etiam infinite parva minor fiat error. Quo posito sequitur non tantum errorem non esse infinite

11 It is not uncontroversial that Leibniz is a considered 'fictionalist' about infinitesimals, either in his early or late in his writings; for a competing view, see Jesseph, 1998.

12 "Since we see the hypothesis of infinites and the infinitely small is splendidly consistent and successful in geometry, this also increases the likelihood that they really exist." (DLC, 51)

parvum, sed omnino esse nullum. Quippe cum nullus assumi possit.[13] (A VI, 3, 434)

There is much to say about this passage, but we shall limit discussion to just a few points. Leibniz's hint toward an argument that might show that the differential is "nothing at all" seems obliquely to invoke *Archimedes' Principle* (due originally to Eudoxus) that for any two numbers x, $y > 0$ such that $x > y$, there is a natural number n such that $ny > x$.[14] For the principle that would naturally justify the step from saying that if the error is smaller than any that can be assumed to the claim it is nothing at all is, in effect, a corollary of *Archimedes' Principle*. (Also, Archimedes is clearly on his mind, as Leibniz mentions him by name in the subsequent lines.) Assuming "trichotomy" for the relevant quantities, i.e. that for any x and y, either $x > y$ or $x = y$ or $y > x$, *Archimedes' Principle* yields the following as a *principle of equality* (PE):

(PE) if, for any $n > 0$, the difference $|x - y|$ is less than $1/n$, then $x = y$.

In later writings Leibniz will sometimes describe this idea by saying that equality is the limit of inequalities or differences (cf. GM IV, 106). In any case, the new principle of equality will come to play a pivotal role in Leibniz's mathematics, and various conceptual extensions of it will emerge in his broader philosophical thought as well. In the present instance, both tendencies are already at work. Let me explain.

The proposed reduction of differentials to "nothing at all" is part of an effort to capture the mathematical device of an infinitely small quantity, such as an infinitesimal interval of a line, while also being able to argue that an infinitely small difference between quantities can be rigorously disregarded. Leibniz does not say here that talk of differentials can be systematically replaced by phrases to the effect that "the error is less than any given error," though he must by now appreciate the force of that style of argument. A par-

13 "We need to see exactly whether it can be demonstrated in quadratures that a differential is nonetheless not infinitely small, but that which is nothing at all. And this will be shown if it is established that a polygon can always be bent inwards to such a degree that even when the differential is assumed infinitely small, the error will be smaller. Granting this, it follows not only that the error is not infinitely small, but that it is nothing at all – since, of course, none can be assumed." (DLC, 65)

14 Archimedes introduces the principle as a postulate about extended quantities: "That among unequal lines, as well as unequal surfaces and unequal solids, the greater exceeds the smaller by such <a difference> that is capable, added itself to itself, of exceeding everything set forth (of those which are in a ratio to one another)". (Archimedes, 2004, 36; see also Netz's discussion, pp. 40f.)

allel pattern of reasoning is clearly intended. The proof sketched in *De infinite parvis* would try show that infinitely small differentials are nothing at all by arguing "ut differentia assumta etiam infinite parva minor fiat error."[15] (A VI, 3, 434) The new principle of equality will certainly yield this result, since any infinitely small difference $|x - y|$ will be less than any given finite ratio $1/n$, and therefore $x - y$, thus making their difference "nothing at all."

But the context presupposed by the sketched proof would seem to be one in which it is *granted* that quantities might differ by infinitely small amounts. Let d be the difference $|x - y|$. The claim of the argument is that even if we suppose the existence of infinitely small differences between quantities, for any given infinitely small value i, it can be shown that d is still less than i. In this context, the new principle of equality would be out of place. For if $|x - y|$ could differ by the infinitely small value d, then it would not automatically be true that $x = y$ if their difference is less than $1/n$ for any n. An infinitely small difference between quantities is precisely one in which, for any n, the difference is less than $1/n$. The finitistic aspect of the new principle of equality thus makes a nonsense of the presupposition of the proof. What is called for in this case, rather, is a 'weaker' principle of equality along the following lines:

if for any $\varepsilon > 0$, the difference $|x - y|$ is less than $1/\varepsilon$, then $x = y$,

where 'ε' is to be interpreted as allowing not only finite values in its range but infinite values as well. At any rate, taking this principle as a premise can cohere with Leibniz's sketched argument for the claim that even if the differential is allowed to be infinitely small (i.e., less than $1/n$ for any n), it can still be shown to be nothing at all if the error is smaller than $1/\varepsilon$ for any ε.

For present purposes we shall not pursue the question whether the argument of *De infinite parvis* can be filled out suitably to show that infinitely small differentials are nothing at all. What matters is simply to observe how Leibniz is taken with the "logic" of the new principle of equality – both for the internal rationale of limit-style argument and the particular idea that equality can be understood as a limit of differences. Still, for all the intriguing hints of *De infinite parvis*, we are left without a clear view of the reason behind Leibniz's change in attitude toward the existence of infinitely small differentials.

Nonetheless, the change is certainly taking place, and within a few short weeks, it's all over for the infinitely small. Leibniz begins confidently de-

---

15 "[...] even when the differential is assumed infinitely small, the error will be smaller." (DLC, 65)

scribing infinitesimals and their ilk as "fictions" and in his philosophical writings, at least, they rapidly fade into the background as entities that becomes less and less worth considering at all. Good-bye to all the wonderful limit entities: good-bye parabolic ellipse with one focus at infinity, good-bye infinilateral polygon, good-bye infinitesimal angles residing within a point, and so on. In a noteworthy piece from 10 April 1676, titled *Numeri infiniti*, Leibniz discusses a number of cases of limit entities – his remarks include a nice series of reflections on the circle taken as an infinilateral polygon, the limit of the series of regular polygons – and notes: "quod etsi non sit in rerum natura, ferri tamen eius expressio potest; compendiosarum enuntiationum causa."[16] (A VI, 3, 498) And further: "Etsi Entia ista sint fictitia, Geometria tamen reales exhibet veritates, quae aliter, et sine ipsis enuntiari possunt, sed Entia illa fictitia praeclara sunt enuntiationum compendia, vel ideo admodum utilia"[17] (A VI, 3, 499). This is starting to become an element in his defense of the use of these fictions in his calculus, a topic to be discussed later. Here it is enough to note that the "fictitious" entities are preserved only as "abbreviations for expressions."

3. Leibniz's *De Quadratura Arithmetica* and the Infinitely Small

As we noted, Leibniz's reasons for abandoning actual infinitesimals in the Spring of 1676 are not immediately evident. From some clues in later writings it can be tempting to think that Leibniz had struck upon some proof of the impossibility of an infinitely small quantity; he mentions to Johann Bernoulli, for instance, that if he were to admit the possibility of infinitesimals, he would then have to accept their existence (cf. GM III, 524 and 551). And it is not hard to imagine how he might have done so, for with his extensive reflections on the concept of the infinite, Leibniz was well supplied with resources for a purely conceptual argument against the existence of infinitely small quantities if he had cared to construct one. Recall, for example, his already-entrenched argument against infinitely large numbers that relies on the "axiom" that the part is less than the whole (cf. A VI, 3, 98 and 168).

---

16 "And even though this ultimate polygon does not exist in the nature of things, one can still give an expression for it, for the sake of abbreviation of expressions." (DLC, 89)

17 "Even though these entities are fictitious, geometry nevertheless exhibits real truths which can also be expressed in other ways without them. But these fictitious entities are excellent abbreviations for expressions, and for this reason extremely useful." (DLC, 89–91)

Consider the infinite number that is the number of all numbers. It would contain as a part the number of all even numbers (imagine assigning a "one" to each natural number to count it: the number of all numbers is the aggregate of all the ones, the number of evens is contained in the total as a sub-aggregate), but a one-one map of each number onto its double establishes the "equality" of the part with the whole, contrary to the dictate of the axiom. Other infinite numbers can be handled likewise, *mutatis mutandis*. If infinitesimals are inverses of infinitely large numbers, as it seems they would be, a simple extension of the same reductio should carry through to refute their reality as well.

Yet no such argument has so far appeared in his writings. Perhaps no disproof is forthcoming because his reasons for rejecting actual infinitesimals are of a different kind. Compared to his writings on the concept of the infinite, which fall recognizably into the tradition of "philosophical foundations" for mathematics and proceed at a high level of generality, Leibniz's dealings with the concept of the infinitely small are more closely interwoven with questions of mathematical practice. Context is important, and the best clues to his new thought about the infinitely small, I think, occur in *De Quadratura Arithmetica* (DQA).

In the opening sections of DQA, Leibniz lays out the pieces from which his calculus will be constructed. Of particular interest for us is Proposition 6 (DQA, 28–33). The demonstration of Prop. 6 articulates a general technique for finding the quadrature of any continuous curve that contains no point of inflection and no point with a vertical tangent (DQA, 29). And of those conditions, only continuity is truly essential, since a curve can always be cut at points of inflection or at "singularities" and the general technique Leibniz produces can then be applied piecewise to the resulting segments. What Leibniz has demonstrated, then, is the integrability of a "huge class of functions."[18] The technique itself is also of interest, for Leibniz's use of "elementary" and "complementary" rectangles very precisely anticipates Riemannian integration.[19] The proof is complex – Leibniz himself describes it as "most thorny" (*spinosissima*) – and other commentators have explained it elegantly and in depth.[20] Here we shall take the liberty of proceeding with a mere impressionistic sketch and then single out a few details for comment.

---

18 Knobloch, 2002, 63.

19 Cf. Knobloch, 2002, and Arthur's contribution in this volume.

20 Including Arthur, see his contribution in this volume.

In the demonstration, Leibniz finds the quadrature of the generic continuous curve by constructing a step space built of up of finite rectangles that approximates the area under the curve. What he proves is that the difference between the step space and the "whole Quadrilineal" (the gradiform space) can always be shown to be smaller than any given finite area. Specifically, Leibniz proves that for any given construction of the step space, the difference between the step space and the Quadrilineal can be shown to be smaller than the area of a finite rectangle whose base is the aggregate of the bases of the rectangles in the step space (and thus fixed below a finite bound in length) and whose height is no greater than the maximum height of any of the rectangles in the step space. Yet it is always possible to refine the step space by increasing the number of rectangles and reducing the maximum height of any rectangle in it, no matter how small the value of the maximum height might be. Thus the maximum height for any rectangle can be made smaller than any given finite quantity. Correspondingly, then, the finite rectangle representing an upper bound on the difference between the area of the step figure and the are of the Quadrilineal can also have its height made less than any given finite quantity, and so its total area can be made less than any given quantity. Therefore, as Leibniz notes expressly at the end of the demonstration: "*Differentia hujus Quadrilinei,* (de quo et propositio loquitor) *et spatii gradiformis data quantitate minor reddi potest. Q.E.D.*"[21] (DQA, 32)

To add a last step reaching the conclusion that the two spaces are therefore equal, one need only advert to the new principle of equality. Leibniz does not do so, perhaps at this point regarding the inference as obvious; the principle goes without saying. Still, if in Prop. 6 he does not explicitly articulate the new principle of equality upon which the argument relies, in follow-up remarks to Prop. 7, he says directly (in words that would equally apply to Prop. 6): "Et proinde si quis assertiones nostras neget facile convinci possit ostendendo errorem quovis assignabili esse minorem, adeoque nullum."[22] (DQA, 39) When the error, or difference, is smaller than any that can be assigned, it is not merely negligible or somehow incomparably small, it is nothing at all. That is, there is no error: the two values are equal.

21 "[...] *the difference between this Quadrilineal* (which is the subject of this proposition) *and the step space can be made smaller than any given quantity. Q.E.D.*"

22 "Therefore, anyone contradicting our assertion [that the area is the same as the sum of the rectangles] could easily be convinced by showing that the error is smaller than any assignable, and therefore null."

Leibniz's demonstration of Prop. 6 is 'rigorous' in the modern sense of involving only finite quantities; it makes no reference to infinite or infinitely small values. And it is specifically the new *Archimedean principle of equality* that allows this. No direct construction of the area of the quadrilineal by means of a single step space would be possible without representing the step space as composed of infinitely many infinitely small (narrow) rectangles. But with the new principle of equality in play, it suffices to show that any given claim of finite inequality between the two areas can be proved false by some particular finite construction, even if there is no single finite construction that at once gives the quadrature of the curve exactly. No 'ultimate construction' lying at the limit is required. Under the aegis of the principle of equality, the system of relations among the series of finite constructions already proves the equality; Leibniz's novel technique of elementary and complementary rectangles thus obviates the need to appeal to infinitely small quantities altogether.

The proof is also notably 'Archimedean' in style in the degree to which its strategy recalls the ancient method of exhaustion. Of course the method of exhaustion proceeded by means of a double-reductio, effecting two different constructions of polygonal spaces, one circumscribing the given gradiform space, the other inscribed within it, to prove that the area of the given space could be neither greater than nor less than a certain quantity. By contrast, as Leibniz points out, his own method requires only a single arm of construction and only a single reductio, making it more natural, direct and transparent than the two-sided classical technique (DQA, 35). Leibniz has, in effect, integrated the two sides of the classical double reductio by fashioning a step figure that neither circumscribes nor is inscribed within the gradiform space but nonetheless converges on it as a limit. The two sides of the underlying logic of the ancient method are correspondingly integrated in the new principle of equality. The method of exhaustion contends that the area given by quadrature is neither greater nor less than that of the given space and must therefore be equal to it. The reasoning is familiar. For any quantity that is given as the amount by which the area of the quadrature exceeds that of the space, it can be shown that any actual difference must be smaller than the given quantity. Likewise for any quantity given as the amount by which the area of the quadrature is supposed to be smaller than the space: by construction it can be shown that the spaces must differ by less than that amount. In Leibniz's hands, both possibilities of error are handled at once under the new principle of equality: if for any given difference (whether by excess or shortfall) the error can be shown to be still smaller (in "absolute value"), then the areas are in fact equal and the error is nothing at all. It goes without saying that his technical accomplishments in quadratures far outstrip the original reaches of the method of exhaustion; the technique of Riemannian integration by itself is an enormous advance, and for Leibniz it is not even particularly a showpiece of DQA (the subsequent infinitesimalist results are touted with greater fanfare). Yet at the level of the basic logic of the proof strategy, Leibniz's reasoning in Prop. 6 very much bears the stamp of Archimedes; perhaps we should call it a neo-Archimedean style of proof.

The special import of Leibniz's achievement for early modern mathematics becomes more vivid when he considers a special case of Prop. 6's general result, one in which the method is restricted to parallel ordinates and the intervals between successive ordinates are always supposed equal. As Leibniz notes, the "common method of indivisibles" was forced to operate under those constraints *securitatis causa* – "for safety's sake" – as was Cavalieri (DQA, 69). This means that these earlier, predecessor techniques (due to Wallis as well as Cavalieri) were considerably less general than Leibniz's new method of DQA; and moreover, the common method of indivisibles could in effect be *modeled* in Leibniz's new approach. Leibniz saw this quite clearly, noting that Prop. 6 "servit tamen ad fundamenta totius Methodi indivisibilium firmissime jacienda" (DQA, 24).[23] That method, suitably interpreted, is nothing more than a special case of a wholly finitary method.

Here I suspect we have the decisive ground for Leibniz's change of mind about the status of infinitesimals. With the mathematical advances of DQA, infinitely small quantities are no longer necessary for finding quadratures, so there is nothing in particular to preclude their being discarded. But, more subtly, the very context in which the infinitesimals had their most significant actual mathematical application – the "common method of indivisibles" – has now been shown to disappear into an entirely finitary method. Unlike the concept of the infinite, which is intellectually attractive in its own right as a subject of study even independently of particular applications, the concept of the infinitely small is of interest only, or mostly, as part of the working conception of a specific mathematical technique. Once that mathematical technique has been absorbed into a more general method that does not posit infinitely small quantities, the question whether the infinitely small might "really" exist becomes idle. No extra argument is required for abandoning the "ontological" conception of the infinitely small. It simply gives up the ghost.

---

[23] "[...] serves to lay the foundations of the whole method of indivisibles in the firmest possible way."

At least two sorts of evidence for this view of the interest of the idea of the infinitely small can be discerned in Leibniz's writings. The first lies in the fact, noted above, that Leibniz appears not to provide abstract conceptual reasons for denying that there are, or could be, infinitely small quantities in nature. Such reasons would not be hard to construct given his views about number, quantity and the infinite. But the case of the infinitely small seems not to engage Leibniz philosophically in the same way; he has very little to say about the ontological issue after the development of DQA other than to refer to the infinitely small as a fiction.

The second strand of evidence for seeing the infinitely small as holding real intellectual interest only in its "working conception" comes from the role that infinitesimals continue to play in DQA (and Leibniz's later mathematical writings). For of course the treatise does not strive to sidestep or eliminate the use of infinitesimals; on the contrary, it is one of the central aims of DQA to *promote* the use of infinitesimals in mathematics, and starting with Prop. 11 infinitesimals are featured prominently in its demonstrations. Despite the fact that the concept of the infinitely small can be bypassed in favor of finitary techniques, and so is not essential as a matter of the "logical foundations" of quadratures, it nonetheless retains a vital heuristic value for the actual practice of mathematics. Thinking of curves or spaces as decomposing into infinitely many infinitely small pieces proves enormously fruitful for the creative work of mathematics; it is perhaps even indispensable from the point of view of discovery. "Cujus specimen totus hic libellus erit," Leibniz writes, "si quis methodi fructum quaerit"[24] (DQA, 69). Leibniz regards his new method not as displacing the mathematical use of infinitesimals but rather as securing and extending it. The calculus of DQA is intended, and understood, to be a more certain, flexible and general technique than Cavalieri's geometry of indivisibles, one that will be far more expansive in its theoretical reach. Leibniz predicts that readers of the DQA,

> sentient autem quantus inveniendi campus pateat, ub hoc unum recte perceperint, figuram curvilineam omnem nihil aliud quam polygonum laterum numero infinitorum, magnitudine infinite parvorum esse. Quod, si Cavalerius, imo ipse Cartesius satis considerassent, majora dedissent aut sperassent.[25] (DQA 69)[26]

[24] "If anyone should question the fruitfulness of this method, the whole of this little book will serve as a specimen of it."

[25] "[...] they will sense just how much the field of discovery has been opened up when they correctly comprehend this one thing, that every curvilinear figure is nothing but a polygon

The texts also indicate that Leibniz sees the role of the new principle of equality in securing the infinitesimal techniques in quadratures. Noting in his prefatory remarks about Prop. 6 that the method will show that the difference between the area of the step space and the area under the curve "differat quantitate minore quavis data"[27] (DQA, 29) he concludes: "Adeoque methodus indivisibilium, quae per summas linearum invenit areas spatiorum, pro demonstrata haberi potest."[28] (ibid.). In Leibniz's new technique, of course, there are no sums of lines, strictly speaking, but only sequences of sums of ever-narrower rectangles. As he notes in the definitions after his comments on Prop. 7, in his method by the phrase "sum of all straight lines" we are to understand the sum of all rectangles, each of which has one side equal to one of the straight lines in question, and the other side equal to a constant interval assumed to be indefinitely small (DQA, 39). 'Indefinitely small'? Any finite size, as small as you like.

Our answer to the question of why Leibniz comes to reject infinitely small quantities by mid 1676 thus involves two conceptions of the infinitely small, or perhaps two perspectives from which the idea might be regarded, and correspondingly two frames of mind about the infinitely small. From an ontological point of view, the infinitesimals of his mathematics are taken merely to be fictions, and the question of their reality is decided in the negative, if, apparently, only by default. From the point of view of mathematical practice, however, infinitesimals are not discarded but retained and actively promulgated.

It should be noted as well that the working conception of the infinitely small is also carefully scrutinized by Leibniz. The "firm foundation" he lays for "the common method of indivisibles" in fact refines a key notion of that method by replacing the idea of an indivisible magnitude with the idea of an infinitely small one that is nonetheless still further divisible (Leibniz notes

---

with an infinite number of sides, of an infinitely small magnitude. And if Cavalieri or even Descartes himself had considered this sufficiently, they would have produced or anticipated more".

26 Leibniz's faith in the fecundity of the infinitesimalist picture of mathematical objects is notable also in *Cum prodiisset* when Leibniz speculates that it was also a secret method of the ancient geometers: "Et certe Archimedem et qui ei praeluxisse videtur, Cononem ope talium notionum sua illa pulcherrima theoremata invenisse credibile est" (H&O, 42). – "Truly it is very likely that Archimedes and one who seems to have surpassed him, Conon, discovered their very beautiful theorems with the help of such ideas" (Child 149). And about Archimedes, at any rate, Leibniz may have guessed right; cf. Dijksterhuis, 1987, 148.

27 "[...] will be less than any given quantity".

28 "Thus the method of indivisibles, which finds the areas of spaces by means of the sums of lines, can be regarded as demonstrated".

"plurimum interest inter indivisibile et infinite parvum"[29] (DQA, 133)). The infinitely small parts of lines, for instance, are themselves lines "neque enim puncta vere indivisibilia"[30] (ibid.). On Leibniz's view, treating infinitesimals as truly indivisible leads into paradox, as he discusses in detail in the scholium to Prop. 22. There he considers the decomposition of a space bounded by a hyperbola of equation $xy = 1$ and the x and y-axes into indivisible lines (the curve's abscissas), and shows that applying the techniques of the common method of indivisibles, it can be proved that a given portion of the space is equal in area to a subspace contained within it – i.e. that the part is equal to the whole, which is absurd (DQA, 67). The solution requires interpreting infinitesimals as infinitely small divisible quantities – in this case, as infinitely small rectangles rather than as indivisible lines – which in effect prevents one from taking a key step in the proof (that of calculating with an infinitely long "last abscissa" to find the sum of lines making up the space). Thus the paradoxical result cannot be derived with the "indivisibles" now suitably reinterpreted.[31] This is a subtle change at the level of practice; in many contexts there would be no reason to consider the difference between understanding infinitesimals as indivisible or divisible quantities. Yet as the case shows, the conceptual distinction is important. Leibniz warns his readers: "Has cautiones nisi quis observet, facile ab indivisibilium [methodo] decipi potest."[32] (DQA, 39).

With all this in view, Leibniz's change of mind about infinitesimals in Spring of 1676 becomes easier to understand. His discovery of the technique of Riemannian integration cut free his mathematics of quadratures from any essential "ontological commitment" to infinitesimal quantities. His interpretation of the infinitesimal as a divisible quantity rather than an indivisible one yielded a new reading of the common method of indivisibles that allowed a resolution to various paradoxical results. And the derivation, or modeling, of the common method of indivisibles in the new method of DQA meant a safe haven for the infinitesimalist techniques within a mathematical framework whose foundations were strictly finitist. Thus the ontology of the infinitely small could be dropped even while the practices that incorporate them could be promoted and extended. And that is precisely

29 "[...] a profound difference between the indivisible and the infinitly small".

30 "[...] not truly indivisible points".

31 For detailed discussion of the paradox see Knobloch, 1990, Knobloch, 1994, and Mancosu, 1996, 128f.

32 "One who does not observe these cautions can easily be deceived by the method of indivisibles."

what Leibniz can be seen to do in 1676 as he advances a revolutionary infinitesimalist mathematics while at the very same time relegating infinitesimals to the status of fictions.

4. What is Leibniz's Fictionalism?

In calling infinitesimals 'fictions' Leibniz signals that he is not endorsing an ontology of actual infinitely small quantities. Still, one might ask just what the fictionalism comes to. In the abstract, three possibilities for interpreting scientific theories come to mind in this connection, each of which can provide a potential understanding of the claim that infinitesimals are fictions.

The first might be termed *reductionism*: The language of infinitesimals as it occurs in Leibniz's mathematics can be systematically translated into a language that involves only finitary terms while preserving the mathematical results. Infinitesimals are then "linguistic fictions": apparent reference to infinitely small quantities is only an artifact of a device of abbreviation that, properly understood, involves no such reference at all. The language of infinitesimals may have some cognitive value as a shorthand or an aid to the imagination, but the form of words is logically dispensable, and what those words say, on analysis, is true.

The second is *pragmatism*: The language of infinitesimals aims not directly at truth but only at a certain form of scientific adequacy in describing the data that the theory – here, the calculus – attempts to organize, explain, predict, etc.[33] The terms in the theory are to be taken at face value, but with indifference to ontological consequences outside of scientific application. The theory is intended to be measured in terms of its scientific success, and it is not put forward to capture truth itself beyond adequacy. If the theory happens not to be true the facts, especially on point of the entities hypostasized in it, then the elements of the theory are fictions in the most straightforward sense: they are merely elements of a story. But since the theory aims no higher than scientific adequacy, the status of infinitesimals as a "useful fiction" is not undermined by the final consilience, or not, of the calculus with reality.

Last is *ideal-theory instrumentalism*: Leibniz's mathematics, or at least that component of it which traffics in the language of infinitesimals, is not

33 This sort of view has been urged for scientific theories generally by Bas van Fraassen (1980), though it has a series of earlier anticipations as well, and the term 'fictionalism' has lately been adopted for it. See Rosen, 2006.

reducible to some entirely factual theory nor it is taken to be a story that is good whether or not it is true. Rather, it is not to be interpreted as meaningful at all but only regarded as an intermediary device for inferring meaningful results from meaningful premises.[34] The intermediary notation might be well-suited for disciplined imaginings or fantasy about infinitely small quantities, areas decomposing into lines, etc., but that is only for heuristic value. Given some background demonstration (or faith) that the whole theory is a conservative extension of its interpreted component, the infinitesimalist techniques are embraced, though now seen only as rules for the manipulation of symbols, while a strictly finitist ontology is retained. (Perhaps the mantra for this view of the infinitesimal calculus: No one shall expel us from this paradise that Leibniz has created!)

Leibniz does not appear to suggest a division of his mathematics into real and ideal components in the manner characteristic of ideal-theory instrumentalism. But it is not hard to detect pragmatist and reductionist elements in Leibniz's writings on infinitesimals, as concerns for both utility and ontology feature in his remarks. Of those two, it is the reductionist model that would appear to jibe best with his overall treatment. The fiction of infinitesimals is a fiction not because the theory aims to be nothing more than a scientifically useful story – though in the DQA Leibniz voices official neutrality about the real existence of infinitesimals, as we shall see in a moment – but because the terms for infinitesimals can be explained away. On the present interpretation, expressions for infinitesimals are syncategorematic: they are not designating terms for infinitely small quantities but rather they are shorthand devices for complex expressions that refer only to finite quantities. Such is the import of the syncategorematic analysis. As we have seen, by Spring of 1676 Leibniz tells his readers how to interpret phrases such as 'the sum of an infinite series' and 'the sum of all straight lines' in rigorously finitary terms. And in DQA itself while discussing the reliance on the ideas of infinite and infinitely small quantities he says expressly: "Nec refert an tales quantitates sint in rerum natura, sufficit enim fictione introduci, cum loquendi cogitandique, ac proinde inveniendi pariter ac demonstrandi compendia praebeant"[35] (DQA, 69). The fiction is preserved for its heuristic value to the mathematical imagination and for its

[34] Obviously this adapts Hilbert's celebrated view of mathematics, announced at the Westphalian Mathematical Society in 1925. Cf. Hilbert, 1983.

[35] "Nor does it matter whether there are such quantities in nature, for it suffices that they be introduced by a fiction, since they allow abbreviations of speech and thought in discovery as well as in demonstration" (DLC, 393, fn 5, Richard Arthur's trans.).

economy of expression. From the point of view of mathematical practice, considerations of utility "justify" the use of infinitesimals in the calculus. From the point of view of foundations, the practice is "justified" by its reducibility to finitary techniques – which is the point of the *spinosissima* demonstration of Prop. 6 by the Riemannian technique and of the subsequent derivation of the (reinterpreted) method of indivisibles as a special case.

Still, the reduction of infinitesimal mathematics to finitist techniques should not be overemphasized in describing Leibniz's view of infinitesimals. As before, the ontological issue is not foremost in his thinking. In fact he views his own demonstration of the method of indivisibles more as a concession to community demands than as an accomplishment to be celebrated in its own right, as he makes clear in a scholium to Prop 6., appended just after that demonstration:

> Hac propositione supersedissem lubens, cum nihil sit magis alienum ab ingenio meo quam scrupulosae quorundam minutiae in quibus plus ostentationis est quam fructus, nam et tempus quibusdam velut caeremoniis consumunt, et plus laboris quam ingenii habent, et inventorum originem caeca nocte involvunt, quae mihi plerumque ipsis inventis videtur praestantior. Quoniam tamen non nego interesse Geometriae ut ipsae methodi ac principia inventorum tum vero theoremata quaedam praestantiora severe demonstrata habeantur, receptis opinionibus aliquid dandum esse putavi.[36] (DQA, 33)

The construction of the common method of indivisibles from finitist foundations ensures reducibility, but its primary role in the treatise is not to stress the eliminability of infinitesimals but to placate potential critics. By offering the 'minutiae' necessary to set aside doubts about the soundness of the basic principles, Prop. 6 then clears the way for the main agenda of DQA, the advancement of infinitesimalist mathematics, which is advertised by Leibniz for its high rewards in mathematical results rather than for its low costs in ontology.

Once the foundations are established in Prop. 6, Leibniz moves ahead in DQA to unlimber the calculus and to display a specimen of its results. The discussion of ontology is essentially over, and the remaining, scattered

[36] "I would gladly have omitted this proposition because nothing is more alien to my mind than those scrupulous minutiae of certain authors in which there is more ostentation than reward, for they consume time as if on certain ceremonies, include more labor than insight, and envelop the origins of discoveries in blind night, which often seems to me more prominent than the discoveries themselves. I do not deny that it is in the interest of geometry to have the very methods and principles of discovery rigorously demonstrated, so I thought I must yield somewhat to received opinions."

philosophical remarks mainly concern epistemic matters in mathematics – stressing the advantages of the infinitesimal methods for directness, lucidity, fruitfulness, etc. He does not take pains to offer a guidebook for recasting infinitesimalist proofs in finite terms, though his handling of infinitary expressions appears to operate within a carefully confined set of procedures and his discussion allows an exacting reconstruction of an 'arithmetic of the infinite' statable in twelve precise rules.[37] These rules themselves can in turn be reduced to principles concerning finite quantities.[38] Thus at least the basic resources for effecting a reduction of infinitesimalist demonstrations are available in DQA. But doing so is no priority, indeed no real concern, of Leibniz, whose eyes are now oriented toward the mathematical frontier.

Two and a half decades later when the public debate about foundations has broken out and he is expressly asked to justify the use of infinite and infinitely small quantities in his calculus, Leibniz's attitude appears to be unchanged. He stresses the practical value of the techniques to mathematics, distances mathematical issues from matters of metaphysics, and says that the disputed quantities can simply be taken as fictions, as is already the case for other common ideas in mathematics such as square roots for negative numbers (cf. GM IV, 91 ff.). He also points to the possibility of reformulating the infinitesimalist procedures in finite terms. He has not forgotten his link with Archimedes. Writing in 1701 to Pinsson, in reply to anonymous criticisms of the calculus published by Abbé Gouye, Leibniz notes:

> Car au lieu de l'infini ou de l'infiniment petit, on prend des quantités aussi grandes et aussi petites qu'il faut pour que l'erreur soit moindre que l'erreur donnée. De sorte qu'on ne differe du style d'Archimede que dans les expressions qui sont plus directes dans nostre Methode, et plus conformes à l'art d'inventer.[39] (A I, 20, 494)

Similarly, in the note on 'the justification of the calculus in terms of ordinary algebra' attached to the his 1702 letter to Varignon, in defending (*inter alia*) the introduction of infinitesimal quantities as limit cases of finite quantities, he writes:

---

[37] See Knobloch, 1994, 273, and Knobloch, 2002, 67 f.

[38] See Arthur's contribution in this volume.

[39] "[...] in place of the infinite or infinitely small one can take quantities as great or small as one needs so that the error be less than any given error, so that one does not differ from Archimedes' style but for the expressions which in our method are more direct and more in accordance with the art of discovery."

> Et si quelqu'un n'en est point content, on peut luy faire voir à la facon d'Archimede, que l'erreur n'est point assignable et ne peut estre donnée par aucune construction. C'est ainsi qu'on a repondu à un Mathematicien tres ingenieux d'ailleurs, lequel, fondé sur des scrupules semblables à ceux qu'on oppose à nostre calcul, trouve à redire à la quadrature de la parabole, car on luy a demandé si par quelque construction il peut assigner une grandeur moindre que la difference qu'il pretend estre entre l'aire parabolique donnée par Archimede et la vertiable, comme on peut tousjours faire lorsqu'une quadrature est fausse.[40] (GM IV, 105–6)

Apart from the vantage point provided by the demonstration of Prop. 6 in DQA, Leibniz's references to recasting infinitesimalist proofs into 'the style of Archimedes' might be taken as a vague suggestion to the effect that the same results could be attained by the method of exhaustion. But with Prop. 6 in view, those remarks can be read more definitely: quadratures described in terms of infinitesimals could alternatively be presented via Leibniz's neo-Archimedean method that progressively constructs a single step space and argues by means of a single-sided "direct" reductio showing that for any given error, the error must be still smaller. And coupled with the new principle of equality, it is thereby proved that there is no error at all. The way of infinitesimals is "more direct" – i.e. it is not forced to proceed by reductio, whether two-sided as in the classical form or one-sided as in Leibniz's innovative proof – and it is "more in accordance with the art of discovery." But for those whose "scruples" are offended by such techniques, the far thornier path of the neo-Archimedean (and proto-Riemannian) approach also remains open.

Even when Leibniz does not mention Archimedes by name, the link is often evident in his characteristic emphasis on the tactic of arguing that the error will be less than any given error, a phrase that, for Leibniz, codes within it the new principle of equality and the prospect of the one-sided reductio. For instance in a 1706 letter to Des Bosses, Leibniz's finitism, his fictionalism and the reference to his neo-Archimedean method are visible all at once:

---

40 "And anyone who is not satisfied with this can be shown in the manner of Archimedes that the error is less than any assignable quantity and cannot be given by any construction. It is in this way that a mathematician, and a very capable one besides, was answered when he criticized the quadrature of the parabola on the basis of scruples similar to those now opposed to our calculus. For he was asked whether he could by means of any construction designate any magnitude that would be smaller than the difference he claimed to exist between the area of the parabola given by Archimedes and its true area, as can always be done when the quadrature is false." (L, 546)

> Ego philosophice loquendo non magis statuo magnitudines infinite parvas quam infinite magnas, seu non magis infinitesimas quam infinituplas. Utrasque enim per modum loquendi compendiosum pro mentis fictionibus habeo, ad calculum aptis, quales etiam sunt radices imaginariae in Algebra. Interim demonstravi, magnum has expressiones usum habere ad compendium cogitandi adeoque ad inventionem, et in errorem ducere non posse, cum pro infinite parvo substituere sufficiat tam parvum quam quis volet, ut error sit minor dato, unde consequitur errorem dari non posse. R. P. Gouye, qui objecit, non satis videtur mea percepisse.[41] (GP II, 305)

Though Archimedes is not named in this passage, I hope it is clear by now that he is nonetheless on Leibniz's mind.

5. *Archimedes' Principle* Again, The *Law of Continuity* and Leibniz's Fictionalisms

Paulo Mancosu has suggested that Leibniz's defense of the calculus involves a theory of "well-founded fictions,"[42] a phrase that Leibniz himself uses on at least a few occasions for infinite and infinitesimal quantities (cf. GM IV, 110: "fictions bien fondées"). And it is clear enough by now that for the use of such quantities in his calculus, the fiction is indeed well-founded and can be rigorously recast in non-fictional terms. But in Leibniz's writings the trope of the useful fiction extends into his mathematical reasoning well beyond manipulations of infinitesimals in quadratures. Alongside the infinitesimal is a netherworld of other fictional entities: the infinite ellipse with one focus at infinity, the unextended angle contained in a point, the point of intersection of parallel lines, the representation of rest as a kind of motion, etc. It may be that these fictions too can be understood to be well-founded in Leibniz's philosophy of mathematics. But if so, it is not at all clear that an accounting similar to that described for infinitesimals can be provided to cover the other cases. The understanding of infinitesimals as

[41] "Philosophically speaking, I hold that there are no more infinitely small magnitudes than infinitely large ones, i.e. that there are no more infinitesimals than infinituples. For I hold both to be fictions of the mind due to an abbreviated manner of speaking, fitting for calculation, as are also imaginary roots in algebra. Meanwhile I have demonstrated that these expressions have a great utility for abbreviating thought and thus for discovery, and cannot lead to error, since it suffices to substitute for the infinitely small something as small as one wishes, so that the error is smaller than any given, whence it follows that there can be no error. R. P. Gouye, who objected, seems to me not to have understood adequately."

[42] Mancosu, 1996, 173.

fictions does not extend in any obvious way to the remaining 'limit entities', for the reason that the mathematical theory of infinitesimals can claim to be modeled in – and so rigorously reducible to – a non-fictional finitist theory. There is not yet any evident counterpart model available for each, or any, of the other limit entities. If they too can be reinterpreted as disguised descriptions of facts, Leibniz does not say what the reductive analysis would be – what the undisguised truth is behind the fiction.

Leibniz does suggest a line of defense for the limit myths based on his *Law of Continuity*, which appears to have been formulated expressly for this purpose – or, at any rate, with the justification of mathematical fictions clearly in mind. Our discussion here must of necessity be brief,[43] but it is worthwhile to consider a precise statement of the Law in mathematical contexts. Here is how Leibniz states it in the 1701 document now called *Cum prodiisset*: "*Proposito quocunque transitu continuo in aliquem terminum desinente, liceat ratiocinationem communem instituere, qua ultimus terminus comprehendatur.*"[44] (H&O, 40) Its application to fictions such as the ellipse with one focus at infinity is clear. The infinite ellipse is equally a parabola – "transitur de Ellipsi in Ellipsin, donec tandem ipse focus evanescat seu fiat impossibilis, quo casu Ellipsis in parabola evanescit" (with ommissions; H&O, 41)[45] – and serves to link the two types of entities together into a single continuum. The principles describing the properties of ellipses will, upon the introduction of the fictional intermediary, translate smoothly to the case of parabolas. "Et ita licet ex nostro postulato parabolam una ratiocinatione cum Ellipsibus complecti" (ibid.).[46] Likewise the idea of the circle as an infinilateral polygon serves to connect "a common reasoning" about polygons with the circle itself by including the circle in the same series. With the *Law of Continuity* in force to uphold the generality of the reasoning, the introduction of the intermediate cases as fictions is then justified.

The precise character of the justification afforded to the use of such fictional entities by the *Law of Continuity* is somewhat more difficult to make out, however. A natural thought would be that the justification is prag-

---

43 For detailed discussions, see Bos, 1974, and Arthur, 2008b.

44 "If any continuous transition is proposed that finishes in a certain limiting case, then it is permissible to formulate a common reasoning which includes that final limiting case." (Child, 147)

45 "[...] we pass from ellipse to ellipse, until at length [...] the focus becomes evanescent or impossible, in which case the ellipse passes into a parabola." (Child, 148)

46 "Hence it is permissible, by our postulate, that the parabola should be considered with the ellipses under a common reasoning" (Child, 148).

matic: imagining the existence of such limit cases, or the projection of properties to them, provides economy in the formulation of principles and serves as a fertile heuristic in the process of discovery. The Law need not be taken strictly as a ("metaphysical") truth in that case, but only as a principle of inquiry or an "architectonic" aspect of mathematical theory-building.

Leibniz sometimes appears to envision a stronger status for the Law, however, and he can occasionally be found writing as if the *lack* of a fictional limit would threaten to *violate* the law. For instance, the 1702 note on the justification of the calculus sent to Varignon has this tone:

> Cependant quoyqu'il ne soit point vray à la rigueur que le repos est une espece de mouvement, ou que l'égalité est une espece de inégalité, comme il n'est point vray non plus que le Cercle est une espece de polygone regulier: neantmoins on peut dire, que le repos, l'égalité, et le cercle terminent les mouvemens, les égalités, et le polygones reguliers, qui par un changement continuel y arrivent en evanouissant. Et quoyque ces terminaisons soyent exclusives, c'est à dire non-comprises à la rigueur dans les varietés qu'elles bornent, neantmoins elles en ont les proprietés, comme si elles y estoient comprises, suivant le langage des infinies ou infinitesimales, qui prend le cercle, par exemple, pour un polygone regulier dont le nombre des costés est infini. Autrement la loy de la continuité seroit violée, c'est à dire puisqu'on passe des polygones au cercle, par un changement continuel et sans faire de saut, il faut aussi qu'il ne se fasse point de saut dans le passage des affections des polygones à celle du cercle.[47] (GM IV, 106)

The reductio here, as stated, is in order simply as an argument. If the *Law of Continuity* implies that the limiting cases be treated as belonging to the series that they limit, to deny that treatment would be absurd. Still, it would seem more plausible for the defense of fictions to invoke the Law as *vindicating* the *introduction* of limiting cases. Perhaps this is only a matter of right emphasis. But it remains perplexing. Notice that the stronger reading,

---

47 "Although it is not at all rigorously true that rest is a kind of motion or that equality is a kind of inequality, any more than it is true that a circle is a kind of regular polygon, it can be said nevertheless that rest, equality and the circle terminate the motions, the inequalities and the regular polygons which arrive at them by a continuous change and vanish in them. And although these terminations are excluded, that is, are not included in any rigorous sense in the variables which they limit, they nevertheless have the same properties as if they were included in the series, in accordance with the language of infinities and infinitesimals, which takes the circle, for example, as a regular polygon with an infinite number of sides. Otherwise the law of continuity would be violated, namely, that since we can move from polygons to a circle by a continuous change and without making a leap, it is also necessary not to make a leap in passing from the properties of polygons to those of the circle." (L 546)

according to which the Law straightforwardly implies that the limiting cases must be treated as belonging to the series they limit, would leave us asking why, in that case, this is a fiction at all rather than a matter of mathematical fact.[48] We need the distinction between fictional and factual consequences of the Law to remain intact; the reductio argument of the letter to Varignon, however, would seem to break it down. At the very least, an explanation is wanted.

I do not mean to suggest that Leibniz cannot construct a satisfactory defense of the use of fictions in his mathematics on the basis of the *Law of Continuity.* On the contrary, it strikes me as a promising resource for such a defense and one that deserves a detailed analysis, though such an analysis must fall outside the scope of the present essay. The point to observe here is simply that the justification based on the Law – whatever, precisely, it should turn out to be – will be quite different in character from the justification developed in DQA specifically for the use of infinite and infinitesimal quantities in the calculus. If we wish to call Leibniz a fictionalist about the whole range of entities and principles that he describes as 'fictions' in his mathematics, we should not be too quick to assume a single, integrated fictionalism in his philosophy equally embracing them all. Perhaps it would be wiser to consider Leibniz's fictionalism as divided into two different branches, one addressing infinite and infinitesimal quantities, the other concerning "intermediate" limit entities and the projection of properties and theorems to limit cases. Whereas the justification for the first will claim both pragmatic and reductionist grounds, the justification for the second will appeal to the *Law of Continuity.*[49]

If this is right, it would then be better to speak of Leibniz's fictionalisms than of a single fictionalist account in his philosophy of mathematics. Yet even if we come to see Leibniz's view as divided into two separate branches, there is a way to view them also as sharing a common root. For the Law of Continuity itself can be understood as a conceptual extension of *Archimedes' Principle.*[50] Recall again the Principle: for any quantities x, $y > 0$, if $x > y$, there is a natural number n such that $ny > x$. And this

48 Thanks to Emily Grosholz for pressing this point.

49 This result accords, at least superficially, with a suggestion of Bos, 1974, that Leibniz's considers two approaches to the justification of the calculus, "one connected with classical methods of proof by 'exhaustion'," the other in connection with a law of continuity" (ibid., 55). Bos's classic paper did not have the benefit of the DQA, however, and does not recognize the reducibility of infinitesimal terms to finite ones.

50 Richard Arthur also has noticed this link (correspondence). I do not claim that he would necessarily agree with the particulars of my presentation.

yielded the new principle of equality as a limit of differences: if for any n, $|x-y| < \frac{1}{n}$, then $x = y$. As we noted above in the discussion of *De infinite parvis*, Leibniz seems already to be extending the principle of equality in one way to consider differences smaller than finite differences by (in the terms of our analysis) allowing the variable for the degree of difference to include not just natural numbers but any value whatever, perhaps even infinite ones. A different sort of extension of the principle of equality would seem to lead to the Law of Continuity. Consider in particular the statement of continuity conditions in a 1688 document setting forth some general principles useful in mathematics and physics: "Cum differentia duorum casuum infra omnem quantitatem datam diminui potest, *in datis* sive positis, necesse est, ut simul diminuatur infra omnem quantitatem *in quaesitis*, sive consequentibus quae ex positis resultant."[51] (A VI, 4, 2032) The familiar thought of differences becoming less than any given difference is evident here already. This can be pressed just a little further. Let x and y be "what is given" or what is "presupposed," and let $f(x)$ and $f(y)$ be "what follows" or "is sought." The Law then says that as the difference $|x-y|$ becomes smaller than $1/\varepsilon$ for any $\varepsilon > 0$, the corresponding difference $|f(x) - f(y)|$ likewise becomes smaller than any given quantity.[52]

Consider for example the circle and the series of regular n-sided polygons. As n increases, the difference between the circle and the polygons becomes smaller without bound: for any given difference, it can always be shown that some polygon differs from the circle by less than the given difference. Likewise for the results of general principles true of polygons and applied to the circle: the differences between the resulting values diminish without bound as the series of polygons is extended. By the *Archimedean principle* of equality as the limit of differences, the difference between the circle and the polygons will then be nothing at all – i.e., the circle will simply *be* a polygon – and likewise the results of applying general principles concerning polygons to the circle will not differ at all – i.e. those principles will be valid for the circle as well. Hence the circle, which is the limit of the series of regular polygons, will be included in the series which it terminates, and "liceat ratiocinationem communem instituere, qua ultimus

[51] "When the difference between two instances in what is given, or is presupposed, can be diminished until it becomes smaller than any given quantity whatever, the corresponding difference in what is sought, or what follows, must of necessity also be diminished or become less than any given quantity whatever."

[52] With the right articulation of 'corresponding', of course, the ε,δ definition of continuity can be elicited here.

terminus comprehendatur."[53] (H&O, 40) And this is precisely what Leibniz enshrines as the Law of Continuity.

Archimedes' Principle runs very deep in Leibniz's thought, and we have seen it surfacing in two key places with respect to the fictions he promulgates in his mathematics. It plays a pivotal role in his finitist foundation for infinitesimalist techniques in DQA. And it appears in the kernel of the Law of Continuity. Those two strands of thought lead in different directions but come back together again in his philosophy to yield two different forms of justification for the use of ideas in mathematics that Leibniz calls fictions. If there is no single across-the-board account of fictions in mathematics that it would be proper to call "Leibniz's fictionalism", nonetheless his fictionalisms can happily be styled Archimedean.[54]

---

53 "[…] it is permissible to formulate a general reasoning which includes that final limiting case." (Child, 147)

54 My thanks to the participants of the 2006 Loemker Conference at Emory University, where an earlier version of this paper was presented, and to the Editors of the present volume. Thanks also to Christie Thomas and Bob Fogelin for discussion, and special thanks to Richard Arthur for suggestions, clarifications, answers to several questions and help with passages from DQA.

O. Bradley Bassler

An Enticing (Im)Possibility: Infinitesimals, Differentials, and the Leibnizian Calculus

1. Introduction: The Argument

In this paper I consider a brief manuscript passage, first published by André Robinet in his volume *Architectonique disjonctive automates systémiques et idéalité transcendenantale dans l'Œuvre de Leibniz* (1986), in which Leibniz claims to prove the impossibility of infinitely small quantities. As Robinet remarks, this passage is crossed out ("barré") and not taken up again later ("non repris") (Robinet, 1986, 292). According to Robinet, the passage occurs in the broader context of Leibniz's correspondence with Varignon, in which Leibniz announces that he believes he has found such a proof (1702).[1] However, in his edition of this and other manuscripts, Enrico Pasini dates the manuscript according to several mutually supporting criteria as coming from the first period of Leibniz's residence in Hannover, hence in the years 1676 and following.[2] Conceptually and textually, it is aligned with Leibniz's exploration of the so-called *tetragonal method* and the issue of whether when the number of inscribed tetragons is allowed to go to infinity the error in the calculation of the area covered goes to zero.[3] Although it seems clear that Leibniz does not further pursue the strategy outlined in this fragment, inspecting this argument will nonetheless give us an appreciation for one line of thought Leibniz entertained about the non-existence of infinitely

---

1 "Je crois qu'il n'y a point de créatures au-dessous de laquelle il n'y ait une infinité de créatures, cependant je ne crois point qu'il y en ait, ni même qu'il y en puisse avoir d'infiniment petites et c'est ce que je crois pouvoir démontrer." (GM IV, 110)

2 Of this and a closely related manuscript, Pasini says: "Questi due manoscritti, risalenti al primo periodo della permanenza hannoveriana di Leibniz, sono qui considerati come facenti parte di un unico testo: a ciò inducono la rispondenza lessicale, la continuità del contenuto, la somiglianza grafica e, infine, l'identità della carta, riscontrabile nella filigrana (rosetta)." (Passini, 1985–1986) I am grateful to Siegmund Probst for bringing Pasini's edition of this manuscript to my attention, and to Tamara Levitz for help with Italian.

3 See Robinet, 1986, footnote 64. See also Knobloch, 2002.

small quantities. My claim will be, in particular, that by uncovering the assumptions and antecedent arguments on the basis of which this line of reasoning proceeds we can come to a better appreciation of the competing constraints which directed Leibniz's thoughts about infinitely small quantities. In particular, I will suggest that it is with respect to the fundamental indeterminacy in Leibniz's conception of the continuum that we may understand both the power and the limitations of Leibniz's treatment of quantity and, in particular, the status of infinitely small quantities.

Here is a translation of the passage under concern as transcribed by Robinet. In the translation, I have not endeavored to be absolutely literal, but rather to make my understanding of the argument as clear as possible. In two cases I have followed Pasini's reading rather than Robinet's, as noted below. In both cases the difference involves an orthographically plausible alternate reading of a letter used as a symbol.

> That, however, infinitely small quantities are fictions I thus /*easily* / prove. Let AB be any infinitely small straight line, and let CD be a normal finite line. Now seek between AB and CD a mean proportional EF. This will either be in infinite proportion to AB or else will be finite relative to AB, contra hypothesis. Now seek for this same EF.CD a third proportional GH, which will be infinite. For it cannot be infinitely small, since then CD would be larger. But it will be greater than anything finite, for if it were finite GH and CD would be of the same level of magnitude. Now seek for the third proportional EF.CD.GH a fourth proportional IK, which will be greater than GH. Now just as greatly infinite as GH (infinite) is to CD (finite), so will IK (which is to GH as GH is to CD) be infinitely greater than GH. I adjoin:
>
> a ______________ b infinitely small
> e* ____________________ f infinitely small
> c __________________________ d common finite
> g __________________________________ h infinite – 1
> i ____ k infinitely infinite.
>
> Transposing IK† into GL so that G is its common beginning, IK or GL itself certainly stretches out to a much greater length than GH, since indeed it is greater, and will have a part HL beyond GH, so that GH is finite; that is, the point H is a common end of GH and that part of GL which extends beyond it. But it is absurd that any line terminating in points G and H itself have an infinite magnitude." * reading 'e' with Pasini for Robinet's 'b'. † reading 'IK' with Pasini for Robinet's 'LK'.[4]

[4] I give here verbatim Robinet's transcription of this textual passage, which appears at Robinet, 1986, 292. My translation follows Robinet's transcription of LH XXV, VIII, f. 37:

Let me begin by attempting to put my basic understanding of the argument into English prose. Leibniz begins by assuming the existence of an infinitely small straight line AB. All quantities which Leibniz discusses are understood to be geometrical; this is important in general terms and will also play a specific and key role in the argument when infinite terminating magnitudes are considered. Next, Leibniz juxtaposes to this infinitely small line *AB* a regularly finite line segment *CD* and tells us to take a mean proportional between the two segments, which he calls *EF*. Here a mean proportional is a (geometric) quantity satisfying the equation:

$$\frac{AB}{EF} = \frac{EF}{CD} \quad . \tag{1}$$

Why Leibniz should think that a mean proportional exists in the case of an infinitely small quantity and a finite quantity is not made clear in this passage, and I will return to this point later; for now, along with Leibniz, I simply assume the existence of such a mean proportional. The mean proportional whose existence Leibniz declares bears some resemblance to broader seventeenth century notions of infinitesimals, which were often understood to hover precariously between something and nothing, and which were sometimes even characterized as magnitudes standing in proportion to a finite quantity as a finite quantity stands in proportion to infinity.[5] Although Leib-

---

Quantitates autem infinite parvas esse fictitias / *facile* / ita ostendi potest: sit vera aliqua recta AB infinite parva, CD vero linea finita communis. Jam inter AB et CD quaeratur media proportionalis EF, ea etiam erit infinite proportionalis AB etiam foret finita communis AB, contra hypothesin jam ipsis EF.CD. inveniatur tertia proportionalis GH ea insit infinita. Nam infinite parva esse non potest, cum ipsa CD finita sit major. Finita autem quavis major est, alioqui si finita communis esset ipsis GH.CD. tertia proportionalis. EF.CD.GH quaeratur dextra proportionalis IK ea erit major ista GH. Imo major infinities quia cum sit GH (infinita) infinities major quam CD (finita) etiam IK (quae est ad GH ut GH ad CD) ipsa GH infinities major erit. Applicetur:

a ____________ b infinite parva
b ____________________ f infinite parva
c ______________________________ d finita communis
g ____ h infinita – 1
i ____ k infinitiis infinitus

LK ipsi GH transponendo eam in GL ita est earum initium commune G necesse est IK vel GL quippe majorem longius praetendi quam GH, cum enim major sit, habebit partem HL ultra GH finita est ergo GH; seu terminum habet H punctum scilicet commune ipsi et excessi majoris GL supra ipsam. Absurdum est autem rectam utriusque terminatur punctis G et H ipse magnitudine infinitam."

[5] Thanks to Doug Jesseph for suggesting this way of putting it.

niz speaks of "infinitely small quantities" rather than "infinitesimals" here, the resemblance to this broader context is significant.

In assuming the existence of a mean proportional, Leibniz is not *directly* assuming the existence of a proportion between AB and CD, but he does go on to declare that EF cannot stand in finite proportion to AB, for then it would stand in finite proportion to AB and infinite proportion to CD. So it must stand in infinite proportion to AB (and although Leibniz does not say so explicitly, in infinitely small proportion to CD as well). Next Leibniz seeks a "third proportional" to EF and CD; in this context, this is a line segment satisfying the equation:

$$\frac{EF}{CD} = \frac{CD}{GH} \quad . \qquad (2)$$

Again, Leibniz assumes the existence of such a proportion, but I will argue that this follows by an argument much like the one which guarantees the existence of the mean proportional EF. Next, Leibniz seeks for the three quantities EF, CD, and GH a "fourth proportional" IK satisfying:

$$\frac{EF}{CD} = \frac{CD}{GH} = \frac{GH}{IK} \quad . \qquad (3)$$

If, informally, we think of GH as the infinitely large quantity corresponding to EF (relative to the finite segment CD) then we may think of IK as the infinitely large quantity corresponding to the original infinitely small quantity AB.

Next, Leibniz asks that we extend the line GH to a line GL which has the same length as the line IK. Then the point H will lie internally upon the line GL which has initial point G. This means, in particular, that the infinitely large quantity GH is identified as a line segment which terminates on both ends, and Leibniz finds this absurd. This is the absurdity which contradicts the original supposition that AB was an infinitely small quantity. Since this is a proof by contradiction, and we have derived the contradiction that an infinitely long terminating line exists, the original supposition that AB is infinitely small must be false.[6] Consequently, infinitely small quantities are not possible.

The proof breaks rather naturally into two parts, and so I will organize my discussion along these lines. In the first part of the proof a series of proportionals is established stretching from infinitely small to infinitely large

---

[6] On the status of proof by contradiction in 17th century debates in the philosophy of mathematics, see (Mancosu, 1996).

quantities. Notice that the largest quantity introduced, IK, corresponds to the smallest quantity, AB, with which we begin. Effectively, we have taken an infinitely small quantity (AB), a finite quantity (CD) and an infinitely large quantity which stands as a third proportional to the first two (IK), and we have interposed mean proportionals between each pair of them. That is,

$$\frac{AB}{CD}=\frac{CD}{IK} \tag{4}$$

and EF and GH stand as mean proportionals between AB and CD and CD and IK, respectively.

In the second part of the proof, Leibniz embeds the line GH in a longer line GL which is quantitatively equivalent to IK and derives a contradiction from this embedding. Substantively, we may say that the proof has two "components": the production of a series of proportionals which stand in quantitative relation and then a geometric argument, the force of which relies on Leibniz's conviction that infinite lines cannot terminate on both ends.

The first part of the argument critically depends on the possibility of finding a mean proportional between two incommensurable quantities, i.e. two quantities which do not stand in finite proportion. Leibniz indirectly addressed the existence of such a mean proportional in a 1695 article in the *Acta Eruditorum*, *Responsio ad nonnullas difficultates a dn. Bernardio Nieuwentijt circa methodum differentialem seu infinitesimalem motas* (GM V, 320–328).[7] In a passage which, as Henk Bos commented, has "repeatedly bewildered historians of mathematics," Leibniz demonstrates the existence of a third proportional for the quantities x and dx, i.e. a finite quantity x and its differential increment dx, which measures the difference between two "successive" values of x. Specifically, Leibniz shows that a third proportional ddx can be found such that

$$\frac{ddx}{dx}=\frac{dx}{x}\ . \tag{5}$$

This makes dx a mean proportional between ddx and x, and so bears indirectly on the existence of mean proportionals between finite and infinitesimal quantities.

In order to understand the demonstration Leibniz provides for the existence of such a third proportional, some background regarding the Leibnizian approach to differentials is required, and on the basis of this back-

[7] A French translation with commentary is given in Leibniz, 1989, 316–337.

ground we will first be in a position to see how issues concerning the *indeterminacy* of geometrical quantity enter into the set of issues surrounding the status of infinitesimals.

2. Differentials

Crucial for understanding Leibniz's production of the third proportional *ddx* in the *Responsio* is the predominance of the notion of *differential* in the Leibnizian calculus. This notion is an infinitary (specifically, infinitesimal) extrapolation[8] of the notion of a finite difference of terms. As Bos remarks: "The usual concept of the differential was connected with the concept of the variable as ranging over an ordered sequence of values; the differential was the infinitesimal difference between two successive values of the variable." (Bos, 1974, 11) As such, the differential is intimately connected with the manner in which we proceed from one value in an ordered sequence to the next. A great part of the flexibility of the Leibnizian version of the calculus lay precisely in the liberty of specifying the "progression of the variable," i.e. the way in which we move from one value to the next, in a way which was advantageous to the solution of any given problem. What we will see, however, in the argument concerning the existence of a third proportional, is that the existence proof which Leibniz provides for this proportional depends on the specification of the progression of the variable as well.

In this argument, Leibniz assumes that the quantity x is in geometric progression, which means that the differences between infinitesimally proximate terms in the progression of x values stand in geometric progression. Thus, if I think of $x_1, x_2, x_3, x_4, \ldots$ as a progression of (infinitely proximate) x values, the claim that x is in geometric progression means that for some finite constant quantities r and c,

$$x_1 = c,\ x_2 = cr,\ \ldots,\ x_n = cr^{(n-1)}.$$[9]

Now we take a second variable, y, and assume that it is in arithmetical progression, so that if $y_1, y_2, y_3, y_4, \ldots$ is a progression of y values, then the claim that y is in arithmetic progression means that for some (infinitesimal) constant quantity s and finite d,

$$y_1 = d,\ y_2 = d + s,\ \ldots,\ y_n = d + (n-1)s,$$

[8] I borrow the notion of extrapolation from Bos, 1974, 13. See also Lavine, 1994, 257–258.

[9] Note that r will be a quantity differing from 1 by an infinitesimal amount.

so that

$$y_1 + s = y_2, y_2 + s = y_3,$$

etc. The fact that y is in arithmetic progression means that all the dy's are equal to each other. Now since dy is constant, choose a particular dx: Leibniz declares that it will stand in relation to dy as x stands in relation to some constant a, i.e.

$$\frac{dx}{dy} = \frac{x}{a} \quad . \tag{6}$$

From a modern perspective, it is easy to see that this is equivalent to letting x and y vary in the relation

$$x = be^{y/a},$$

since then

$$\frac{dx}{dy} = \frac{be^{y/a}}{a} = \frac{x}{a} \quad .$$

To see Leibniz's claim directly in terms of the condition on the progressions of the x and y variables, note that if the variable x progresses geometrically, then so do the differentials dx. (Cf. Leibniz, 1989, 333, n.50) For if $x_1 = c$, $x_2 = cr, \ldots, x_n = cr^{(n-1)}$, then $dx_1 = x_2 - x_1 = cr - c = (r-1)c$, and in general, $dx_n = x_n - x_{n-1} = cr^{(n-1)} - cr^{(n-2)} = (r-1)cr^{(n-2)}$. This means that dx varies with respect to the constant dy as x varies with respect to some constant a, as was to be shown. Multiplying both sides of equation (6) by dy, we have:

$$dx = \frac{xdy}{a} \quad . \tag{7}$$

Then, since dx is also a quantity we may look at its differences, which we write ddx. Thus, for example, $ddx_1 = dx_2 - dx_1$, etc. On the left hand side of the equation, we will have simply ddx, but what about on the right hand side? Consider, for example,

$$d(xdy)_1 = (x_2[y_3 - y_2]) - (x_1[y_2 - y_1]).$$

Since the differences in y are some constant factor s, this amounts to $s(x_2 - x_1) = sdx_1$, and in general, noting that $dy = s$, we will have that $ddx = \frac{dxdy}{a}$. Using (7) to substitute in for $\frac{dy}{a}$, we may now rewrite this in the form Leibniz gives, i.e., $xddx = dxdx$, or, finally,

$$\frac{x}{dx} = \frac{dx}{ddx} \quad . \tag{8}$$

Hence *ddx* stands in the relation of a third proportional to *x* and *dx*, so that *dx* is a mean proportional between *x* and *ddx*.

3. Evaluating the Argument

As remarked above, this argument has fared rather badly among historians of mathematics, and even Bos would find it at fault were Leibniz to claim that *ddx* always stands in the relation of third proportional to *x* and *dx*. But as Bos insists, "[i]t is, however, a perfectly acceptable argument, if one bears in mind that Leibniz does not claim that *ddx* is always the third proportional of *x* and *dx* but rather gives an example in which such is the case." (Bos, 1974, 24)[10] That is, we must find a curve such that when the *x* variable proceeds geometrically the *y* variable proceeds arithmetically (and vice versa). Such is the case for the logarithmic curve, which in Leibniz's time was usually defined precisely in terms of the condition just stated on the progression of variables.

Next, I must make some remarks about the extent to which this argument, even properly understood, helps us with the main assumption in the first half of Leibniz's argument against the possibility of infinitely small quantities. First, we should note that the argument of the *Responsio* assumes the existence of infinitely small quantities such as the differentials *dx* and *dy*, or, more accurately, specific differentials such as dx_1, dx_2, etc.[11] But,

---

10 Even more recently, Marc Parmentier has remarked: "L'argument de Leibniz est peu convaincant car il repose sur une généralisation un peu brutale B partir d'un cas particulier ad hoc qui ne concerne que des séries numériques et non des fonctions" (Leibniz, 1989, 332–333, n. 49). I am not entirely sure what Parmentier's point here is, since Leibniz frequently moves back and forth between the consideration of series and (not functions but) relations between variables; the whole attitude surrounding the specification of variables relies upon it. Further, if we adopt a function theoretic point of view then Leibniz's point can be made with respect to the exponential function of *y*. The fact that Leibniz treats a particular case here does not bother me any more than it does Bos, and should not bother Parmentier.

11 We should not think of differentials such as *dx* or *dy* as quantities but rather as variables, just as *x* and *y* are thought of as variables. It is particular *instances* of these variables which are themselves quantities. On the differential as variable, see Bos, 1974, 17. Herbert Breger asserts that Bos' identification of inconsistencies in the Leibnizian calculus is a result of Bos' treating differentials as fixed infinitesimals, but if so, then the problem would be one of an

then, so does the argument at hand, since it has the form of a proof by contradiction. We should not, in any case, expect the argument from the *Responsio* to shed any direct light on the existential status of infinitesimals. Rather, we should look to it to understand how Leibniz thinks about obtaining proportionals between infinitely small and finite quantities.

This brings us to an obvious, second point. The argument in the *Responsio* is about finding a third proportional, not a mean proportional. It does, however, establish dx as a mean proportional between ddx and x. Does this help us at all?

I think it does, but only if we make explicit a Leibnizian commitment that he does not clearly articulate. For Leibniz's argument to go through, we must assume that *no* infinitely small quantity can lie at a level of the infinitely small which is closest to the level of the finite. Further, there is no level of infinity from which the differential dx need necessarily be drawn. Bos makes this latter point explicitly, but also notes that the early practitioners of the Leibnizian calculus seemed not to notice this (Bos, 1974, 24).[12] Leibniz does not make this point explicitly. However, I would argue that if Leibniz is even willing to consider the possibility that a mean proportional could exist between an arbitrary infinitesimal quantity AB and a finite quantity CD, he must recognize the former, much weaker, point in some way. At least by implication, it seems that Leibniz is (consistently) committed to the position that an infinitely small quantity cannot stand in "closest proximity" to the finite quantities if it is to behave like a quantity at all (and hence stand in proportional relation to quantities at other levels).

With this point in hand, the procedure that Leibniz employs in the *Responsio* argument is "invertible" in such a fashion as to allow for the construction of a mean proportional between AB and CD. In particular, let x be a variable progressing geometrically, and y be a second variable progressing arithmetically, such that for a particular value of the variable x, $\frac{AB}{ddx}=\frac{CD}{x}$. Then for that value of x, dx will satisfy $\frac{AB}{ddx}=\frac{EF}{dx}=\frac{CD}{x}$, and so

---

internal inconsistency in Bos' own treatment of the Leibnizian differential issue. See Breger's paper in this volume.

12 This "indiscernibility" of the level from which dx is drawn should be compared with Jan Mycielski's indiscernibility axioms in his "analysis without actual infinity," which seems to play an analogous role; see Mycielski, 1981, and for a more accessible presentation, Lavine, 1994, esp. 278–288. This would be particularly useful for an attempt to follow out the strategy of legitimating Leibniz's approach to the calculus by way of a grounding of it in something like the Greek method of exhaustion; see Bos, 1975, 55.

we have effectively constructed EF. I suggest that Leibniz may have had something like this in mind when he assumed the existence of a mean proportional between AB and CD. The further proportionals Leibniz goes on to construct in the first half of the argument all take the form of third proportionals, so their construction can be modeled even more directly on the argument of the *Responsio*.

Let me turn, then, to the second half of the argument. Here, I will suggest, Leibniz is involved in a quandary which may be traced back to various aspects of the Leibnizian calculus. On the one hand, the calculus as Leibniz pursues it naturally invokes the concept of infinitely large quantities, and given that Leibniz conceives of quantities geometrically, this seems to point, in particular, to the existence of infinite lines. On the other hand, by virtue of the fact that Leibniz does consider quantity geometrically, he expresses discomfort at various points throughout his career about the idea that there could be lines infinite in length but terminating at both ends. It is a major point of discussion in the correspondence with Johann Bernoulli in the mid-1690's, for example, where Leibniz expresses worries about the status of infinitely large and small geometric quantities,[13] as he does in the relevant letter to Varignon in a bracketed passage not included in the letter as sent. Here Leibniz recognizes that there is a strong tension between the notion of actually existing infinite quantities terminating at both ends and his tendency to identify the source of the infinite in the unterminated ("interminé"), and hence the indefinite.[14] In short, there seems to be an incompatibility between Leibniz's idea of the mathematical infinite as indefinite, his idea of quantity as geometric, and the inverse relation in which (existent) infinitely small quantities stand to infinitely large ones. I will argue that the best way for Leibniz to circumvent the discomfort he feels in holding all three of these commitments simultaneously is to dispense with the idea of there really being infinitely small quantities, i.e. to treat them as fictions. This strategy is still not entirely comfortable, for there are ways in which Leibniz's "fictionalism" leads him to hedge on all three of these commitments in various ways as well.

---

13 For a discussion, see Bassler, 1998a, 860.

14 "[...] it is unnecessary to make mathematical analysis depend on metaphysical controversies or to make sure that there are lines in nature which are infinitely small in a rigorous sense in contrast to our ordinary lines, or as a result, that there are lines infinitely greater than our ordinary ones [yet with ends; this is important inasmuch as it has seemed to me that the infinite, taken in a rigorous sense, must have its source in the unterminated; otherwise I see no way of finding an adequate ground for distinguishing it from the finite]". (GM IV, 91) The translation is taken from (Leibniz, 1969, 543).

In any case, I will argue that the impossibility proof be seen along these lines. In the first half of the proof, Leibniz commits to the existence of infinite quantities, which he produces, in the first instance, as a third proportional with respect to infinitely small and finite quantities and, in the second instance, with respect to a finite and an infinite quantity. As such, infinitely small and infinitely large quantities stand in an inverse relation about a finite quantity which itself stands as their mean proportional. In particular, keeping the example from the *Responsio* in mind, we may see these quantities as individual instances of variable quantities subject to the inverse operations of differentiation and integration (summation) in the process of generating proportional quantities.[15]

Nonetheless, Leibniz's conception of quantities, whether fixed (e.g., AB, or dx_1) or variable (ddx, dx, x) is consistently geometric, and from the perspective of geometry the problems accruing to the notion of the infinitely large must be distinguished from those associated with the infinitely small, at least psychologically and arguably conceptually as well. That is, while it is difficult if not impossible to imagine a line starting at a particular point, going on forever, and then terminating at another point, the idea that two points on a line could be infinitely close to each other seems considerably more palatable. It is, I believe, the impalatability of the former idea – of an infinite line terminating at both ends – which leads Leibniz to the declaration of absurdity in the impossibility proof. Further, if we include the fact that for Leibniz, the notion of the mathematical infinite is the notion of the indefinite, this makes it all the more difficult to imagine an infinite line terminating on both ends, for this would be an infinite (i.e. indefinite) extension which terminates. At any rate, this is much different from Leibniz's conception that the model for the mathematical infinite is the sequence of natural numbers, which has a beginning but no end.

15 There is a critical point in the background which I am not arguing for explicitly here. It is that, insofar as infinitesimal and infinite quantities are at issue, the conceptual analysis should go, e.g., from the concept of differential to the concept of infinitesimal and not vice versa. This point of view is supported in detail by Bos and, in a very different and extremely illuminating way, by Herbert Breger (Cf. Breger, 1990a, 56–7). Breger's point, which strengthens Bos's approach, involves noticing a variety of contexts in 17th century mathematics in which the concept of motion made possible "pioneering achievements" in mathematics, achievements that were later taken over by appeal to the mathematical structure of the continuum. This point goes some way toward explaining why Leibniz's attitude toward the famous *labyrinthus de compositione continui* largely privileges the metaphysical foundations of physics over those of mathematics.

Still, even in the face of all this discomfort we must recognize that Leibniz struck the impossibility argument and did not return to it. Why? Any answer seems likely to remain mere speculation, but I would suggest that, on the one hand, Leibniz did not feel he needed to demonstrate the inexistence of infinitely large and small quantities – their existence would not threaten the consistency of his calculus in any obvious way any more than their inexistence[16] – but, on the other, that Leibniz was all too aware that the calculus traded on a formal analogy between infinitely small and large quantities that conceptually did not sit well with the existence of such quantities. My evidence for this latter claim is that, although the Leibnizian calculus as, for example, Bos presents it, puts infinitely small and infinitely large quantities on a par, as Bos also points out as a matter of fact infinitely large quantities rarely appear in Leibniz's own work (Bos, 1974, 78–80). Rather than treating sums of finite quantities as infinite, which seems on the face of it the most natural approach, Leibniz usually evaluates such sums relative to a differential so that the quantity involved is finite rather than infinite. On the other hand, he does not manifest such scruples when infinitesimal quantities are involved. Perhaps, as the abandoned impossibility proof suggests, it was not directly the existence of infinitesimal quantities which troubled Leibniz so much as the implication that should they exist then so would infinitely large ones.

I think there is a more general reason why Leibniz was generally unsympathetic to the existence of infinitely small quantities, but it is also more difficult to articulate, both because the point lies at a deeper conceptual level of the Leibnizian calculus and also because outside of this technical context there is no comparable set of philosophical terminology for considering it directly.[17] We can begin to see this point, however, by con-

---

16 As Leibniz presents it, from a "modern" perspective his calculus *is* inconsistent, at least according to Bos, but Bos also stresses that we must undertake to explain why a demonstrably inconsistent calculus was capable of being so productive; see Bos, 1974, 12–13. This, however, is not the sort of inconsistency I have in mind when I say Leibniz's calculus was not threatened either by the existence or inexistence of infinitesimal quantities. What I mean, which is more pedestrian, is that neither the existence nor the inexistence of infinitesimals seemed to be an impediment to the manipulation of the calculus and the use of it to solve mathematical problems. This is a point that Leibniz himself makes as early as the 1676 manuscript, *De quadratura arithmetica circuli, ellipseos et hyperbolae cuius corollarium est Trigonometria sine tabulis*, in the Scholium to Proposition 23; see DQA, 69.

17 Indeed, the closest analogue seems to lie in Leibniz's metaphysical foundations of physics, and in particular in that dimension of it which concerns the sprecification of a physical quantity progressing with respect to space or time. Here see especially Kangro, 1969.

sidering the conceptual grounds for Leibniz's turn away from the Cavalierian theory of indivisibles and toward his own differential approach to the calculus. The point, briefly, is that when Cavalieri considers areas, for example, as surfaces enclosing "all the lines," the collection of lines (which need not actually compose the surface, though Leibniz seems to read Cavalieri this way) is *naturally* infinite. When Leibniz turns from Cavalieri's method to his own, he eliminates this infinite collection, but not simply because he finds infinite collections undesirable – although this may very well be a part of the motivation. Rather, when Leibniz multiplies such infinite sums by a differential and thereby returns them to the domain of finite quantity, he also produces a mathematical object which is *independent of the specification of the progression of the variable.* By this method, Leibniz says, he can "das dx expliciren und die gegebene quadratur in andere infinitis modis transformiren und also eine vermittelst der anderen finden."[18] (Leibniz to Bodenhausen, GM VII, 387) Leibniz views this benefit in terms of his capacity to measure the infinitly small: "Elementa infinite parva sunt mesuranda". (Leibniz, 1875, 597) But this was different from Cavalieri's method: "Ea vero infinite parvorum aestimatio Cavalierianae methodi vires excedebat".[19] Whether there is an essential conceptual connection between Leibniz's innovations regarding the "measurement of the infinitely small" and his evaluation of finite areas as opposed to infinite sums is not entirely clear to me, but it is clear that his focus on the measurement of the infinitely small dovetails with his tendency to think of infinitely small quantities as terminating geometrical line segments and areas as terminating two-dimensional figures,[20] which is not possible in the case of infinite lines.

The chief technical benefit of Leibniz's differential geometrical approach to the calculus is, we may say, that it preserves symmetries between the roles of the variables: we are not required to take x as an independent variable and hence y as a function of it, or vice versa. Indeed, the analogous procedure in the context of the Leibnizian calculus (which Bos argues, I think rightly, is not function based) is to specify the x variable as constantly progressing, hence making the y variable effectively a function of it; thus

18 "[...] separate the dx and transform a given quadature into other infinite modes, and thus find the one by means of the other."

19 "The infinitely small elements are to the measured. [...] But this measurement of the infinitely small was beyond the power of the Cavalierian method."

20 Not all areas are terminating figures, as is witnessed by the example of the Torricelli "tube." On the evaluation of such unbounded areas, see especially Knobloch, 35–37, Knobloch, 1994, esp. 276–277, and, most recently, Knobloch, 2006.

the symmetry between variables is broken in terms of the particular specification of the progression of variables. But this general preservation of symmetry is achieved at the expense of various indeterminacies associated with the Leibnizian calculus. There is freedom in the specification of the progression of the variable, but even more dramatically the differential need not even be drawn from any particular level of the infinitely small. More generally still, it is not the differentials themselves which are determinate but the relations of differentials. This point alone already suggests the irreality of differentials as (variable) infinitesimal magnitudes outside of the relations in which they stand to each other. Beyond this there is the thorny question of the even more radical indeterminacy of higher-order differentials, which I leave aside for the purposes of this paper.[21] And finally, there is the growing conviction in Leibniz's philosophical development that the source of the mathematical infinite must be located in the indeterminate, that is, the indefinite. Ultimately, I believe it is on the basis of this growing conviction that Leibniz finds the sort of impossibility proof offered here unnecessary, inferring the impossibility of infinitesimal and infinite magnitudes directly from the conception of the indefinite as infinite.[22]

The more basic symmetry, which we may say Leibniz preserves in one regard and breaks in another, is the symmetry between the infinitely small and the infinitely large. On the one hand, Leibniz requires that differentiation and integration stand as inverse operators, and this requires a formal analogy between the roles played by the infinitely small and the infinitely large. On the other hand, the desire alone to preserve the freedom of specification of the progression of variables leads Leibniz to treat the domain of the infinitely small differently than the domain of the infinitely large; then, in addition to this there are the problems associated with infinitely large geometric quantities terminating on both ends. The differential has an "edge" over the integral in the Leibnizian calculus. Ironically, it may be just this edge, along with the psychologically greater plausibility of infinitely small quantities, which made actually existing infinitesimal quantities such an enticing (im)possibility.

---

[21] See Bos, 1974, esp. 26–30.

[22] See Bassler, 1998a.

4. Comparison Texts and Conclusion

In the final portion of this essay, I turn to the necessarily more technical issue concerning the place of the impossibility proof within the context of Leibniz's repeated investigations of the status of infinitesimal quantities. Given Pasini's thorough dating of the manuscript passage, it seems most natural to take the argument as dating from Leibniz's first Hannover residency, but whichever of the two datings is accepted, the manuscript derives from Leibniz's "mathematical maturity." In what follows, I would like to consider a short census of other passages in which discussions of the status of infinitesimals occur. First, I offer a comparison of the argument with other authoritatively dated passages in which Leibniz discusses the status of infinitesimals and infinite lines, and second, arguments which show that the conceptual issues in the argument I have considered remain vitally at play up through the late 1680's. If we include the 1702 letter to Varignon itself, this provides a skeleton of passages indicating Leibniz's interest in the status of infinitesimals over the majority of his mathematically mature career.

There are a number of passages in the Leibnizian corpus in which Leibniz discusses the status of infinitely small quantities, and also passages in which he discusses the status of infinite lines, whether these be conceived of as terminating or "interminate." I have discussed, in particular, the status of infinitely small quantities in Leibniz's pre-Paris writings in a series of articles,[23] and I do not find anything like the argument considered here in that period, so I will confine myself to a consideration of writings from the Paris period and after. First, there is a significant short manuscript dating from 26 March 1676, *On the infinitely small*, in which Leibniz discusses the status of infinitesimal quantities quite explicitly; this manuscript has been commented on rather extensively by Richard Arthur in his Yale Leibniz edition of Leibniz's writings on the labyrinth of the continuum (DLC). Here is the most relevant paragraph from the manuscript:

> Videndum exacte an demonstrari possit in quadraturis, quod differentia non tamen sit infinite parva, sed quod omnino nulla, quod ostendetur, si constet eousque inflecti semper posse polygonum, ut differentia assumta etiam infinite parva minor fiat error. Quo posito sequitur non tantum errorem non esse infinite parvum, sed omnino esse nullum. Quippe cum nullus assumi possit.[24] (A IV, 3, 52)

[23] See Bassler, 1998b; Bassler, 1999; and Bassler, 2002.

[24] "We need to see exactly whether it can be demonstrated in quadratures that a differential is nonetheless not infinitely small, but that which is nothing at all. And this will be shown if it

Leibniz's argument in this passage is conditional in nature: he asserts that if a polygon can approximate to such a fineness that when the differential is infinitely small the error is smaller, then the error is nothing at all. Then, Leibniz asserts, the differential will be nothing at all. Leibniz does not assert directly that the differential is nothing at all. But Arthur has argued that in fact by a tetragonal approximation, the error term can be made smaller than the differential.[25] Then, according to Leibniz, it follows that the differential is nothing at all. Granting Arthur's point that the error term can be made smaller than the differential, we still need to ask about the logical structure of the hypothetical argument. It seems Leibniz makes what, from a later vantage in his career, would be an error. For assuming that the error is smaller than the differential need not imply that the error is zero, only that the error *goes to zero* (Arthur concedes this is a possible objection as well).[26]

In any case, later in his career, Leibniz certainly does not say that the differential is nothing at all, and in fact he explicitly criticizes Nieuwentijt on exactly this point: although incomparably small quantities can be neglected, if we replace them by 0 we retain an equation which is true, but identical, and comes to nothing ("non prodest").[27] The law of homogeneity, to which Leibniz alludes in his response to Nieuwentijt and which he makes explicit in the "Symbolas memorabilis calculi algebraici et infinitesimalis in comparatione potentiarum et differentiarum, et de lege homogeneorum transcendentali,"[28] (GM V, 377–382) allows Leibniz to neglect incomparable quantities while not setting them equal to zero.

By way of contrast, the argument I have considered in this paper makes no such mistake. Further, it relies on issues which involve distinguishing between levels of infinitesimals which were not explicitly on the table in the passage which Arthur cites. Indeed, with respect to this set of issues the argument from 1676 seems still quite naive. There is no (explicit) recognition that the error term will be of the same infinitesimal order as the differential, and the argument as stated seems to preclude the existence of infinitesimals of different orders. The argument I have considered, in contrast, relies on just this distinction.

---

is established that a polygon can always be bent inwards to such a degree that even when the differential is assumed infinitely small, the error will be smaller. Granting this, it follows not only that the error is not infinitely small, but that it is nothing at all – since, of course, none can be assumed." (DLC, 64–5)

25 See DLC, Arthur's introduction, lv-lvi, and his notes to *On the infinitely small*, 392–3.

26 In footnote 71 of his introduction, given at DLC, 372–373.

27 GM V, 324; Leibniz, 1989, 331.

28 Leibniz, 1989, 409–421.

Around the same time, Leibniz also drafts several arguments concerning the status of interminate lines ("linea interminata").[29] However, these passages, which have frequently been cited in the secondary literature, are concerned with problems associated with accepting the status on non-terminating infinite lines. As such, they bear on the set of issues at hand, if at all, in the pressure they exert *against* committing to the status of the infinite as indefinite in the domain of quantity. As a matter of fact, I think the arguments are neutral so far as the status of infinite *terminating* lines is concerned, but it is enough for my point here to see that they could not be used directly to support arguments against them.

Finally, there is a very interesting passage, tentatively dated in the Akademie Edition from the summer of 1689, in which Leibniz discusses infinite terminating lines explicitly in a note written during his reading of Thomas White's *Euclides Physicus*.[30] To those who would commit to infinite terminating lines, Leibniz responds that they do not provide an essential mark by which to distinguish the finite from the infinite.[31] I take this passage to be an indication of Leibniz's sense that a (definite) distinction between the finite and the infinite can only be drawn on the basis of the distinction between the finite as definite, that is, terminating, and the infinite as indefinite, that is, interminate. This passage is further indication that the issues Leibniz discusses in the letter to Varignon and the manuscript passage I have considered were of interest to Leibniz in the late 1680's.

In sum, the best reading of the situation seems to be that the manuscript dates from the years following Leibniz's return from Paris, but that the issues it raises continue to be of interest to Leibniz throughout the rest of his philosophical and mathematical career. In any case the argument points toward commitments Leibniz was only to make fully explicit during his later years.

---

29 *Linea Infinita est Immobilis*, 3 January 1676 (A VI, 3, 471); *Linea Interminata*, April 1676 (A VI, 3, 485–89); *Extensio Interminata*, April 1676 (?) (A, VI, 3, 489–90).

30 "Aus und zu Thomas White's *Euclides Physicus*" (A VI, 4, 2088ff.), here in particular, 2092–2093.

31 "Cui respondens numerus [infinitus] palmorum cadet in lineam terminatam, sed infinitam; id est quae est major quavis data, sed semper mihi haeret scrupulus, quod non datur mihi nota essentialis discernendi finitam ab infinita […]." (A VI, 4, 2093)

Emily Grosholz

Productive Ambiguity in Leibniz's Representation of Infinitesimals[1]

In this essay, I argue that Leibniz believed that mathematics is best investigated by means of a variety of modes of representation, often stemming from a variety of traditions of research, like our investigations of the natural world and of the moral law. I expound this belief with respect to two of his great metaphysical principles, the *Principle of Perfection* and the *Principle of Continuity*, both versions of the *Principle of Sufficient Reason*; the tension between the latter and the *Principle of Contradiction* is what keeps Leibniz's metaphysics from triviality. I then illustrate my exposition with two case studies from Leibniz's mathematical research, his development of the infinitesimal calculus, and his investigations of transcendental curves.

1. The Principle of Continuity

Leibniz wrote a public letter to Christian Wolff, written in response to a controversy over the reality of certain mathematical items sparked by Guido Grandi; it was published in the Supplementa to the *Acta Eruditorum* in 1713 under the title *Epistola ad V. Cl. Christianum Wolfium, Professorem Matheseos Halensem, circa Scientiam Infiniti* (AE Supplementa 1713 = GM V, 382–387). Towards the end, he presents a diagram (discussed below in Section 2) and concludes:

[1] I would like to thank the National Endowment for the Humanities and the Pennsylvania State University for supporting my sabbatical year research in Paris during 2004–2005, and the research group REHSEIS (Equipe Recherches Epistémologiques et Historiques sur les Sciences Exactes et les Institutions Scientifiques), University of Paris 7 et Centre National de la Recherche Scientifique, and its Director Karine Chemla, who welcomed me as a visiting scholar.

> Atque hoc consentaneum est Legi Continuitatis, a me olim in Novellis Literariis Baylianis primus propositae, et Legibus Motus applicatae:[2] unde fit, *ut in continuis extremum exclusivum tractari posit ut inclusivum*, et ita ultimus casus, licet tota natura diversus, lateat in generali lege caeterorum (GM V, 385).[3]

He cites as illustration the relation of rest to motion and of the point to the line: rest can be treated as if it were evanescent motion and the point as if it were an evanescent line, an infinitely small line. Indeed, Leibniz gives as another formulation of the *Principle of Continuity* the claim that "l'egalité peut estre considerée comme une inegalité infinement petite".[4] (*Lettre de M.L. sur un principe generàle utile*, 1687 = GP VII, 53)

The *Principle of Continuity*, he notes, is very useful for the art of invention: it brings the fictive and imaginary (in particular, the infinitely small) into rational relation with the real, and allows us to treat them with a kind of rationally motivated tolerance. For Leibniz, the infinitely small cannot be accorded the intelligible reality we attribute to finite mathematical entities because of its indeterminacy; yet it is undeniably a useful tool for engaging the continuum, and continuous items and procedures, mathematically. *The Principle of Continuity* gives us a way to shepherd the infinitely small, despite its indeterminacy, into the fold of the rational. It is useful in another sense as well: not only geometry but also nature proceeds in a continuous fashion, so the *Principle of Continuity* guides the development of mathematical mechanics.

But how can we make sense of a rule that holds radically unlike (or, to use Leibniz's word, heterogeneous) terms together in intelligible relation? I want to argue that two conditions are needed. First, Leibniz must preserve and exploit the distinction between ratios and fractions, because the classical notion of ratio presupposes that while ratios link homogeneous things, proportions may hold together inhomogeneous ratios in a relation of analogy that is not an equation. This allowance for heterogeneity disappears with the replacement of ratios by fractions: numerator, denominator, and fraction all become numbers, and the analogy of the proportion collapses

---

2 Réplique à l'abbé D.C. sous forme de letter à Bayle (AE Feb. 1687 = GP III, 45).

3 "All this accords with the *Law of Continuity*, which I first proposed in Bayle's *Nouvelles de la République des Lettres* and applied to the laws of motion. It entails that with respect to continuous things, one can treat an external extremum as if it were internal, so that the last case or instance, even if it is of a nature completely different, is subsumed under the general law governing the others."

4 "[...] the equation can be treated as an infinitesimally small inequality."

into an equation between numbers.[5] However, Leibniz's application of the *Principle of Continuity* is more strenuous than the mere discernment of analogy: the relation between 3 and 4 is analogous to the relation between the legs of a certain finite right triangle. But the relation between the legs of a finite and those of an infinitesimal 3–4–5 right triangle is not mere analogy; the analogy holds not only because the triangles are similar but also because of the additional assumption that as we allow the 3–4–5 right triangle to become smaller and finally evanescent, "the last case or instance, even if it is of a nature completely different, is subsumed under the general law governing the others." (cf. above, fn.2)

Thus, the notation of proportions must co-exist beside the notation of equations; but even that combination will not be sufficient to express the force of the *Principle of Continuity.* The expression and application of the principle requires as a second condition the adjunction of geometrical diagrams. They are not, however, Euclidean diagrams, but have been transformed by the *Principle of Continuity* into productively ambiguous diagrams whose significance is then explicated by algebraic equations, differential equations, proportions, and/or infinite series, and the links among them in turn explicated by natural language. In these diagrams, the configuration can be read as finite or as infinitesimal (and sometimes infinitary), depending on the demands of the argument; and their productive ambiguity, which is not eliminated but made meaningful by its employment in problem-solving, exhibits what it means for a rule to hold radically unlike things together.

This is a pattern of reasoning, constant throughout Leibniz's career as a mathematician, which the Logicists who appropriated Leibniz following Louis Couturat and Bertrand Russell could not discern, much less appreciate. As Herbert Breger argues in his essay "Weyl, Leibniz und das Kontinuum," the *Principle of Continuity* and indeed Leibniz's conception of the continuum – indebted to Aristotle on the one hand, and seminal for Her-

[5] Some commentators have been puzzled by Leibniz's allegiance to the notion of ratio and proportion. Marc Parmentier, for example, writes, "nous devons nous rappeler que les mathématiques de l'époque n'ont pas encore laïcisé les antiques connotations que recouvre le mot ratio. A cette notion s'attache un archaïsme, auquel l'esprit de Leibniz par ailleurs si novateur, acquitte ici une sorte de tribute, en s'obstinant dans une position indéfensable. La ratio constitue à ses yeux une entité séparée, indépendante de la fraction qui l'exprime ou plus exactement, la mesure. En ce domaine l'algèbre n'a pas encore appliqué le rasoir d'Occam. La preuve en est que la ratio était encore le support de la relation d'analogie, equivalence de deux rapports, toute différente de la simple égalité des produits des extremes et des moyens dans les fractions." (Leibniz, 1989, 42)

mann Weyl, Friedrich Kaulbach and G.-G. Granger on the other – is inconsistent with the Logicist program, even the moderate logicism espoused by Leibniz himself, not to mention the more radical versions popular in the twentieth century. The intuition [*Anschauung*] of the continuous, as Leibniz understood it, and the methods of his mathematical problem-solving, cannot be subsumed under the aegis of logical identity. Breger adds:

> Ich kann dieser Vermutung hier nicht nachgehen und möchte mich mit der Feststellung begnügen, dass Leibniz zwar ein dem Logizismus entsprechendes philosophisches Programm vertreten hat, dass er aber durch seine Mathematik selbst sich weit von diesem Programm entfernt hat.[6] (Breger, 1986)

In the two sections that complete this essay, I will show that this pattern of reasoning characterizes Leibniz's thinking about, and way of handling, non-finite magnitudes throughout his active life as a mathematician.

2. Studies for the Infinitesimal Calculus

In 1674, Leibniz wrote a draft entitled *De la Methode de l'Universalité*, (C 97–122) in which he examines the use of a combination of algebraic, geometric and arithmetic notations, and defends a striking form of ambiguity in the notations as necessary for the 'harmonization' of various mathematical results, once treated separately but now unified by his new method. He discusses two different kinds of ambiguity, the first dealing with signs and the second with letters.

The simplest case he treats is represented this way:

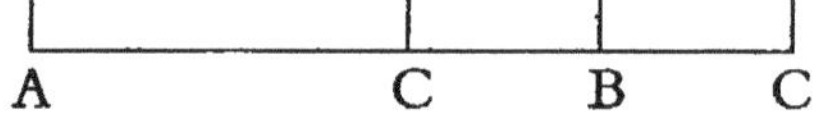

The point of the array is to represent a situation where A and B are fixed points on a line; this means that if the line segment AC may be determined by means of the line segment AB and a fixed line segment BC=CB, there is an ambiguity: the point C may logically have two possible locations, one on each side of B. Leibniz proposes to represent this situation by a sole

[6] "I can't go into this conjecture here, and would like simply to assert that although Leibniz did advocate a philosophical program corresponding to Logicism, he also distanced himself a great deal from it in his mathematical practice."

equation, which however involves a new kind of notation. He writes it this way:

$$AC=AB\mp BC$$

and goes on to suggest a series of new operations, corresponding to cases where there are three, four, or more fixed points to begin with. He generates the new symbols for operations by a line underneath (which negates the operation) or by juxtaposing symbols (One sees some nascent group theory here.) (C 100).

Re-expressing the same point in algebraic notation, he writes that

$$\mp a+b, \text{ or } +a\mp b=c$$

means that

$$+a + b, \text{ or } - a + b, \text{ or } + a + b, \text{ or } + a - b$$

is equal to *c* and goes on to give a more complex classification for ambiguous signs. The important point, however, is that the ambiguous signs can be written as a finite number of cases involving unambiguous signs. (C 102)

The treatment of ambiguous letters, however, is more complex, truly ambiguous, and fruitful. He illustrates his point with a bit of smoothly curved line AB(B)C intersected at the two points B and (B) by a bit of straight line DB(B)E. The notation AB(B)C and DB(B)E is ambiguous in two different senses, he observes. On the one hand, the concatenated letters may stand for a line, or they may stand for a number, "puisque les nombres se representent par les divisions du continu en parties egales"[7] (C 105), and because, by implication, Descartes has shown us how to understand products, quotients, and n^{th} roots of line segments as line segments. On the other hand, and this is a second kind of ambiguity, lines may be read as finite, as infinitely large, or as infinitely small. The mathematical context will tell us how to read the diagram, and he offers the diagram just described as an example: "donc pour concevoir que la ligne DE est la touchante, il faut seulement d'imaginer que la ligne B(B) ou la distance des deux points ou elle coupe est infiniment petite: et cela suffit pour trouver les tangentes."[8] (C 105)

---

7 "[...] because the numbers are represented through division of the continuum into equal parts".

8 "[...] thus in order to understand that the ligne DE is the tangent, one has only to imagine that the line B(B) or the distance between the two points where it intersects the curve is infinitely small: and this is sufficient for finding the tangents."

In this configuration, reading B(B) as finite so that the straight line is a secant, and as infinitesimal so that the straight line is a tangent, is essential to viewing the 'harmony' among the cases, or, to put it another way, to viewing the situation as an application of the Principle of Continuity. The fact that they are all represented by the same configuration, supposing that B(B) may be read as ambiguous, exhibits the important fact that the tangent is a limit case subject to the same structural constraints as the series of secants that approach it. And this is the key to the method of determining tangents. A good characteristic allows us to discern the harmony of cases, which is the key to the discovery of general methods; but such a characteristic must then be ambiguous.

To further develop the point, Leibniz returns to his original example, adumbrated.

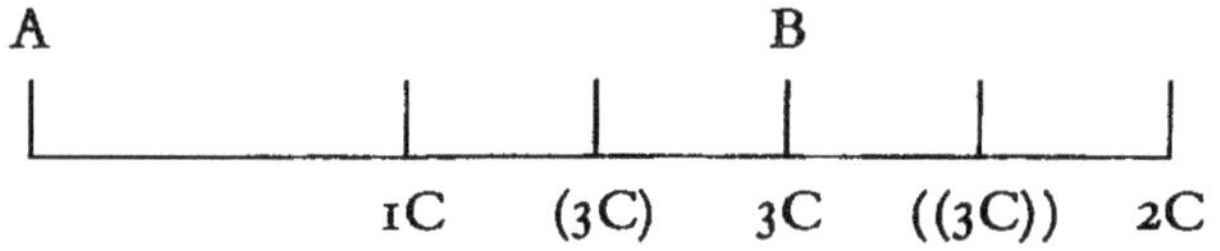

Once again, A and B are fixed points on a line. When we set out the conditions of the problem where a line segment AC is determined by two others, AB and BC, the point C may fall not only to the left or right of B, but directly on B: "le point C qui est ambulatoire peut tomber dans le point B."[9] (C106) Since we want the equation

$$AC=+AB\mp BC$$

to remain always true, we must be sure to include the case where B and C coincide, that is, where BC is infinitely small, "afin que l'equation ne contradise pas l'egalité entre AC et AB."[10] (C 106) In other words, the equality AC=AB is a limit case of the equation just given. In order to exhibit its status as a limit case, or (to use Leibniz's vocabulary) to exhibit the harmony among these arithmetic facts and thus the full scope of the equation, we must allow that BC may be infinitely small.

Here, Leibniz observes, the ambiguity of the sign $\mp$ is beside the point and doesn't matter; but the ambiguity of the letters is essential for the application of the principle of continuity, and thus cannot be resolved but must be preserved.

---

9 "[...] the point C which is moveable may fall on the point B."

10 "[...] so that the equation may not contradict the equality between AC and AB."

> Puisque on peut placer 3C, non seulement directement sous B, pour faire AC=AB et BC egale à rien, mais on le peut aussi placer en deça entre A, et B en (3C) ou au dela de B, en ((3C)) pour verifier par l'une des positions l'Equation AC=+AB–BC et par l'autre l'Equation AC=+AB+BC. pourveu que la ligne (3C)B ou ((3C))B soit conceüe infiniment petite. Voilà comment cette observation peut servir à la methode de l'universalité pour appliquer une formule generale à un cas particulier.[11] (C 106)

In the diagram, (3C) and 3C, or 3C and ((3C)) may be identified when AC=AB, as B and (B) are in the preceding diagram when the secant becomes the tangent.

Leibniz's intention to represent series or ranges of cases so as to include boundary cases and maximally exhibit the rational interconnections among them all depends on the tolerance of an ineluctable ambiguity in the characteristic. Some of the boundary cases involve the infinitesimal, but some involve the infinitary. Scholars often say that while Aristotle abhorred the infinite and set up his conceptual schemata so as to exclude and circumvent it, Leibniz embraced it and chose conceptual schemata that could give it rational expression. This is true, and accounts for the way in which Leibniz devises and elaborates his characteristics in order to include infinitary as well as infinitesimalistic cases; but it has not been noticed that this use renders his characteristic essentially ambiguous. And he says as much.

He notes that the use of ambiguously finite / infinitesimal lines had been invoked by Guldin, Gregory of St. Vincent and Cavalieri, while the use of ambiguously finite / infinite lines was much less frequent, though not unknown:

> Car il y a longtemps qu'on a observé les admirable proprietez des lignes Asymptotes de l'Hyperbole, de la Conchoeide, de la Cissoeide, et de plusieurs autres, et les Geometres n'ignorent pas qu'on peut dire en quelque façon que l'Asymptote de l'Hyperbole, ou la touchante menée du centre à la courbe est une ligne infinie egale à un rectangle fini [...] Et pour ne pas prevenir mal à propos l'exemple dont nous nous servirons pour donner un essay de cette methode, on trouvera dans la suite, que latus transversum de la parabole doit estre conceu d'une longueur infinie.[12] (C 106–107)

---

11 "Since one may place 3C, not only directly under B, in order to make AC = AB and BC equal to zero, but over towards A at (3C), or over on the other side of B at ((3C)) in order to make the equation AC = + AB – BC true on the one hand or on the other to make the equation AC = + AB + BC true, provided that the line (3C)B or ((3C))B be conceived as infinitely small. You see how this observation can serve the method of universality in order to apply a general formula to a particular case."

12 "For long ago people noticed the admirable properties of the asymptotes of the hyperbola, the conchoid, the cissoid, and many others, and the geometers knew that one could say in

Leibniz alludes to the fact that if we examine a hyperbola (or rather, one side of one of its branches) and the corresponding asymptote, the drawing must indicate both that the hyperbola continues ad infinitum, as does the asymptote, and that they will meet at the ideal point of infinity; moreover, a rule for calculating the area between the hyperbola and the asymptote (identified with the x-axis) can be given. The two lines may both be infinite, but their relation can be represented in terms of a finite (though ambiguous) notation – involving both letters and curves – and can play a determinate role in problems of quadrature. In the spring of 1673, Leibniz had traveled to London, where John Pell referred him to Nicolaus Mercator's *Logarithmotechnia*, in which Leibniz discovered Mercator's series. Taking his lead from the result of Gregory of St. Vincent, that the area under the hyperbola $y = \frac{1}{(1+t)}$ from $t = 0$ to $t = x$ is what we now call $\ln(1 + x)$, Mercator represented the latter by the series

$$\frac{x}{1} - \frac{x^2}{2} + \frac{x^3}{3} - \frac{x^4}{4} + \ldots$$

The more important example is that of the parabola; at stake are its relations to the other conic sections. Leibniz gives the following account of how to find a 'universal equation' that will unify and exhibit the relations among a series of cases. He offers as an illustration the conic sections, and what he writes is an implied criticism of Descartes' presentation of them in the *Geometry*, which does not sufficiently exhibit their harmony:

> La formation d'une Equation Universelle qui doit comprendre quantité de cas particuliers se trouvera en dressant une *liste de tous les cas particuliers*. Or pour faire cette *liste* il faut reduire tout à une ligne, ou grandeur, dont la valeur est requise, et qui se doit determiner par le moyen de quelques autres lignes ou grandeurs adjoustées ou soubstraites, par consequent il faut qu'il y ait certains *points fixes*, ou pris pour fixes, [...] et d'autres *ambulatoires*, dont les endroits possibles differents nous donnent le catalogue de tous les cas possibles [...]. Ayant trouvé cette liste, il faut songer à reduire à une formule generale tous les cas possibles, par le moyen de signs ambigus, et des lettres dont la valeur est tantost ordinaire,

a certain manner that the asymptote of the hyperbola, or the tangent drawn from the center to that curve, is an infinite line equal to a finite rectangle [...] and in order to avoid trouble apropos the example we are using in order to try out this method, we will find in what follows that the *latus tranversum* of the parabola must be conceived as an infinite length."

> tantost infiniment grande ou petite. J'ose dire qu'il n'y a rien de si brouillé, et different qu'on ne puisse reduire en harmonie par ce moyen [...].[13] (C 114–115)

He gives a diagram, with a bit of curved line representing an arbitrary conic section descending to the right from the point A, ABYE; a vertical axis AXDC descending straight down from A; and perpendicular to that axis at X another axis XY which meets the curve in Y; the line DE is drawn parallel to XY. Two given line segments *a* and *q* represent the parameters of the conic section. Leibniz asserts that the general equation for all the cases, where AX=*x* and XY=*y*, must then be,

$$2ax \mp \left(\frac{a}{q}\right)x^2 - y^2 = 0$$

When *a* and *q* are equal and $\mp$ is explicated as –, we have the circle of radius *a*=*q*; when *a* and *q* may be equal or unequal and $\mp$ is explicated as –, we have an ellipse where *a* is the *latus rectum* and *q* is the *latus transversum*; when $\mp$ is explicated as +, the conic section is the hyperbola.

However, in order to include further both the parabola and the straight line as cases of the conic section, Leibniz asserts, one must make use of infinite or infinitely small lines.

> Or posons que la ligne, *q*, ou le latus transversum de la Parabole soit d'une longueur infinite il est manifeste, que l'Equation $2axq \mp ax^2 = qy^2$, sera equivalente à celle cy: $2axq = qy^2$ (qui est celle de la Parabole) parce que le terme de l'Equation ax^2, est infiniment petit, à l'egard des autres $2axq$, et qy^2 [...].[14] (C 116)

And with respect to the straight line, he asserts, we must take both *a* and *q* as being infinitely small, that is, infinitesimal.

---

13 "The formation of a universal equation which must comprehend a number of particular cases will be found by setting up a *list of all the particular cases*. Now in order to make this list we must reduce everything to a line segment or magnitude, whose value is sought, and which must be determined by means of certain other line segments or magnitudes, added or subtracted; consequently there must be certain fixed points, or points taken as fixed, and others which move, whose possible different locations give us the catalogue of all the possible cases [...] having found this list, we must try to reduce all the possible cases to a general formula, by means of ambiguous signs, and of letters whose values are sometimes finite, sometimes infinitely large or small. I dare to claim that there is nothing so mixed up or ill-assorted that can't be reduced to harmony by this means."

14 "Now supposing that the line *q*, or the *latus transversum* of the parabola be of infinite length, it is clear that the equation $2axq+ax^2=qy^2$, will be equivalent to this one: $2axq=qy^2$ (which is that of the parabola) because the term ax^2 of the equation is infinitely small compared to the others $2axq$, et qy^2".

> Par consequent dans l'Equation: $2ax \mp (a/q)x^2 = y^2$, le terme $2ax$ evanouira comme infiniment petit, à l'egard de $(a/q)x^2$ et y^2, et ce qui restera sera $+(a/q)x^2 = y^2$ le signe $\mp$ estant change en + or la raison de deux lignes infiniment petites peut estre la mesme avec celle de deux lignes ordinaires et mesme de deux quarrez ou rectangles soit donc la raison a/q egale à la raison e^2/d^2 et nous aurons $(e^2 / d^2)x^2 = y^2$ ou $(e/d)x = y$ dont le lieu tombe dans une droite.[15] (C 116)

Leibniz concludes that this equation, by exhibiting the conic sections as limit cases of one general equation, not only displays their mutual relations as a coherent system, but also explains many peculiar features of the special cases: why only the hyperbola has asymptotes, why the parabola and the straight line do not have a center while the others do, and so forth.

At the end of the essay, Leibniz notes that we must distinguish between ambiguity which is an equivocation, and ambiguity which is a 'univocation.' The ambiguity of the sign $\mp$ is an example of equivocation which must be eliminated each time we determine the general equation with respect to the special cases. But the ambiguity of the *letters* must be retained; it is the way the characteristic expresses the *Principle of Continuity*, for Leibniz believed that the infinitesimal, the finite, and the infinite are all subject to the same rational constraints. One rule will embrace them, but it must be written in an irreducibly ambiguous idiom.

> A l'egard des signes, l'interpretation doit delivrer la formule de toute l'equivocation. Car il faut considerer que l'ambiguité qui vient des lettres donne une Univocation ou Universalite mais celle qui vient des signes produit une veritable equivocation de sorte qu'une formule qui n'a que des lettres ambigues, donne un theoreme veritablement general [...]. La première sorte d'interpretation est sans aucune façon ni difficulté, mais l'autre est aussy subtile qu'importante, car elle nous donne le moyen de faire des theorems, et des constructions absolument universelles, et de trouver des proprietez generales, et mesme des definitions ou genres subalterns communs à toutes sortes de choses qui semble bien éloignées l'une de l'autres [...] celle-cy donne des lumieres considerables pour l'harmonie des choses.[16] (C 119)

---

15 "Consequently, in the equation: $2ax+(a/q)x^2=y^2$, the term $2ax$ will vanish as it is infinitely small compared to $(a/q)x^2$ et y^2, and that which remains will be $+(a/q)x^2=y^2$ with the sign + changed into +. Now the ratio of two infinitely small lines may be the same as that of two finite lines and even of two squares or of two rectangles; thus let the ratio a/q be equal to the ratio e^2/d^2 and we will have $(e^2/d^2)x^2=y^2$ or (e/d)x=y whose locus is the straight line."

16 "With respect to signs [for operations], the interpretation must free the formula from all equivocation. For we must consider the ambiguity that comes from letters as giving a 'univocation' or universality but that which comes from signs as producing a true equivocation,

We should not think that Leibniz wrote this only in the first flush of his mathematical discoveries, and that the more sophisticated notations and more accurate problem-solving methods which he was on the threshold of discovering would dispel this enthusiasm for productive ambiguity. A look at two of his most celebrated investigations of transcendental curves by means of his new notation will prove my point.

3. The Principle of Perfection

Leibniz's definition of perfection is the greatest variety with the greatest order, a marriage of diversity and unity. He compares the harmonious diversity and unity among monads as knowers to different representations or drawings of a city from a multiplicity of different perspectives, and it is often acknowledged that this metaphor supports an extension to geographically distinct cultural groups of people who generate diverse accounts of the natural world, which might then profitably be shared. However, it is less widely recognized that this metaphor concerns not only knowledge of the contingent truths of nature but also moral and mathematical truths, necessary truths. As Frank Perkins argues at length in Chapter 2 of his *Leibniz and China: A Commerce of Light*, the human expression of necessary ideas is conditioned (both enhanced and limited) by cultural experience and embodiment, and in particular by the fact that we reason with other people with whom we share systems of signs, since for Leibniz all human thought requires signs. Mathematics, for example, is carried out within traditions that are defined by various modes of representation, in terms of which problems and methods are articulated.

After having set out his textual support for the claim that on Leibniz's account our monadic expressions of God's ideas and of the created world must mutually condition each other, Perkins sums up his conclusions thus:

> We have seen [...] that in its dependence on signs, its dependence on an order of discovery, and its competition with the demands of embodied experience, our expression of [necessary] ideas is conditioned by our culturally limited ex-

---

so that a formula that only contains ambiguous letters gives a truly general theorem [...] The first kind of interpretation is without difficulty, but the other is as subtle as it is important, for it gives us the means to create theorems and absolutely universal constructions, and to find general properties, and even definitions and subaltern kinds common to all sorts of things which seem at first to be very distant from each other [...] it throws considerable illumination on the harmony of things."

> pression of the universe. We can see now the complicated relationship between the human mind and God. The human mind is an image of God in that both hold ideas of possibles and that these ideas maintain set relationships among themselves in both. Nonetheless, the experience of reasoning is distinctively human, because we always express God's mind in a particular embodied experience of the universe. The human experience of reason is embodied, temporal, and cultural, unlike reason in the mind of God. (Perkins, 2004, 96–97)

Innate ideas come into our apperception through conscious experience, and must be shaped by it.

With this view of human knowledge, marked by a sense of both the infinitude of what we try to know and the finitude of our resources for knowing, Leibniz could not have held that there is one correct ideal language. And Leibniz's practice as a mathematician confirms this: his mathematical *Nachlass* is a composite of geometrical diagrams, algebraic equations taken singly or in two-dimensional arrays, tables, differential equations, mechanical schemata, and a plethora of experimental notations. Indeed, it was in virtue of his composite representation of problems of quadrature in number theoretic, algebraic and geometrical terms that Leibniz was able to formulate the infinitesimal calculus and the differential equations associated with it, as well as to initiate the systematic investigation of transcendental curves (See Grosholz, 1992). Leibniz was certainly fascinated by logic, and sought to improve and algebraize logical notation, but he regarded it as one formal language among many others, irreducibly many. Once we admit, with Leibniz, that expressive means that are adequate to the task of advancing and consolidating mathematical knowledge must include a variety of modes of representation, we can better appreciate his investigation of transcendental curves, and see why and how he went beyond Descartes.

4. Transcendental Curves: The Isochrone and the Tractrix

Leibniz's study of curves begins in the early 1670's when he is a Parisian for four short years. He takes up Cartesian analytic geometry (modified and extended by two generations of Dutch geometers including Van Schooten, Sluse, Hudde, and Huygens) and develops it into something much more comprehensive, analysis in the broad 18th century sense of that term. Launched by Leibniz, the Bernoullis, L'Hôpital and Euler, analysis becomes the study of algebraic and transcendental functions and the operations of differentiation and integration upon them, the solution of differential equations, and the investigation of infinite sequences and series. It also plays a major role in the development of post-Newtonian mechanics.

The intelligibility of geometrical objects is thrown into question for Leibniz in the particular form of (plane) transcendental curves: the term is in fact coined by Leibniz. These are curves that, unlike those studied by Descartes, are not algebraic, that is, they are not the solution to a polynomial equation of finite degree. They arise as isolated curiosities in antiquity (for example, the spiral and the quadratrix), but only during the seventeenth century do they move into the center of a research program that can promise important results. Descartes wants to exclude them from geometry precisely because they are not tractable to his method, but Leibniz argues for their admission to mathematics on a variety of grounds, and over a long period of time. This claim, of course, requires some accompanying reflection on their conditions of intelligibility.

For Leibniz, the key to a curve's intelligibility is its hybrid nature, the way it allows us to explore numerical patterns and natural forms as well as geometrical patterns on the other; he is as keen a student of Wallis and Huygens as he is of Descartes. These patterns are variously explored by counting and by calculation, by observation and tracing, and by construction using the language of ratios and proportions. To think them all together in the way that interests Leibniz requires the new algebra as an *ars inveniendi*. The excellence of a characteristic for Leibniz consists in its ability to reveal structural similarities. What Leibniz discovers is that this 'thinking-together' of number patterns, natural forms, and figures, where his powerful and original insights into analogies pertaining to curves considered as hybrids can emerge, rebounds upon the algebra that allows the thinking-together and changes it. The addition of the new operators d and ∫, the introduction of variables as exponents, changes in the meaning of the variables, and the entertaining of polynomials with an infinite number of terms are examples of this. Indeed, the names of certain canonical transcendental curves (log, sin, sinh, etc.) become part of the standard vocabulary of algebra.

This habit of mind is evident throughout Volume I of the VII series (*Mathematische Schriften*) of Leibniz's works in the Akademie-Ausgabe, devoted to the period 1672–1676. As Marc Parmentier admirably displays in his translation and edition *Naissance du calcul différentiel, 26 articles des* Acta eruditorum (Leibniz, 1989), the papers in the *Acta Eruditorum* taken together constitute a record of Leibniz's discovery and presentation of the infinitesimal calculus. They can be read not just as the exposition of a new method, but as the investigation of a family of related problematic things, that is, algebraic and transcendental curves. In these pages, sequences of numbers alternate with geometrical diagrams accompanied by ratios and proportions, and with arrays of derivations carried out in Cartesian algebra augmented by new concepts and symbols. For example, *De vera proportione circuli ad quadratrum circumscriptum in numeris rationalibus expressa* (AE February, 1682 = GM V, 118–122) which treats the ancient problem of the squaring of the circle, moves through a consideration of the series

$$\frac{\pi}{4} = 1 - \frac{1}{3} + \frac{1}{5} - \frac{1}{7} + \frac{1}{9} - \ldots,$$

to a number line designed to exhibit the finite limit of an infinite sum. Various features of infinite sums are set forth, and then the result is generalized from the case of the circle to that of the hyperbola, whose regularities are discussed in turn. The numerical meditation culminates in a diagram that illustrates the reduction: in a circle with an inscribed square, one vertex of the square is the point of intersection of two perpendicular asymptotes of one branch of a hyperbola whose point of inflection intersects the opposing vertex of the square. The diagram also illustrates the fact that the integral of the hyperbola is the logarithm. Integration takes us from the domain of algebraic functions to that of transcendental functions; this means both that the operation of integration extends its own domain of application (and so is more difficult to formalize than differentiation), and that it brings the algebraic and transcendental into rational relation.

During the 1690s, Leibniz investigates mathematics in relation to mechanics, deepening his command of the meaning and uses of differential equations, transcendental curves and infinite secycloidries. In this section I will discuss two of these curves, the isochrone and the tractrix. The isochrone is the line of descent along which a body will descend at a constant velocity. Leibniz publishes his result in the *Acta eruditorum* in 1689 under the title, *De linea isochrona, in qua grave sine acceleratione descendit, et de controversia cum Dn. Abbate de Conti* (AE April, 1689 = GM V, 234–237). However, the real analysis of the problem is found in a manuscript published by Gerhardt (GM V, 241–243), and accompanied by two diagrams (in the appendix): the first, reversed, is incorporated in the second.

On the first page of this text, the diagramm labeled 119 is read as infinitesimal. It begins:

> Quaeritur Linea descensoria isochrona YYEF (fig. 119), in qua grave inclinate descendens isochrone seu uniformiter plano horizontali appropinquet, ita nempe ut aequalibus temporibus, quibus percurrantur arcus BE, EF, aequales sint descensus BR, RS, in perpendiculari sumti. Sit linea quaesita YY, cujus recta

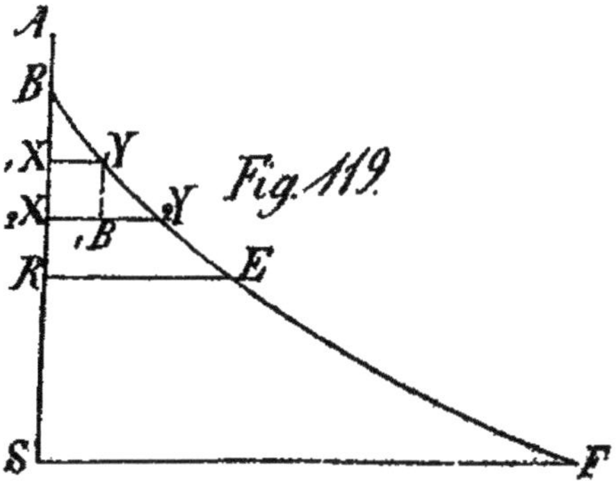

Directrix, in qua ascensus perpendiculares metiemur, sit AXX; abscissa AX vocetur y, et ${}_1X_2X$ seu ${}_1Y_1D$ erit d*x* et ${}_1D_2Y$ vocetur d*y*.[17] (GM V, 241).

The details of the analysis are interesting, as Leibniz works out a differential equation for the curve and proves by means of it what was in fact already known, that the curve is a quadrato-cubic paraboloid. However, what matters for my argument here is that we are asked to read the diagram as infinitesimalistic, since ${}_1X_2X$, ${}_1Y_1D$, and ${}_1D_2Y$ are identified as differentials.

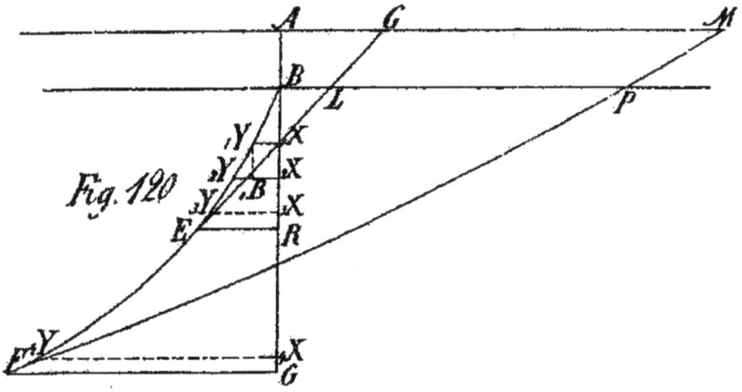

[17] "The line of descent called the isochrone YYEF is sought, in which a heavy body descending on an incline approaches the plane of the horizon uniformly or isochronously, that is, so that the times are equal, in which the body traverses BE, EF, the perpendicular descents BR, RS being assumed equal. Let YY be the line sought, for which AXX is the straight line directrix, on which we erect perpendiculars; let us call x the abscissa AX, and let us call y the ordinate XY, and ${}_1X_2X$ or ${}_1Y_1D$ will be dx and let ${}_1D_2Y$ be called dy." Note that in figure 119, ${}_1D$ is misprinted as ${}_1B$ (and this misprint continues in figures 120).

Immediately afterwards, in the section labeled "Problema, Lineam Descenscoriam isochronam invenire", exactly the same diagram is used, but reversed, incorporated into a larger diagram, and with some changes in the labeling. Here, by contrast, the diagram labeled 120 is meant to be read as a finite configuration; but it intended to be the same diagram. Note how Leibniz begins: "Sit linea BYYEF (fig.120) paraboliformis quadrato-cubica, cujus vertex B, axis BXXRS [...]"[18] (GM V, 242). There is no S in Figure 120; but the argument that follows makes sense if we suppose that 'G' ought to be 'S' as it is in Figure 119. Leibniz shows, using a purely geometrical argument cast in the idiom of proportions, that if the curve is the quadrato-cubic paraboloid, then it must be the isochrone. A heavy object falling from B along the line BYY, given its peculiar properties, must fall in an isochronous manner:

> Nempe tempus quo grave ex B in linea BYY decurret ad E, erit ad tempus quo ex E decurret ad F, ut BR ad RS, ac proinde si BR et RS sint aequales, etiam temporis intervalla, quibus ex B descenditur in E et E in F, erunt aequalia.[19] (GM V, 242)

What we find here is the same diagram employed in two different arguments that require it to be read in different ways; what a diagram means depends on its context of use. We might say that in the second use here, the diagram is iconic, because it resembles the situation it represents directly, but in the first use it is symbolic, because it cannot directly represent an infinitesimalistic situation. Yet the sameness of shape of the curve links the two employments, and holds them in rational relation.

We can find other situations in which the same diagram is read in two ways within the same argument. The tractrix is the path of an object dragged along a horizontal plane by a string of constant length when the end of the string not joined to the object moves along a straight line in the plane; you might think of someone walking down a sidewalk while trying to pull a recalcitrant small dog off the lawn by its leash. In fact, in German the tractrix is called the *Hundkurve*. The Parisian doctor Claude Perrault (who introduces the curve to Leibniz) uses as an example a pocket watch attached to a chain, being pulled across a table as its other end is drawn

---

18 "Let the line BYYEF be a quadrato-cubic paraboloid, whose vertex is B and whose axis is BXXRS [...]".

19 "Namely, the ratio between the time in which the heavy object runs down along line BYY from B to E, and the time in which it runs down from E to F, will be [the same as] the ratio of BR to RS; and then if BR and RS are equal, so also the intervals of time, in which it descends from B to E and from E to F, will be equal."

along a ruler. The key insight is that the string or chain is always tangent to the curve being traced out; the tractrix is also sometimes called the 'equi-tangential curve' because the length of a tangent from its point of contact with the curve to an asymptote of the curve is constant. The evolute of the tractrix is the catenary, which thus relates it to the quadrature of the hyperbola and logarithms.[20] So the tractrix is, as one might say, well-connected.

Fig. 139

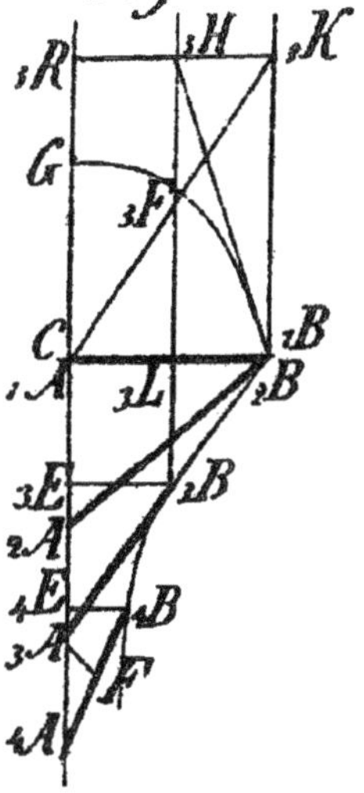

Leibniz constructs this curve in an essay that tries out a general method of geometrical-mechanical construction, *Supplementum geometriae dimensoriae seu generalissima omnium tetragonismorum effectio per motum: similiterque multiplex constructio lineae ex data tangentium conditione*, published in the *Acta Eruditorum* in September, 1693 (GM V, 294–301). His diagram, like the re-casting of Kepler's *Law of Areas* in Proposition I, Book I, of Newton's *Principia*, represents a curve that is also an infinite-sided polygon, and a situation where a continuously acting force is re-conceptualized as a series of impulses that deflect the course of something moving in a trajectory. The diagram labeled 139 must thus be read in two ways, as a finite *and* as an infinitesimal configuration. Here is the accompanying demonstration:

20 The evolute of a given curve is the locus of centres of curvature of that curve. It is also the envelope of normals to the curve; the normal to a curve is the line perpendicular to its tangent, and the envelope is a curve or surface that touches every member of a family of lines or curves (in this case, the family of normals).

> Centro $_3B$ et filo $_3A_3B$ tanquam radio describatur arcus circuli utcunque parvus $_3AF$, inde filum $_3BF$, apprehensum in F, directe seu per sua propria vestigia trahatur usque ad $_4A$, ita ut ex $_3BF$ transferatur in $_4B_4A$; itaque si ponatur similiter fuisse processum ad puncta $_1B$, $_2B$, ut ad punctum $_3B$, utique punctum B descripsisset polygonum $_1B_2B_3B$ etc. cujus latera semper incident in filum, unde imminuto indefinite arcu, qualis erat $_3AF$, ac tandem evanescente, quod fit in motu tractionis continuae, qualis est nostrae descriptionis, ubi continua, sed semper *inassignabilis* fit circumactio fili, manifestum est, polygonum abire in curvam, cujus tangens est filum.[21] (GM V, 296)

Up to the last sentence, we can read the diagram as the icon of a finite configuration; in the last sentence, where the diagram becomes truly dynamical in its meaning, we are required to read it as the symbol of an infinitesimalistic configuration, a symbol that nonetheless reliably exhibits the structure of the item represented. (A polynomial is also a symbol that reliably exhibits the structure of the item it represents.) After Leibniz invents the dx and ∫ notation, his extended algebra can no longer represent mathematical items in an ambiguous way that moves among the finite, infinitesimal, and infinitary; thus, he must employ diagrams to do this kind of bridging for him. In the foregoing argument, and in many others like it, we find Leibniz exploiting the productive ambiguity of diagrams that link the finite and the infinitesimal in order to link the geometrical and dynamical aspects of the problem.

---

21 "We trace an arbitrarily small arc of a circle $_3AF$, with center $_3B$, whose radius is the string $_3A_3B$. We then pull on the string $_3BF$ at F, directly, in other words along its own direction towards $_4A$, so that from position $_3BF$ it moves to $_4B_4A$. Supposing that we have proceeded from the points $_1B$ and $_2B$ in the same fashion as from $_3B$, the trace will have described a polygon $_1B_2B_3B$ and so forth, whose sides always fall on the string. From this stage on, as the arc $_3AF$ is indefinitely diminished and finally allowed to vanish – which is produced in the continuous tractional motion of our trace, where the lateral displacement of the string is continuous but always *unassignable* – it is clear that the polygon is transformed into a curve having the string as its tangent."

Eberhard Knobloch

Generality and Infinitely Small Quantities in Leibniz's Mathematics – The Case of his Arithmetical Quadrature of Conic Sections and Related Curves

The so-called *Fields Medal* takes the place of the non-existent Nobel Prize for mathematics.* Up until 2006 it had been awarded 48 times. In August, 2006, at the International Mathematical Union Congress in Madrid, it was awarded to Andrej Okounkow (Princeton), Terence Tao (Los Angeles), Wendelin Werner (Paris), and Grigori Perelman (St. Petersburg) (who, however, rejected it). The head on the obverse represents Archimedes facing right (Knobloch, 2005):

The sculptor Robert Tait McKenzie designed the medal in 1933. The date is written in Roman numerals: MCNXXXIII. There is an N instead of an M. When I wrote about this to the Fields medallist Sir Michael Atiyah, he sent me the following reply, dated the 16th of July 2000:

* The picture can be found on the website of the International Mathematical Union: www.mathunion.org.

> I had a look at my Fields Medal. It took me a little time to find the date you mentioned. I believe you are correct in saying that the second M appears as an N and is therefore a mistake. However, it is in very small characters and the difference at that scale between M and N is almost invisible to the naked eye. Michael Atiyah.

The question immediately arises whether very small or at least infinitely small errors are permissible in mathematics. Further, what are infinitely small quantities?

This paper studies Leibniz's different answers to these questions in their historical context. I proceed in the following steps:
1. Ancient models: Aristotle, Archimedes, Euclid
2. Rigor: Archimedes, Cavalieri
3. Predecessors-successors: Leibniz in 1673
4. Methods and principles: Leibniz in 1675/76
5. Generality
6. Epilogue

1. Ancient models: Aristotle, Archimedes, Euclid

In order to understand Leibniz' use of quantities, we have to grasp the ancient models, starting with the Aristotelian theory of quantities. In his *Metaphysics* Aristotle defines: "Quantity [ποσόν] is what is divisible [διαιρετόν] into the parts being in it". There are two kinds of quantities: "A quantity, then, is a plurality [πλῆθος] if it can be counted [ἀριθμητόν]; and a magnitude [μέγεθος], if it can be measured [μετρητόν]." (Book V, 13) All three definitions are based on actions in the mode of possibility – operations one could possibly perform on an object. It is in this Aristotelian tradition that Archimedes formulated what nowadays is called *Archimedean axiom*. He himself called it only an "assumption" [λαμβανόμενον] in his treatise *On the Sphere and Cylinder*:

> Further, of unequal lines and unequal surfaces and unequal solids, the larger exceeds the less by so much as, when added to itself, can be made [δυνατόν ἐστιν] to exceed any assigned [παντός τοῦ προτεθέντος] [sc. magnitude] among those which are comparable with one another. (Book I, Assumption 5)

In other words: certain quantities are given; then something can be done with them. In his so-called proofs by exhaustion Archimedes did not use the multiplicative form of his assumption (or axiom in modern terms) but the divisive form as demonstrated by Euclid in his *Elements*. Euclid writes:

> Two unequal magnitudes being set out [ἐκκειμένων], if from the greater there be subtracted a magnitude greater than its half, and from that which is left a magnitude greater than its half, and if this be repeated continually[ἐὰν … καὶ τοῦτο ἀεὶ γίγνηται], there will be left some magnitude which will be less than the lesser magnitude set out [ἐκκειμένου]. (Book X, Theorem 1)

Here, we encounter the same situation as in the case of Archimedes. The first proposition of his *Measurement of a Circle* might serve as an example of such a situation: "The area of any circle is equal to a right-angled triangle in which one of the sides about the right angle is equal to the radius r and the other to the circumference, of the circle."

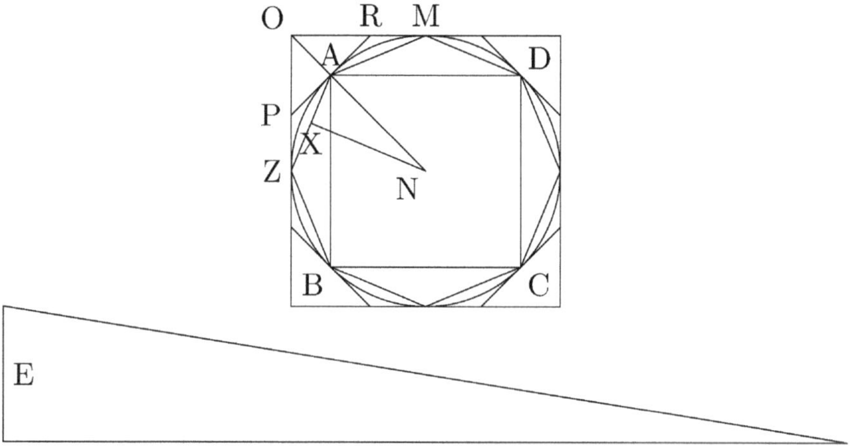

Figure 1.

Let c be the area of the circle ABC etc., t be the area of the triangle described, p be the area of the polygon ABC, etc.

Suppose $c \neq t$. Then $c = t + e$ (for example). According to Euclid $c - p$ can be made smaller than e. Hence we get

$$(t + e) - p < e \text{ or } p > t \quad .$$

On the other hand

$$XN < r \quad .$$

So, the perimeter of the polygon is smaller than the perimeter of the circle. Hence we get

$$p < t$$

and thus a contradiction.

2. Rigor: Archimedes – Cavalieri

Archimedes became the model of a rigorous mathematician, his demonstrations models of rigorous demonstrations. Yet, surprisingly, he himself defended a more modest standpoint, appealing simply to the authority of the ancient geometers to justify his method (Knobloch, 2000). In the preface to his *Quadrature of the Parabola* he justified the application of his fifth assumption by saying: "However, earlier geometers did also use this lemma." Obviously, this was a weak argument in order to justify his own mathematical practice. Yet he even added:

> But it appears that each of these theorems mentioned above inspired a level of confidence that is less than that of none of those [in other words: that is at least as great as the confidence inspired by every theorem] proved without this lemma. Yet, it suffices that the theorems that are published by us are pursued to a level of confidence (πίστις) similar to that of these theorems [sc. which are proved without the lemma].

It might be called an irony of the history of science that Archimedes, the very model of rigor, avowed that he was content with "confidence" instead of "unshakable certainty" such as was aimed at by Bonaventura Cavalieri in 1635:

> Illa [structura geometriae] quidem adeo firma, atque inconcussa, esse decuit, ut velut adamantina summorum ingeniorum tamquam arietum ictibus pulsata ne minimum quidem nutantia [fundamenta] agnoscerentur: Hoc enim Mathematicarum dignitati, ac summae certitudini, quam prae omnibus aliis humanis scientiis, nemine philosophorum reclamante, ipsa sibi vindicarunt, maxime convenire manifestum est. (Cavalieri, 1635, Book VII, 1 f.)[1]

It is worth keeping in mind that Cavalieri was one of the authors cited again and again by Leibniz when he developed the foundations of his infinitesimal mathematics.

[1] "That (the structure of geometry) should be so fine and unshakable that one recognizes that the so to say steely foundations, hit by the impacts of the greatest minds as if it were of battering rams do not waver in the least. Obviously this corresponds most of all to the dignity and highest certainty of the mathematical (sciences), which they themselves claimed for them more than all other human sciences without that any philosopher contradicted."

3. Predecessors and Successors: Leibniz in 1673

In the spring of 1673, Leibniz produced a series of voluminous studies, the so-called *Collectio mathematica* (Mathematical Collection). This collection constitutes numbers 9, 10, 12, 14, 15, 16, 17 of volume VII, 4 of the academy edition of Leibniz's Complete Writings and Letters (in press). After reading Blaise Pascal's *Lettres de A. Dettonville contenant quelques-unes de ses inventions de géométrie* (dating from 1659) Leibniz set about finding an appropriate meaning for the notion of "indivisibles" that Pascal had explained as "infinitely small quantities."

In n. 12, Leibniz uses the notions "pars minor assignabili" ("part smaller than an assignable (part)"), "differentia minor qualibet data" ("difference smaller than an arbitrary given (quantity)"), and "portio infinite parva" ("infinitely small portion") without explicitly equating these *non-equivalent* notions. Somewhat later, in n. 16,1 he declares: "Nota: Indivisibilia definienda sunt infinite parva, seu quorum ratio ad quantitatem sensibilem (vel differentia) infinita est."[2]

Pasini's Italian translation, "Nota: gli *indivisibili* devono essere definiti infinitamente piccoli, ossia il cui rapporto alla quantità sensibile è infinito", left out the passage "(vel *differentia*)" added by Leibniz (Pasini, 1993, 56). Obviously, Leibniz had only shifted the problem. When he postulated that indivisibles have to be defined as infinitely small, he still had to reply to the question: What does it mean to be infinitely small?

Leibniz provides two answers which are not equivalent. Such an impression might be induced by the word "vel". The first explanation can be translated into the equation:

$$\text{finite} : \text{infinitely small} = \text{infinite}$$

The meaning of this equation presupposes an understanding of the notion of "infinite." Thus, what does 'to be infinite' mean for Leibniz at this time? On the same page he declares: "Certum est enim istam summam $\frac{1}{1}\ \frac{1}{2}\ \frac{1}{3}$ etc. esse maiorem quolibet numero finito assignabili."[3] In other words, "infinite quantity" means a quantity which is larger than any assignable

[2] "Note: Indivisibles have to be defined as infinitely small or whose ratio (or difference) to a perceivable quantity is infinite."

[3] "Because it is certain that that sum $\frac{1}{1}+\frac{1}{2}+\frac{1}{3}$ a.s.o. is larger than any finite assignable number."

number. Leibniz does not say "assignato," "assigned." It is worth mentioning that such a definition excludes any potential or variable infinity. It necessarily implies a cardinality beyond all natural numbers, a real infinity. The ancient conception of potential infinity no longer suffices, but Leibniz neither mentions nor discusses this crucial implication in the manuscript. I would like to emphasize, though, that Leibniz replaces this early definition of infinity with another definition in the *Arithmetical Quadrature of the Circle* (DQA). In this later work, Leibniz defines infinity as "larger than any assigned quantity".

In Leibniz's *Arithmetical Quadrature of the Circle*, the first explanation of "infinitely small" becomes a consequence of a new definition of "infinitely small" (Knobloch, 2002, 70, rule 11).

The second explanation can be translated into the equation:

finite – infinitely small = infinite.

This explanation leads astray and was never – as far as I can see – repeated by Leibniz: he added it too quickly.

In no. 16,4 (A VII, 4), which was also written in the spring of 1673, he still adhered to the notions of "assignable" and "unassignable" using expressions like:

a) pars inassignabilis: unassignable part
b) aliquota inassignabilis: unassignable portion
c) magnitudo inassignabilis: unassignable magnitude
d) punctum seu quantitas inassignabilis : point or unassignable quantity
e) differentia minor assignabili quavis: difference smaller than any assignable (difference)
f) differentia erit nulla vel quod idem est assignabili qualibet minor: the difference will be zero or, equivalently, smaller than any assignable (difference)
g) differentia inter duas minimas applicatas minor est qualibet recta quae non dicam cogitari, sed fingi posset: the difference between the two smallest ordinates is smaller than any straight line that could, I would not like to say, be thought of but could be imagined.

It is worth emphasizing that in f) Leibniz drew the correct conclusion: A quantity that is smaller than any assignable quantity is equal to zero. Yet such a conclusion necessarily led to new problems. For this reason, Leibniz kept searching for a better definition of "infinitely small." By speaking of the "smallest ordinates" (g), he used Kepler's undefined, and worse, undefinable terminology: there are no "smallest ordinates" (Knobloch, 2000, 90). Example d) was no less contradictory. According to Euclid (*Elements*

I, def. 1), a point is something having no parts and thus – according to the above-mentioned Aristotelian definition – a non-quantity. Leibniz did not adhere to this terminology (stemming from Nicholas of Cusa and Galileo (Knobloch, 1999, 89)) but called a point an "unassignable quantity," thus expressing a clear contradiction between his own conception and that of Aristotle and Euclid.

For the time being two results can be taken to the records:

1. From his earliest studies, dating from 1673, Leibniz combined the discussion of infinite and indivisible with that of infinitely small.
2. Occasionally Leibniz used the operational definition of "infinitely small" ("smaller than any given"), but he used the static definition ("smaller than any assignable") far more frequently. In the summer of 1673, Leibniz explicitly spoke of the "trigonometria inassignabilium," of the "trigonometry of unassignables" (A VII, 4, no. 26) in order to describe his conception of the characteristic triangle consisting of infinitely small sides.

It is of the highest interest and importance that Leonhard Euler used Leibniz's inappropriate terminology f) dating from the spring 1673 when he explained the notion of "infinitely small" in his *Institutiones Calculi Differentialis*:

> Vocantur itaque differentialia, quae, cum quantitate destituantur, infinite parva quoque dicuntur, quae igitur sua natura ita sunt interpretanda, ut omnino nulla seu nihilo aequalia reputentur. (Euler, 1755, 5)[4]

And somewhat later:

> Nullum autem est dubium, quin omnis quantitas eousque diminui queat, quoad penitus evanescat atque in nihilum abeat. Sed quantitas infinite parva nil aliud est nisi quantitas evanescens ideoque revera erit = 0. Consentit quoque ea infinite parvorum definitio, qua dicuntur omni quantitate assignabili minora. (Euler, 1755, 69)[5]

Euler himself adds a rigorous demonstration of this conclusion. "Namque nisi esset = 0, quantitas assignari posset ipsi aequalis, quod est contra hy-

[4] "Hence, they are called differentials that are also called infinitely small because they lack a quantity which for that reason have to be interpreted by their nature in such a way that they are certainly pondered to be nothing or equal to zero."

[5] "But there is no doubt that every quantity can be diminished to such a degree that it completely vanishes and becomes zero. This definition of infinitely small quantities also accords with that definition by which they are said to be smaller than any assignable quantity."

pothesin."[6] I would like to emphasize this crucial Eulerian insight. The "0" is the real number zero and nothing else. Immediately after these explanations Euler deduces the equation

$$n : 1 = 0 : 0$$

from

$$n \cdot 0 = 0$$

in order to argue that the third proportional is n times greater than the fourth proportional, that arithmetical and geometrical equalities do not coincide. The division by zero, the expression 0 : 0, is not questioned. There are various attempts to save Euler's calculation with zeros and to eliminate Euler's inconsistencies by means of modern mathematical notions and tools like hyperreal numbers or equivalence relations (Laugwitz, 1983; Keisler, 2002), by assuming that Euler's infinitely small quantities are susceptible to increase or to decrease (Suisky, 2006). The disadvantage of this assumption is that not Euler, but Leibniz developed such a notion of "infinitely small" (see next section), whereas Euler explicitly said the contrary:

> Quare illa objectio, qua analysis infinitorum rigorem geometricum negligere arguitur, sponte cadit, cum nil aliud reiiciatur, nisi quod revera sit nihil. Ac propterea iure affirmare licet in hac sublimiori scientia rigorem geometricum summum, qui in veterum libris deprehenditur, aeque diligenter observari. (Euler, 1755, 71)[7]

4. Methods and Principles: Leibniz in 1675/76

In 1675, Leibniz elaborated the longest mathematical treatise he ever wrote, *On the Arithmetical Quadrature of the Circle, the Ellipse, and the Hyperbola. A Corollary is a Trigonometry without Tables*. It was first published in 1993 (DQA). In 2004, a bilingual Latin-French edition appeared (Leibniz, 2004). In this treatise, Leibniz laid a rigorous foundation for the theory of infinitely small and infinite quantities or, in other words, for the theory of quantified indivisibles.

---

6 "Because if it were not equal to zero, a quantity could be assigned that was equal to it. This contradicts the assumption."

7 "Hence that objection collapses by itself by which the analysis of infinites is accused of neglecting geometrical rigor, because nothing else is omitted but what is in truth nothing. And for that reason one can justly claim that in this more elaborated science the highest geometrical rigor which occurs in the books of the ancients is equally diligently observed."

Even in the preface, Leibniz stressed the importance of the sixth theorem, in which he introduced "Riemann sums" in order to demonstrate the integrability of continuous functions:

> Prop. 6. est spinosissima in qua morose demonstrantur certa quaedam spatia rectilinea gradiformia itemque polygona eousque continuari posse, ut inter se vel a curvis differant quantitate minore quavis data, quod ab aliis plerumque assumi solet. Praeteriri initio ejus lectio potest, servit tamen ad fundamenta totius Metodi indivisibilium firmissime jacienda. (DQA, 24)[8]

Leibniz partially repeated these words immediately after he explained the theorem itself:

> Hujus propositionis lectio omitti potest, si quis in demonstranda prop. 7. summum rigorem non desideret. Ac satius erit eam praeteriri ab initio, reque tota intellecta tum demum legi, ne ejus scrupulositas fatigatam immature mentem a reliquis longe amoenioribus absterreat. Hoc unum enim tantum conficit duo spatia, quorum unum in alteriun desinit si in infinitum inscribendo progrediare; etiam numero inscriptionum manente finito tantum, ad differentiam assignata quavis minorem sibi appropinquare. Quod plerumque etiam illi sumere pro confesso solent, qui severas demonstrationes afferre profitentur. (DQA, 28)[9]

Two things are worth emphasizing:

1. Leibniz implicitly refers to Archimedes as his model of rigor. Later on in the same treatise, he mentions him explicitly by name several times (DQA, 34, 78, 80). As we saw, Leibniz was completely justified in doing this.

---

[8] "Proposition 6 is most thorny. Therein it is overly carefully demonstrated that the procedure of constructing certain rectilinear step spaces and in equal fashion polygons can be continued to such a degree that they differ from each other or from curves by a quantity which is smaller than any arbitrary given quantity. This is usually assumed by others in most cases. Its reading can be passed over at the beginning. It serves, however, to lay the foundations of the whole method of indivisibles in the soundest way possible." (Knobloch 2002, 61 f.)

[9] "The reading of this proposition can be omitted if one does not want supreme rigor in demonstrating Proposition 7. And it is better that it is disregarded at the beginning and that it will be read only after the whole subject has been understood, in order that its excessive exactness does not discourage the mind from other, by far more agreeable things by making it become weary prematurely. For it brings about only this, that two spaces, of which one passes into the other if we progress infinitely, approach each other to a difference which is smaller than any arbitrarily assigned difference, even when the number of inscriptions remains only finite. This is generally taken for granted by those who declare to give rigorous demonstrations."

2. In this treatise Leibniz uses the operational, modern meaning of "infinitely small" from the very beginning.

The mathematical details of his integration theory that remind us of a Weierstrassian approach, that is of an ε-δ language, can be found in (Knobloch, 2002). Here it might suffice to compile a few examples of this language: "ad differentiam assignata quavis minorem sibi appropinquare" (DQA, 28)[10], "differre quantitate minor quavis data" (DQA, 29)[11], "differentia dato aliquo spatio minor reddi (assumi, sumi) potest (fit)" (DQA, 32)[12]. In all cases, Leibniz speaks in the Aristotelian or Archimedean mode of possibility. The choice of the size of an "infinitely small" quantity depends on the given quantity. Hence an infinitely small quantity is necessarily a variable quantity so that Leibniz rightly states: "Intervallum indefinitae parvitatis assumtum" (DQA, 39).[13] The assignable error made in such an approach is the infinitely small quantity. Thus Leibniz concludes on the meta-level of this analysis: "error erit minor quovis errore assignabili" (DQA, 33, 39).[14] We might sum up the crucial difference between Euler's and Leibniz's definitions of infinitely small quantities in the following way:

Let i be an infinitely small quantity, gq a given quantity, aq an assignable quantity.

Leibniz: For all $gq > 0$ there is an $i(gq) > 0$ so that $i(gq) < gq$
$\Rightarrow i(gq)$ is a variable quantity.

Euler: For all i and for all $aq > 0$: $i < aq$
$\Rightarrow i = 0$.

With regard to the method of indivisibles, Leibniz states: "Adeoque methodus indivisibilium, quae per summas linearum invenit areas spatiorum, pro demonstrata haberi potest".[15] (DQA, 29) About 25 years later, Leibniz defended his differential calculus in exactly the same fashion, that

10 "To approach one another to a difference which is smaller than any arbitrary assigned (difference)".

11 "To differ by a quantity smaller than any arbitrary given quantity".

12 "The difference can be made (assumed, taken) (becomes) smaller than any arbitrary given space".

13 "An interval being assumed of indefinite smallness".

14 "The error will be smaller than any arbitrary assignable error".

15 "Hence the method of indivisibles, which finds the areas of spaces by means of sums of lines, can be regarded as proven."

is, by referring to Archimedes and to his operational definition of infinitely small quantities:

> L'Auteur de ces réflexions semble trouver le chemin par l'infini et l'infini de l'infini pas assez sûr et trop éloigné de la méthode des anciens. Mais [...] on prend des quantités aussi grandes et aussi petites qu'il faut pour que l'erreur soit moindre que l'erreur donnée, de sorte qu'on ne diffère du stile d'Archimède que dans les expressions.[16] (Leibniz, 1701, 270–1=GM V, 350)

Leibniz was correct in asserting this, but his treatise on the quadrature of the circle was not published in those days. For this reason, Marc Parmentier justly stated in 2004: "Celle-ci éditée, le nouveau calcul aurait-il rencontré tant d'incompréhensions?"[17] (Leibniz, 2004, 32)

5. Generality

According to Leibniz, mathematics reflects the order and the harmony of the world which ideally exists in God. Every harmony implies generality, while generality implies beauty, conciseness, simplicity, usefulness, fecundity (Knobloch, 2006b). This statement especially applies to the mathematics of infinitely small quantities. In his treatise on the quadrature of the circle, Leibniz praised the fecundity of those principles that made him continue related studies: "Ridiculum enim videbatur casus singulares efferre ac demonstrare velle; cum eadem opera iisdem pene verbis generalissima theoremata condi possent."[18] (DQA, 71) The transmutation theorem was one such theorem in his eyes. He said: "Quod ad ipsam attinet propositionem, arbitror unam esse ex generalissimis, atque utilissimis, quae extant in Geometria [...]. Sed et inter fecundissima Geometriae theoremata haberi potest".[19] (DQA, 70)

---

16 "Appearently, the way which the author of these reflections finds through the infinite and the infinite of the infinite is not sufficiently certain and to far away from the method of the ancients. However, [...] one takes the quantities as large and as small as needed in order to keep the error smaller than any given error, in such a way that one does not differ neither from the style nor from the expressions of Archimedes."

17 "If this had been published, would the new calculus have faced such incomprehensions?"

18 "For it seemed to be ridiculous to present and to demonstrate single cases even though most general theorems could be established by the same work and nearly the same words."

19 "As far as the proposition itself is concerned, I believe that it is one of the most general and most useful that exists in geometry [...]. But it can also be considered as one of the most fecund theorems of geometry".

This theorem allowed Leibniz to resolve the area of a curvilinear figure into triangles using convergent ordinates instead of parallelograms. Its proof is based on the above mentioned thorny theorem 6 (section 4).

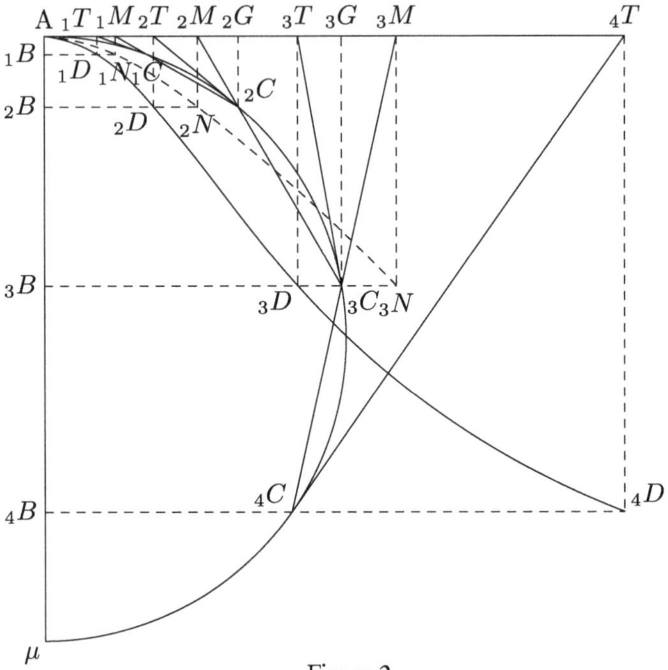

Figure 2
Based on Leibniz, DQA

Let $A_1C_2C_3C$ etc. be a given curve. Leibniz constructs the points of intersection of the tangents in C with the y-axis $A_1T_1M_2T_1G_2M$ etc. The segments AnT are transferred to the ordinates nBnC. The points nD form a new curve. The transmutation theorem reads:

Let Q be the so-called section figure $_1D_1B_3B_3D_2D_1D$, let T be the sector $_1CA_3C_2C_1C$. Then

$$Q = 2T \ .$$

The complete indirect demonstration can be found in (DQA). It consists of five steps:

1. Inscription of a polygon P in the sector T.
2. Inscription of a step figure H in the section figure Q.
3. Application of an auxiliary theorem (proposition 1 of the treatise):

$$H = 2P$$

4. Application of two inequalities. Let us assume that

$$|Q-2T=Z| \ .$$

According to Archimedes and to theorem 6 it is possible to choose H and P in such a way that

$$|T-P|<\frac{1}{4}Z \text{ and } |Q-H|<\frac{1}{4}Z \ .$$

5. Application of the triangular inequality

$$|Q-2T| \leq |Q-2P| + |2P-2T| < \frac{3}{4}Z \ .$$

This contradicts the assumption in step 4.

In other words, in order to lay the foundations of his integration theory in the soundest way possible, Leibniz used Archimedean methods. From them on, he used only ostensive proofs thanks to his "infinitely small quantities."

6. Epilogue

In his *Arithmetical Quadrature of the Circle*, Leibniz defined indivisibles as infinitely small quantities, that is, as quantities which are smaller than any given quantity. Leibniz then combined the ideas of Archimedes and Kepler (Kepler, 1615; Knobloch, 2000). Archimedes' method of proof served as a model of rigor in order to introduce well-defined new quantities that are called "infinitely small." Thus, Leibniz combined exactness with fruitfulness, replacing Archimedean indirect proofs by ostensive proofs. He was well aware of the dangers of this new method, "quam lubrica sit ratiocinatio circa infinita, nisi demonstrationis filo regatur" (DQA, 67).[20]

20 "[...] how slippery the calculation regarding the infinite is if it is not guided by the thread of a demonstration".

Herbert Breger

Leibniz's Calculation with Compendia

It has often been noted that Leibniz's verbal descriptions of infinitesimal magnitudes vary or even appear incoherent (Cf. e.g. Boyer, 1959, 207–221; Earman, 1975, 236–251). But in his use of them Leibniz is in fact being quite clear and explicit; his view of infinitesimals appears not to have altered since the beginning of his Hannover period or a few years later.[1] It is not sufficient to study Leibniz's verbal descriptions of infinitesimal magnitudes in isolation; they need to be interpreted in connection with their mathematical usage. According to Leibniz's own statement (GM V, 257, 398, 399; A III, 2, 931–933; GM III, 71–73), which has been confirmed by research in the history of mathematics (Gerhardt, 1891, 1053–1068; Mahnke, 1926, 5; Scholtz, 1934, 26; Hofmann, 1974, 74), on his path to the infinitesimal calculus Leibniz was influenced by Pascal and Huygens in particular. I would therefore like to turn first to these two mathematicians.

1. The State of the Art I: Pascal

On several occasions, Leibniz reported how he stumbled across the characteristic triangle so important in devising infinitesimal calculus: on reading Pascal something dawned upon him that not even Pascal had noticed (A III, 2, 931–933; A III, 6, 255; GM III, 72–73; GM V, 399). In the proof that inspired Leibniz, Pascal was certainly not speaking of infinitely small magnitudes; the text and drawing are evidence of just the opposite.[2] Pascal states explicitly that one can take point *E* to be on the tangent "où l'on vou-

[1] Admittedly we will only gain a complete image after Series VII of the Academy edition has been completed. Here I am going to ignore the conceptual attempts made by the young Leibniz and am only concerned with the ideas of the mature Leibniz.

[2] This remains the case, even if one agrees with Mahnke (1926, 37–39) that Leibniz was referring to figure 16: Pascal, 1965, IX, 67.

dra," (Pascal, 1965, IX, 61) and he determines as a lemma that the triangles *DIA* and *EKE* are similar.

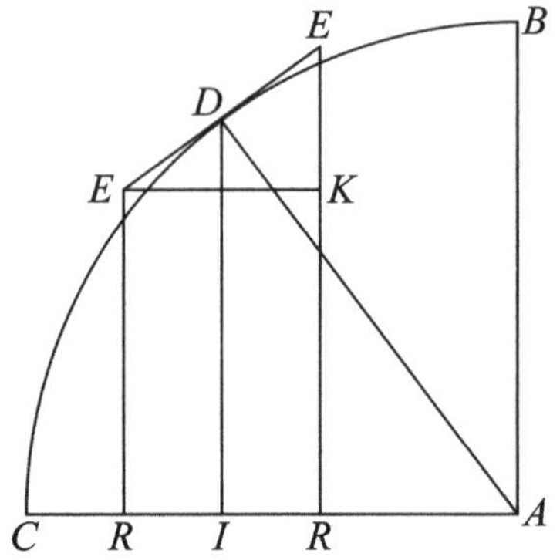

Figure 1.

By way of proof for proposition 1, immediately following, which is concerned with a statement about segments of the arc of a circle, Pascal divides the arc of a circle and a line "en un nombre indefiny de parties." (Pascal, 1965, IX, 63) He then refers to the lemma and concludes with the statement. But that is certainly not correct. Pascal therefore inserts a note: they cannot of course be equal, if the division is finite. But: "l'égalité est veritable quand la multitude est indefinie; parce qu'alors la somme de toutes les touchantes egales entr'elles, *EE*, ne differe [...] de la somme de tous les arcs egaux *DD*, que d'une quantité moindre qu'aucune donnée"[3] (Pascal, 1965, IX, 65). In other words: because the lemma applies to each division, one could offer a correct proof using the apagogic method of Archimedes, by demonstrating that the error is smaller than any positive quantity, however small this may be, and that the error thus equals zero.

Pascal's method of attaching a comment to a false proof, in which he maintains that one can also conduct the proof correctly, may amaze the non-mathematician, but for the mathematician it immediately makes sense. One knows at once that it is possible to conduct the proof using Archimedes' method as well as how this is to be done, and the reader is also grateful to Pascal that he has spared him the long-winded demonstration of Archimedes' method of proof.

Pascal also proceeds in a similar manner elsewhere. In the face of possible objections he argues that one could show that the error is smaller than

[3] "[...] the equality is true when the multitude is indefinite, because then the sum of all the equal tangents EE only differs from the sum of all the equal arcs DD by a quantity less than any given."

any given (positive) quantity and justifies this in turn by arguing that the number of subdivisions is indefinite. What is demonstrated with the method of indivisibles (this is how Pascal calls infinitely small magnitudes or infinitely narrow rectangles) can also be shown in a strict manner and in accordance with Greek mathematics. Pascal continues: both methods differ only in the words they use; for rational people it is sufficient just to point out how this is meant (Pascal, 1965, VIII, 351, 352).

It is not difficult to find further places where Pascal talks of an indefinite division or a division "jusqu'à l'infiny." (Pascal, 1965, IX, 25, 68, 85, 86, 105, 190, 191) If we attempt to sketch out Pascal's method of procedure in the *Lettres de Dettonville*, we have to appreciate that nowhere does Pascal introduce new mathematical magnitudes with a fixed, though infinitely small value. Strictly speaking, a division "jusqu'à l'infiny" is of course impossible. In truth he really means an abbreviation of the method of proof. Pascal is saying, so to speak: take a look at the procedure and make it clear to yourself that the proposed relationships are valid for every other division, however small; in other words, an apagogic proof is possible. I would like to add in passing that in a manuscript, in which he discusses what he had learned from Pascal, the young Leibniz talks of a division "in partes indefinitas."[4]

It has been proposed that Pascal had a strong influence on Leibniz in that Leibniz adopted the neglect of quantities from Pascal (cf. Boyer, 1959, 150). This proposal, however, is out of the question. Boredom at the long-windedness of the apagogic proof is not only typical of Pascal, but also of Fermat, Wallis, Huygens, Leibniz and others.[5] Even if the apagogic proof remains the model and ultimate foundation, in the second half of the 17th century mathematicians were interested in finding a solution, i.e. in the analysis. The connection between infinitesimals and what we now call epsilontics was obvious enough for 17th-century mathematicians. Accordingly, Leibniz often emphasized the relationship between his infinitesimal calculus and Archimedes (GM V, 322), whereby he also underlines the fact

---

4 Mahnke, 1926, 35. The notes ("Mathematicae Collectionis Plagulae Seiunctae") are due to be published in 2008 in: A VII, 4, N. 17.

5 Whiteside, 1960/1962, 331–348. On pp. 331 and 347, Whiteside criticises many 17th-century mathematicians from the vantage point of the higher level of abstraction reached in later mathematics e.g. Leibniz in his *Quadratura circuli*. The mathematicians certainly considered it a banal exercise to provide a complete description of the apagogic proof and saw it as dispensable; cf. Breger, 1994, 214–216 (The italicising of the apagogic proof in Fermat's treatise on rectification is missing in Fermat, 1891); Wallis, 1695, 646, cf. also Boyer, 1959, 171; Scholtz 1934, 33–34.

that Archimedes had provided no formal calculus. To reinforce this emphasis, Leibniz compares his relation to Archimedes with that of Descartes to Apollonius or Euclid.[6] Leibniz's justification of infinitesimal calculus by way of epsilontics has, with few exceptions,[7] not been taken seriously. Leibniz's remark has also been taken to mean that he had thought of replacing an infinitesimal magnitude with a small positive epsilon (Bos, 1974, 55–56) – a procedure that does not in every case allow easy transformation of the analysis into proof and that in addition is not applicable to differentials of a higher order – instead of an apagogic proof, which would demonstrate in retrospect the correctness of the result found in every single case,[8] if one considers it worthwhile.

Taking Pascal as a point of departure, I would now like to turn briefly to Leibniz's first publication of his infinitesimal calculus from 1684. It has been said that Leibniz introduced infinitesimals here as finite magnitudes (Boyer, 1959, 210; Bos, 1974, 19, 62–64). This is not wrong, but it is misleading. Leibniz in fact explains that one can choose any *dx* you like, and he then defines *dy* as the magnitude that has the same relation to *dx* as the ordinate to the subtangent. This definition does not initially explain how one arrives at the tangent. However, further on in the treatise Leibniz explains the tangent as being the line connecting two points separated by an infinitely small distance and the curve as being equivalent to an infinitely angled polygon (GM V, 223; likewise L'Hôpital 1696, 3, 11). This means of proceeding is by no means contradictory; it is the logical continuation of Pascal's method: the sides of the characteristic triangle are assumed to be finite; they can be chosen in any manner whatsoever, thus also as small as one would wish. The infinitely small magnitude is the abbreviation suitable in the context of discovery for a train of thought that the competent mathematician "sees," one that in the context of justification could be justified in an awkward fashion by means of apagogic proof. Whoever is interested in the provability rather than in the art of finding should not stare at the infinitely small magnitude like a rabbit at the snake; he should take a closer look at the process of ever-decreasing divisions. The infinitesimal is therefore not only very small; it has also absorbed, if this casual expression is

---

[6] Cf. Mahnke, 1926, 61; A III, 5, 68, 90; A VI, 4: 431; GM II, 123; GM IV, 54; GM V, 393–394; GM VII, 15; GP IV, 277.

[7] Mahnke, 1926; Scholtz, 1934; Breger, 1992; Knobloch, 2002, 59–73; Bos, 1974, 55, has already pointed out that Lucie Scholtz's work has not received the interest it deserves.

[8] GM IV, 106; GM V, 240 last sentence; GP II, 305. Epsilon should not be taken for the infinitely small magnitude; it stands for the variance from the correct result as assumed in the indirect proof.

allowed, a logical quantifier ("for all positive epsilons ...") so to speak. The meaning and quantitative value of the infinitesimal are dependent on the mathematical context. Or to express it differently: infinitesimals require instructions for use, and we all know that verbal instructions for use are always confusing to those who lack experience in how to use them.

2. The State of the Art II: Huygens

Let us now take a glance at a passage in Huygens's *Horologium oscillatorium*; Leibniz studied this piece of writing thoroughly:

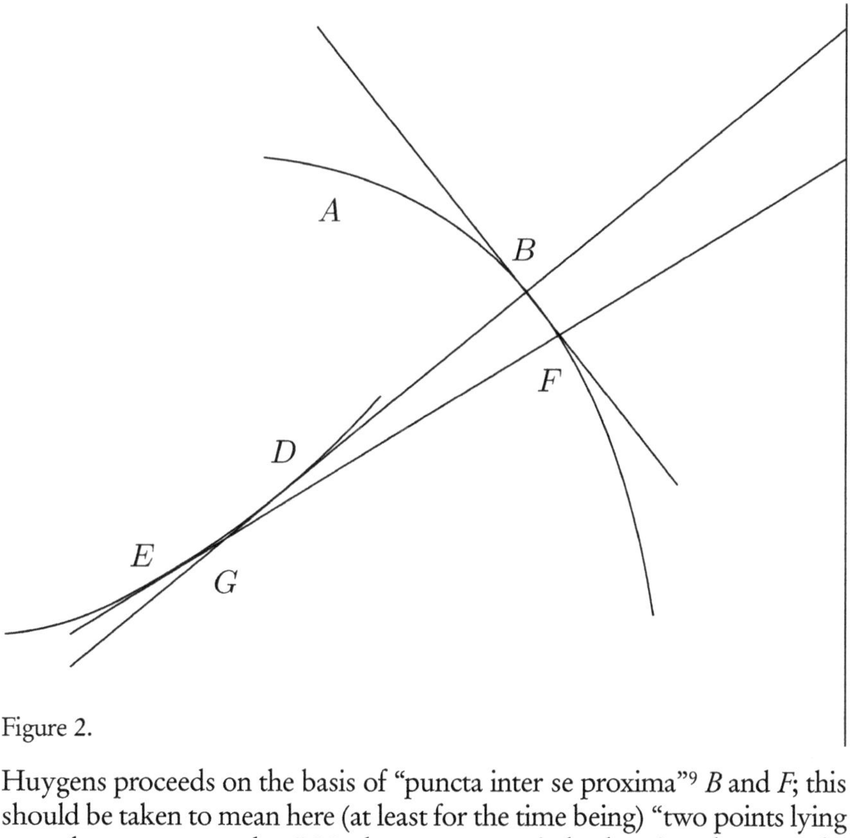

Figure 2.

Huygens proceeds on the basis of "puncta inter se proxima"[9] *B* and *F*; this should be taken to mean here (at least for the time being) "two points lying very close to one another." He then constructs in both points the normals,

[9] Huygens, 1673, 82 (also in Huygens, 1934, XVIII, 225); Cantor, 1901, 142; Yoder, 1988, 89–91. The expression quoted in the text is also to be found in Huygens, 1673, 62.

which are tangents of the curve sought after in two new points *E* and *D*. He then argues: the closer the original points are to one another, the closer the new points created by the construction are to each other. If now the original points – Huygens continues – are separated by an infinitely small distance, then the newly created points will coincide. In fact, one would expect that the newly created points are also separated by an infinitely small distance, but such a statement would not be of any use to Huygens. If one translates his geometric line of reasoning into modern notation, then one would obtain $f(x + dx) = f(x)$. When I encountered a similar line of reasoning a very long time ago in a text written by Tschirnhaus and then again somewhat later in a text by Leibniz (A III, 3, 612; GM V, 267, 281), I was irritated and thought that arguments with infinitely small magnitudes clearly should not be taken seriously. I am now of another opinion; the argument is stringent and correct, assuming that one understands what an infinitely small magnitude is. As long as the two points considered by Huygens are truly different from one another and are thus separated by an small positive distance, Huygens' argument is only an approximation. If the two points coincide, the reasoning seems to lack any basis. The argument only functions with the fiction of two points separated by an infinitely small distance, i.e. with two immediately adjacent points. Expressed differently, only this fiction delivers an exact solution. So we are dealing with a trick or an abbreviated way of saying that the error is smaller than any given magnitude, provided that one selects two original points that are truly different, but that are separated by a sufficiently small distance.

For a justification of this conclusion, one can refer to the continuity of the construction symbolized here by f (or to Leibniz's continuity principle). It is evident that the construction is continuous, but to write this down explicitly would involve a disproportionate amount of effort. In other words, something that is obvious and can be grasped intuitively can be formulated briefly and convincingly by means of an infinitely small magnitude; to write the same line of argument without using infinitely small magnitudes would require no small amount of pen work and would obscure the mathematical gist of the argument. In literature on the history of mathematics it is generally agreed that Huygens was a strict adherent of the stringent Greek mathematics and employed infinitesimals only seldom and even then only if strict proof would have been too boring.[10]

---

10 Zeuthen 1903, 343, Baron, 1987, 222–223; Bos, 1980, 132, 136–137. Cf. also Leibniz's comments on Huygens's *Horologium* in: H&O, 43.

A few lines later we find a further example: Huygens argues that the tangent in *B* is at the same time the tangent in *F*, if *B* and *F* are separated by an infinitely small distance. This is of course fictitious again; it is an abbreviated way of saying something that we today would describe with words referring to a process: "the tangent is the limit of the secant." If Huygens's remark is taken at face value, as if there really were two such points on the curve, then this would clearly produce nonsense. No formulated theory of limits and of sequences etc. can be attributed to Huygens and Leibniz; nor do the mathematicians consider developing such a theory necessary (because they have apagogic proofs in the back of their minds). The verbal apparatus which the mathematician has at his disposal is insufficient (Breger, 1990b); the mathematicians of the 17th and of the early 18th century know this, but they are not interested in this any further, because the mathematical facts are evident and it would be possible to conduct an apagogical proof at any time.

Since Huygens had a strong influence on the young Leibniz, we need to glance briefly at his biography and career. Because Huygens stuck to Greek stringency, proofs by means of moving points or of Cavalierian indivisibles were unacceptable for him (Huygens, 1888, I, 524; Huygens, 1920, XIV, 337), although there is no denying that these unconventional methods had lead to new (and correct) results. Because it was "tediosum" to always have to conduct an apagogic proof, Huygens opted for a compromise: Since finding was more interesting than proving, he wanted to employ infinitely small magnitudes on occasion, but still wished to provide the basis for a stringent proof, so that the expert was in no doubt that a stringent proof was possible. As Huygens says, it would suffice to do this on a couple of occasions, because one then knew how one should proceed in other cases. Thus the author is spared the work of writing the proof down and the reader the work of reading it. Otherwise mathematicians – Huygens continues to argue – would not find enough time to keep abreast of mathematical literature, which in recent times had been appearing so profusely.[11] When the young Leibniz found in Huygens his mathematical mentor, he must at once have absorbed this atmosphere of a new era in which solutions were appearing at such a great pace and in which the proof for each solution, once one had the solution, was so easy for the expert.[12]

11 When publishing *Horologium oscillatorium* Huygens produced a few results without proof; these results had meanwhile been published by Wallis, cf. Huygens, 1916, XIII, 753; Huygens, 1920, XIV, 37, 190–192; Baron, 1987, 221–223.

12 In this connection it is important to note that Leibniz calls the new method of calculating 'analysis', that is the art of finding (cf. Breger, 1992)

The manner in which Pascal and Huygens proceeded to use abbreviated proofs may appear shocking for a logicist philosopher; among mathematicians it has remained normal, simply a matter of course, right down to the present. To choose a primitive example: from $(x + 1)^2 = 36$ everyone immediately deduces $x=5$, although conducting a proof on the basis of the Peano axioms would presumably take two pages. Every mathematical treatise contains a large number of gaps of this nature.[13] But among mathematicians this would only be regarded as a gap in the proof, if an expert were unable to fill it in after brief reflection.

Huygens is interesting here for a further reason. While the criticism of infinitesimal calculus expressed by Nieuwentijt and Berkeley derives from their lack of mathematical knowledge, Huygens's reserve towards infinitesimal calculus, which he upheld for a long time, is of a different nature. What were the reasons for this reserve? I am not aware of any thorough investigation into this question. However much Huygens adhered to classical Greek stringency, he did not, as far as I know, accuse the new manner of calculating of lacking sound foundations. We have seen that Huygens was reasoning with infinitely small magnitudes before Leibniz. It is easy enough to find further examples of this; for example, Huygens explained Fermat's purely algebraic method of determining extreme values as one employing infinitely small magnitudes (Huygens, 1940, XX, 231, 249; cf. also M. Cantor, 1901, 144–145). Huygens clearly had no difficulty in understanding infinitely small magnitudes and in accepting them in some cases as a means of finding something and of abbreviating an exposition. His reservations towards the new method of calculation show us that we miss an opportunity to understand the specific contribution made by Leibniz (and Newton) if we concentrate exclusively on infinitely small magnitudes. This, by the way, is what Leibniz himself emphasized: if using infinitely small magnitudes in itself were to make someone the inventor of the infinitesimal calculus, Leibniz writes, then Huygens and others would already have been the inventors of infinitesimal calculus, but the new calculation method first arose when formulating an algorithm, i.e. when explicitly formulating calculation rules for sums, products etc (GM V, 393). Leibniz did indeed calculate with infinitely small magnitudes for a full two years after discovering the infinitely small triangle and wrote hundreds of pages[14] before he formulated the new method of calculation (Mahnke, 1926, 38; GM III, 73). These

[13] Cf. Huygens' remark: "evidentius est quam ut demonstratione indigeat" (Huygens, 1673, 62).

[14] Cf. the volumes A VII, 4, and A VII, 5, which will appear in 2008.

two years were anything but wasted time; it was only by proving many theorems and gaining experience with the new material that Leibniz arrived at the higher level of abstraction from which he was able to recognize and explicitly formulate the rules of calculus. As long as geometric infinitesimals are being used, the 17th-century mathematician has the appropriate intuition: he "sees" immediately whether an epsilon argument works or not. But as soon as one starts to calculate with infinitesimals in an algebraic manner, the geometric intuition fades away and one has to acquire a completely new kind of intuition.

Huygens did not reach this higher level of abstraction, or if he did, then only at a late date; maybe this was due to his age (though until his death Huygens naturally remained one of the best mathematicians in Europe). In the last years of his life, Huygens realized how useful the infinitesimal calculus really was and made an effort to understand it. In fact, it is even said that he learned the infinitesimal calculus (Bos, 1980, 143; Bos, 1972, 600; Yoder, 1988, 62; cf. also A III, 6, 417). Nevertheless, rather than using the infinitesimal calculus, Huygens employed the geometric thinking he had mastered so skilfully, using it to solve problems at the beginning of the 1690s, of which that of the catenary is the most well-known. The path trodden from employing infinitesimals geometrically to the new method of calculation was certainly a long one, even for such an excellent mathematician as Huygens.

3. Aspects of Leibniz's Concept of the Compendia

To cut a long story short, I would like to argue that we should abandon the prejudices of the second half of the 19th and of the 20th century. As far as Leibniz's infinitesimal calculus (and Newton's fluxional calculus) is concerned, there was no foundational problem (though this situation had changed by the early 19th century at the latest). What was really new and what posed the actual problem of understanding the new method of calculation was the higher level of abstraction; and this is precisely what in the course of development in mathematics (which naturally leads to higher levels of abstraction) has become self-evident and for us is thus invisible so to speak.

What was new in Leibniz's infinitesimal calculus was that calculations became somewhat independent of geometry and that algebraic calculation with infinitesimals was thereby constructed. Before Leibniz, the infinitesimal was a geometric line *AB* or *KL* in a certain geometric constellation.

Leibniz was the first to regard the infinitesimals of functional dependencies such as *dx*, *dy*. Infinitesimals that are not differentials do not occur as far as I know in Leibniz's mature mathematical reasoning (as opposed to Cauchy). The differentials are also more abstract than the traditional geometric infinitesimals, in as much as they themselves vary. The differentials of a higher order are also new. One can define

$$dy = y(x + dx) - y(x)$$

and then continue with a further definition:

$$d^2y = dy(x + dx) - dy(x).$$

In this definition *dx* is taken to be infinitesimal. If one wished to define the differentials in the way Leibniz did in his treatise of 1684, then one needs to refer in the definition to the subtangent of the first derivative.

Nieuwentijt had disputed the existence of second-order differentials; his arguments were very formal. We therefore find in Leibniz's answer to Nieuwentijt the most formal expressions of opinion on infinitesimals that Leibniz to my knowledge ever made.[15] Leibniz first explains that the differentials are to be viewed against the backdrop of a process; even the product of an infinitely small magnitude with an infinite magnitude is to be understood in the context of a process. In his defense of the second-order differentials, Leibniz remarks in passing that one can also calculate with the square of a differential.[16] According to Leibniz second-order differentials are infinitely small compared with first-order differentials. It is important to grasp the meaning of this claim. If first-order differentials have absorbed a logical quantifier, second-order differentials have absorbed two logical quantifiers. A second-order differential is a process that operates on processes. This interpretation fits in well, by the way, with Leibniz's general claim that for him in mathematics there is only a potential infinite,[17] although he certainly talks of the existence of an infinite number of monads in metaphysics.

The calculating rules for sums, products, quotients, powers and roots are of particular importance (GM V, 220–222). In this respect, too, calculus

15 GM V, 320–328. Other places in which Leibniz talks of various sizes of the infinite must presumably also be understood as different speeds of growth; the mature Leibniz at least refused to accept the assumption of an infinitely large number (GP III, 592; GM IV, 218).

16 GM V, 322. Leibniz's calculation also differs from the approach used by John Bell, for whom the squares of differentials are equal to zero.

17 From about 1680 on.

departs somewhat from its geometric foundation: it is no longer the curves, but rather the individual algebraic operations in their equations that are the object of calculus; calculating becomes much easier from this higher vantage point.[18]

It has been said that Leibniz's attention focused on the differential rather than the derived function. In contrast to later developments that is of course correct, but we should note that differentials firstly occur in pairs and secondly, as a means of analysis, they no longer occur in the final solution. If *dy/dx* or *∫ydx* occur in the final result, then these are quantities which are only written with differentials for the sake of convenience; a single *dx* or a $x+dx$ cannot occur in the final solution. It follows from this that *dy/dx* in practice plays a fairly prominent role right from the start. In addition, Leibniz obtains the simple calculation rules by using the tangent gradient (and not the subtangent as had previously been normal practice) for characterizing the tangent (GM V, 223). To be sure, the calculation rules are formulated in the 1684 treatise for the differentials, not for the derivatives.

Thus, prior to Leibniz the infinitely small magnitude was an abbreviated way of speaking of a process: "compendium ratiocinandi" or "per modum loquendi compendiosum;"[19] by inventing infinitesimal calculus, an abbreviated way of speaking of a geometric process becomes an object of calculation on a higher level of abstraction. But the object of calculation does not thereby lose its geometric roots; it remains an object of calculation dependent on its context, which at times is different from zero and at times is equal to zero. There had been no such abstract and strange object of calculation in the whole history of mathematics. One can well understand that Huygens, who was so attuned to Greek stringency, was capable of adopting without difficulty all that shocks us today in the infinitesimals, but that he was incapable of fathoming the transition to a more abstract calculus with abbreviated processes until at least 1691.

Johann Bernoulli, almost 40 years younger than Huygens, was able to cope with the higher level of abstraction with a certain amount of pragmatism. Leibniz was too much a philosopher to be satisfied with such pragmatism; again and again he expressly called the infinitesimal magnitudes

---

18 GM V, 220 (see title), 223.

19 H&O, 43; GP II, 305. Bernoulli (1691, 290) also expresses himself in this manner. In DQA, 69, we even find: "loquendi cogitandique, ac proinde inveniendi pariter ac demonstrandi compendia". On the question raised here by the young Leibniz of how to produce proof by means of infinitely small magnitudes cf. Breger, 1999.

fictitious.[20] If one misunderstands the differentials as "genuine mathematical entities" and as "fixed, but infinitely small," the infinitesimal calculus naturally appears to have "inconsistencies" and an "insecure foundation" (Bos, 1974, 12, 13). The continuity principle expressed as $x+dx=x$ is only valid for fictions.[21] One can only regard the tangents as connecting two points of a curve at an infinitely small distance and the curve as an infinitely angled polygon in a fictitious context (GM V, 223; cf. also GM V, 126); otherwise one would deduce that the whole curve had the same tangent gradient everywhere. As an expedient in analysis, these fictions are comparable with imaginary numbers in the 'casus irreducibilis' of the cubic equation or with a fictive infinitely distant point (GM IV, 92–93; H & O, 40–41, 42). Every number and every geometric line is finite and determinable; the unassignable magnitudes are fictitious, they cannot be determined by any construction (GP VI, 90; GM V, 322).

Leibniz possibly saw calculating with compendia in close connection with his theory of signs. Every instance of human reasoning, in Leibniz's opinion, requires a sign that is employed to abbreviate ("compendii causa") the things themselves or the ideas of the things. If, whenever the mathematician talks of a hyperbola, Leibniz continues, he wanted to envisage its definition and the definition of the terms occurring in this definition, he would only proceed and find new things at a very slow pace. Once one has become familiar with the things, one can calculate with their signs (A VI, 4, 918).

Clearly, over the course of time an imperceptible process was initiated, in which infinitely small magnitudes came to be taken for granted. When Leibniz states that the differential *dx* is the distance "inter duas proximas *x*" or that the distance of the "lineae proximae" of a family of curves is a differential (GM VII, p. 222; GM V, 267), then this is obviously a fiction: if x is a point, then $x+dx$ is a fictitious point. But for Cauchy $x+dx$ was also a real point; in connection with the Fourier series it then became necessary to find an explicit theory of the continuum, thus banning infinitesimals from mathematics for a hundred years. In the various versions of non-standard analysis, $x+dx$ is likewise an actual point (Schmieden/Laugwitz, 1958; Ab-

20 GM III, 524; GM IV, 110; GP II, 305; GP VI, 629; H&O, 43; GM V, 385. As early as 1675 Leibniz regarded infinitesimals as fictitious, cf. DQA, 35, p. 69.

21 "[...] aequalitas considerari potest ut inaequalitas infinite parva" (GM VI, 130). Cf. also L'Hôpital 1696, 2–3. In general the validity of the continuity principle arises from the fact that Leibniz adopted the Aristotelian theory of the continuum: The boundary of a continuum belongs to the continuum (cf. Breger, 1992).

raham Robinson, 1966). In Cauchy's times there was already a mathematical theory of sequences. Cauchy defines infinitesimals as sequences converging to zero. This definition implies likewise the existence of a quantifier in the definition of infinitesimals. The same is true, for example, of the non-standard analysis in the version of Schmieden and Laugwitz (Schmieden/Laugwitz, 1958; modified by Laugwitz, 1986): infinitesimals are defined as the equivalence classes of particular sequences. In this respect there is a similiarity to Leibniz: the infinitesimal is an abbreviation for a process.

After infinitesimals had been strictly rejected for a long time, they were rehabilitated by the non-standard analysis. The work of Henk Bos (Bos, 1974) is a reaction to this new situation; it has the merit of being the first serious attempt to come to grips with Leibniz's infinitesimals. To be sure, some of Bos' work has to be corrected in the light of present-day insights. Comments were made on this earlier in this essay, but we should take a closer look at two aspects.

Firstly, according to Bos there were two strategies by which to justify Leibniz's recourse to infinitesimals: epsilontics and the principle of continuity (Bos, 1974, 55–57). This distinction appears artificial, for the principle of continuity is of course also founded on epsilontics: two magnitudes are equal if their difference is smaller than any magnitude that can possibly be expressed. So the continuity principle is valid in the expression "the equality is an infinitely small inequality" (GM VI, 130) or: "the rule for equality is a special case of the rule for inequality." (GM VII, 25) In either case the processual nature is the decisive point; it is of no great import whether the process is described by means of epsilontics or with reference to the principle of continuity.

The second remark concerns the so-called "indeterminacy of differentials." (Bos, 1974, 24–25) Bos expressed this idea in a somewhat paradoxical manner, for a dx does not become indeterminate because, for example, the magnitudes $0.3dx$ or $(dx)^2$ can be used. A limit process can naturally be shaped in various ways: the process represented by $(dx)^2$ runs faster than the one represented by dx, and the process represented by $0.3dx$ is a step ahead compared with the process represented by dx. But it is somewhat confusing to call this idea an "indeterminacy of differentials".

I would like to make one final remark on infinitesimals in Leibniz's physics. There, too, infinitesimals are fictitious. Leibniz calls the state of rest an infinitely small motion (GM VI, 130). The infinitely small magnitude denotes the state of disappearing or of commencing (GP VI, 90; GM IV, 105). Of course, Leibniz adds, it is not strictly true that rest is a type of movement, but rest terminates continuous motion and in a certain manner one can think of it as still belonging to motion, because it has a certain characteristic in common with motion, just as one can regard a circle as a regular polygon with infinitely many angles (GM IV, 106). The same can be said for living force; dead force is a nisus, an infinitely small living force (GM VI, 238).

We should remember here that Leibniz adopted the Aristotelian theory of the continuum (Aristotle: *Physics*, book 6; Breger 1992). In this continuum points are not parts of the continuum: they constitute the boundary of partial continua, whereby the boundary always belongs to a partial continuum. Aristotle shows that there is a point in time in which a process of change is finished, but that there is no point in time for the beginning of a process. If a body rests in time *AB* and moves in the time *BC*, then it cannot have had any velocity at the point *B*; but in every point after *B* it is already in motion. Leibniz expresses this same idea in an intuitively plausible manner by already assigning to the body an infinitely small speed and an infinitely small living force at the point *B*.

It has been the intention of this essay to give a clearer view of Leibniz's specific contribution to the development of infinitesimal calculation. Leibniz "sees" in a text by Pascal on finite magnitudes the infinitely small triangle, because for him infinitely small magnitudes were the compendia for a process. For Huygens, too, infinitely small magnitudes were compendia for a process, and he certainly uses them as such. If he nevertheless had considerable difficulty understanding Leibniz's infinitesimal calculus, then it was clearly because for him the infinitely small magnitudes were abbreviations of geometric ideas and not fictitious objects in an algebraic calculus on a higher level of abstraction.

Fritz Nagel

Nieuwentijt, Leibniz, and Jacob Hermann on Infinitesimals

When Leibniz published his crucial paper on a "new method for determining maxima and minima" (Leibniz, 1684, 467–73) in 1684, it was Jacob Bernoulli and his brother Johann who were the first mathematicians outside of the circle of Leibniz's personal friends to get an intimate knowledge of the new calculus. Soon both Bernoulli brothers could handle the new method. But of course they were primarily interested in solving a large number of the most difficult problems in mathematics and physics, and not in meditating on the basic notions of the calculus and on their rigorous foundation.

For example, if we consider Johann Bernoulli's lessons, which he presented to L'Hôpital in Paris in 1691/92, as the first comprehensive and accessible text presenting the rules of the calculus to a larger public, we still cannot find much to explain the meaning of a basic notion like "infinitesimally small quantity."

Before starting with the rules of the calculus, Bernoulli places three postulates at the beginning of his text, of which I will only quote two: "1. Quantitas diminuta vel aucta quantitate infinities minore neque diminuitur neque augetur. 2. Quaevis Linea Curva constat ex infinitis rectis iisque infinite parvis."[1] (J. I Bernoulli, 1691/92) If we compare these postulates of Johann Bernoulli with the "demande ou supposition I et II" in L'Hôpital's *Analyse des infinimens petits* (L'Hôpital, 1696) we can observe a small but important difference. In L'Hôpital's *Analyse*, we read:

> I. On demande [...] qu'une quantité qui n'est augmentée ou diminuée que d'une autre quantité infiniment moindre qu'elle, puisse etre considérée comme demeurant la meme.

---

[1] "1. A quantity diminished or augmented by a quantity which is infinitely less does not become diminished or augmented. 2. Any curved line consists of an infinite number of straight lines, and these are infinitely small." (Johann I Bernoulli, De calculo differentialis, Ms UB Basel L I a 6, p. 1 (UB Basel = Öffentliche Bibliothek der Universität Basel).

II. On demande qu'une ligne courbe puisse etre considérée comme l'assemblage d'une infinité de lignes droites chacune infiniment petite.[2] (L'Hôpital, 1696, 3)

Whereas Bernoulli says that a quantity "does not become" diminished or augmented and that a curved line "consists" of straight lines, L'Hôpital says with more carefully chosen words, that a quantity "*could* be considered" as remaining the same and a curved line "*could* be considered" as an assembly of infinitely small straight lines. Bernoulli uses the indicative mood whereas L'Hôpital prefers the subjunctive, as if he wanted to denote something not as fact but as contingent or possible. Thus the difference could be interpreted as if Bernoulli treated the entities to which the notions of the calculus refer as real objects, whereas L'Hôpital speaks of them hypothetically, as if he were using the notions of the calculus as a mere manner of speaking. This leads us to the question: what happened between the first publication of the calculus by Leibniz, the first lessons on it by Johann Bernoulli, and the first printed textbook by L'Hôpital?

1. Leibniz's Dispute with Nieuwentijt

After Leibniz's calculus had been taken up by the scientific community, the first critics appeared and began to question the foundations of the calculus, which seemed to them to be weak. One of these critics was Bernhard Nieuwentijt. He was born on August 10, 1654 in Westgraftdijk (North Holland) and died in Purmerend near Amsterdam on May 30, 1718. He studied medicine and law at Leiden and Utrecht, then settled at Purmerend as a medical practitioner, where he became member of the city council and mayor. He became well known for several books on philosophical and theological subjects and for three treatises concerning the infinitesimal calculus. We will now briefly consider the main arguments in these papers, through which Nieuwentijt provoked a series of discussions in the Leibniz-Bernoulli circle.

In 1694 Nieuwentijt published his *Considerationes circa Analyseos ad quantitates infinite parvas applicatae Principia, & calculi differentialis usum in resolvendis problematibus Geometricis* (Nieuwentijt, 1694). In this treatise, Nieuwentijt concedes on the one hand that mathematics has made great

[2] "I. We request [...] that a quantity which is only augmented or diminished by another quantity being infinitely smaller than the former, *could be considered* as remaining the same." And: "II. We request that a curved line *could be considered* as an assembly of an infinity of straight lines, each of which is infinitely small [...]".

progress by using the new method of calculus founded by Barrow and Newton and then augmented by Leibniz (as he says), and he praises the brilliant results achieved, for example, by the Bernoullis. On the other hand, he reproaches the founders and propagators of the calculus for not having provided a rigorous demonstration of their methods. He says:

> [...] monuero me praelaudatorum modo Virorum ratiocinia hunc in finem prae caeteris elegisse, cum [...] omnes quidem, quantum mihi cognitum est, vel ea potius utantur, quam ex professo demonstrent[3] (Nieuwentijt, 1694, 5).

Nieuwentijt repeated this accusation a year later in a book entitled *Analysis infinitorum, seu Curvilineorum proprietates ex polygonarum natura deductae* (Nieuwentijt, 1695).[4] In our context, three fundamental arguments of Nieuwentijt are of special interest because of the reactions they provoked. The first reaction came from Leibniz himself and the others from Jacob Hermann, a member of the Bernoulli circle in Basel.

Nieuwentijt's first argument concerns the fact that in practice the new calculus discards certain quantities, sometimes being regarded as zero, sometimes as infinitely small, but without providing rigorous criteria to choose between the two options. He criticizes Leibniz for not having clarified why and when infinitesimally small quantities may be regarded as zero and neglected. The second argument concerns the products of infinitesimally small quantities. Nieuwentijt tries to demonstrate that even if we accept regular infinitesimals, we should not accept the infinitesimals of higher order, such as dx^2, dx^3, dx^4 and so on. Nieuwentijt's "demonstration" runs like this: He first provides a definition of an infinitesimal quantity. He says: "Quantitatem qualibet data minorem, compendii gratia, infinitesimam; majorem, infinitam appellare liceat."[5] (Nieuwentijt, 1695, Def. 1, 1) He then introduces an axiom which reads as follows: "Quicquid toties sumi, hoc est per tantum numerum multiplicari non potest, ut datam ullam quantitatem, ut ut exiguam, magnitudine sua aequare valeat, quantitas non est, sed in re

[3] "I will note that for this purpose I have chosen the considerations of the men quoted just above, especially because it seems [...] that they all, as far as I know have rather made use of it than consciously demonstrated it".

[4] Leibniz informed Johann Bernoulli about Nieuwentijt's publications and the reproaches concerning especially the existence of higher order differentials in a letter from June 24, 1695, GM III, 195.

[5] "A quantity, which is less than any given one, can be called, for abbreviation, infinitesimal; which is greater than [any given one], infinite." (Nieuwentijt, 1695, 1, Def. 3, p. 1.)

geometrica merum nihil."[6] (Nieuwentijt, 1695, 2, Axioma 1, 1). From this axiom Nieuwentijt concludes that there are no higher order infinitesimals. He thus argues in the following way (Nieuwentijt, 1695, 4).[7] He supposes an infinitely large number *m* (quantitas infinita), then divides a finite number by *m* and calls the result an "infinitesimum". This infinitesimum is not yet nothing, but if we divide it again by *m*, we do not get a quantity, because, multiplied by *m*, we do not get a finite quantity but only another infinitesimum.

Of course Nieuwentijt's "demonstration" is not very convincing. But for us it is important that, by applying his result to first order infinitesimals, Nieuwentijt concludes that infinitesimals of second or higher order are impossible. This means, with regard to Leibniz, that Nieuwentijt accepts first order differentials and uses them for calculation, but rejects differentials of higher order and objects to their use. Nieuwentijt's second argument against the differential calculus concerns the treatment of exponential functions. Using the Leibnizian concept of differentials (which, if combined with finite quantities, have to be regarded as zero), Nieuwentijt tries to show that we cannot discover the differential quotient of an exponential function like $y^x = z$ (Nieuwentijt, 1695, 62, 280). By treating this equation according to the principles of the calculus and discarding all terms in which products of differentials occur, he gets the following equations:

$$dz = (y + dy)^{x + dx} - y^x = y^{x + dx} + xy^{x + dx - 1}dy - y^x \ .$$

But if now in these terms all the differentials have to disappear, we do not get an expression for the differential quotient of the exponential function but only the identity $0 = y^x - y^x$. This is apparently correct, but does not allow us to find the relation between dy and dx, which is the differential quotient of the exponential function.

Shortly after Leibniz had read Nieuwentijt's publications, he replied to his objections in an article published in the *Acta Eruditorum* of July 1695 (Leibniz, 1695, 310–316). Against Nieuwentijt's first argument, namely the ambiguity in treating differentials sometimes as zero and sometimes as infinitely small, Leibniz defends his definition of equality by saying: "Caeterum aequalia esse puto, non tantum quorum differentia est omnino nulla,

---

6 "What cannot be so often multiplied, that is [multiplied] by such a great number, that it becomes equal to any given quantity, however small it may be, is not a quantity, and therefore a mere nothing in geometry."

7 Nieuwentijt, 1695, 4, Lemma 10.

sed et quorum differentia est incomparabiliter parva."[8] (Leibniz, 1695, 311) But the struggle over these two definitions of equality is really only a struggle over words. For, as he says: "Et si quis talem aequalitatis definitionem rejicit, de nomine disputat. Sufficit enim intelligiblem esse, et ad inveniendum utilem."[9] (Ibid.)

Leibniz of course rejects Nieuwentijt's argument against the existence of higher order differentials. First, he rejects Nieuwentijt's demonstration concerning the quotients of infinitesimal quantities done by means of an infinite number *m*. But Leibniz only indicates that, if one accepts different grades of infinity, then Niewentijt's demonstration does not lead to a contradiction. Leibniz then provides an example to show that second order differentials can be understood and make sense (Leibniz, 1695, 312–3): Let us take x as increasing according to a geometrical progression and y increasing according to an arithmetical progression, thereby keeping dy constant. That means

$$x = e^{y/a} \qquad \text{and} \qquad dx = \left(\frac{1}{\mathrm{a}}\right) e^{y/a} dy\mathrm{a}$$

where a is a constant. Thus

$$\frac{dx}{dy} = \frac{x}{a}.$$

Therefore we have

$$dx = x\frac{dy}{a} \qquad (1)$$

or

$$\frac{dy}{a} = \frac{dx}{x} \qquad (2).$$

If we now differentiate equation (1) on both sides we get

$$ddx = \frac{dxdy}{a} \qquad (3).$$

We now can substitute dy/a in equation (3) by dx/x according to equation (2) and thus we get

---

[8] "I regard objects as equal, not only when their difference is totally zero, but also when their difference is incomparably small."

[9] "Whoever rejects such a definition of equality disputes about names. It is sufficient that [a definition] is intelligible and useful for making discoveries."

$$ddx = \frac{dxdx}{x} \qquad \text{or} \qquad \frac{ddx}{dx} = \frac{dx}{x} \qquad (4).$$

Equation (4) says that the second order differential has the same relation to the first order differential as the first order differential to x. In other words, ddx compared to dx is infinitely small in the same way that dx is infinitely small when compared to the finite quantity x. In addition to this proof, Leibniz illustrates his argument by saying that the ordinary function, the first order differential and the second order differential have the same relation as motion, velocity and acceleration, which is an element of velocity. Motion describes a line, velocity an element of the line, and acceleration an element of an element.[10] Apparently Leibniz thought he had convinced Nieuwentijt that the rejection of higher order differentials is wrong.

To refute Nieuwentijt's assertion that exponential functions could not be treated by the differential calculus, Leibniz confesses first that by treating the function $b^x=y$ he too had come to the same disappointing identity as Nieuwentijt. But then he claims that such an identity could be avoided by treating the exponential function with the right method of his calculus. Leibniz refers to his own contributions in his article on the numerical quadrature of the circle (Leibniz, 1682, 41–6) in the *Acta Eruditorum* of February 1682 and to his correspondence with Huygens. But he adds that he has published only little on this subject because such exponential expressions would rarely occur. Then he continues:

> Nec quisquam mihi notum est, praeter ingeniossium Bernoullium, qui proprio Marte, me non monente, et ipse in calculo differentiali huc pervenerit, atque haec penetrarit, quae Hugenius per jocum hypertranscendentia appelabat.[11] (Leibniz, 1695, 314)

Then Leibniz shows how one can find the differential of the exponential function $x^v=y$ by using the rules of his calculus. In the same way as Bernoulli, he takes the logarithm of both sides of the equation and gets

$$v \log x = \log y.$$

---

10 "[...] quantitas ordinaria, quantitas infinitesima prima seu differentialis et quantitas differentio-differentialis, vel infinitesima secunda, sese habent ut motus et celeritas et solicitatio, quae est elementun celeritatis. Motu describitur linea, velocitate elementum lineae, solicitatione elementum elementi." (Leibniz, 1695, 315)

11 "No one is known to me, except the most ingenious Bernoulli, who, by his own initiative and not induced by me, has forced his way in the field of what Huygens jokingly called the hypertranscendent curves".

Because

$$\log x = \int \frac{dx}{x} \qquad \text{and} \qquad \log y = \int \frac{dy}{y}$$

he gets

$$v \int \frac{dx}{x} = \int \frac{dy}{y}.$$

By differentiation he gets

$$\frac{dy}{y} = dv \log x + v\frac{dx}{x}$$

and therefore he would get:

$$dy = x^v dv \log x + v x^{v-1} dx$$

which means

$$dy = d(x^v) = x^v dv \log x + v x^{v-1} dx.$$

Unfortunately the equation printed in his article in the *Acta Eruditorum* differs from this result and is wrong. But when Leibniz checks his result several lines later by applying his formula to the special case of a constant v, that is, with $dv=0$, he reaches the correct result $d(x^v) = vx^{v-1}dx$. This would not result from the printed formula. So it is evident that in this case Leibniz has not miscalculated, but that a misprint has corrupted his correctly calculated formula (Leibniz, 1695, 314).[12]

If Leibniz had thought that Nieuwentijt would be satisfied by his answer to the objections, he would soon be disappointed. Not even a full year had passed before Nieuwentijt published a reply to Leibniz's article. His booklet bears the title *Considerationes secundae circa calculi differentialis principia et responsio ad Virum Nobilissimum G. G. Leibnitium* (Nieuwentijt, 1696). Nieuwentijt emphasizes once more that he does not intend to fight against the infinitesimal calculus in general or to provoke those prominent men who use the calculus with great success. His sole aim is to free the foundations of the calculus from error and secure the calculus for those who want to go forward to a new kind of mathematics which moves beyond the bounds set by the ancients (Nieuwentijt, 1696, 2). But when Leibniz wrote to Jacob Bernoulli that he wished Nieuwentijt would produce something new (Bernoulli, 1993, 88), he was once again left disappointed. For in his *Considerationes secundae* Nieuwentijt mainly repeats the five points where he thought he had found difficulties unsolved by Leibniz and his adherents.

---

12 Hermann indicates the misprint to Nieuwentijt, cf. Hermann, 1700, 54.

2. Jacob Hermann's Defence of Leibniz's Reply to Nieuwentijt

Leibniz was not very motivated to respond, yet again, to Nieuwentijt's *Considerationes secundae*. So he was glad to discover that a disciple of Jacob Bernoulli at Basel would take over this task. This disciple was Jacob Hermann, born in 1678 in Basel, who had studied mathematics with Jacob Bernoulli and had just passed his M.A. exam before starting his study of theology (Nagel, 2005, 55–75).[13] In 1700 he published a booklet of 62 pages entitled *Responsio ad Clarissimi Viri Bernh. Nieuwentiit Considerationes Secundas circa calculi differentialis principia* (Hermann, 1700). As Hermann tells us in the preface, he wrote this little dissertation for two reasons: first, there was no hope that Leibniz himself would answer Nieuwentijt's renewed attacks. Second, he wanted to show that the Leibnizian party had enough grounds to refute each of the objections. Hermann is right, by the way, when he apologizes for the inelegance of his prose and expresses his confidences that the solidity of his demonstrations and arguments will compensate for this lack of style (Hermann, 1700, praefatio).

Hermann divided his booklet into six chapters, five of which are dedicated to Nieuwentijt's objections, whereas the final chapter presents, according to Hermann, the "cardinal propositions" of the differential calculus. In the following we will focus only on several of the points made in Hermann's expansive and detailed discussion of Nieuwentijt's objections (especially those points made in the first, fourth and fifth chapters).

In the first chapter, Hermann responds to Nieuwentijt's demand that Leibniz might explain and justify the treatment of infinitely small quantities, when combined with finite quantities as nothing and consequently allowing their neglect. According to Hermann the whole difficulty is based on an ambiguous use of the two notions "aequalis" and "incomparabilis". With regard to equality, he reiterates Leibniz's definition, formulating it like this: "quaecunque data quavis minore differentia differunt, aequalia esse." (Hermann, 1700, § 3, 5).[14] Hermann considers this definition to be a basic assumption, which Leibniz himself had thought does not require any demonstration. Hermann underlines the basic role of this assumption by showing that it goes back to Archimedes and by noting that many mathematicians have successfully made use of it when they applied the method of exhaustion in order to find quadratures of curves:

---

[13] For a list of the works and manuscripts of Jacob Hermann cf. Nagel, 1991, 36–54.

[14] "Whatever differs by a difference smaller than any given quantity is equal."

> [...] ceu videre est ex Archimedis scriptis, praecipue de dimensione circuli et Parabolae: Plerique Geometrae qui de Spatiorum quadraturis scripserunt hoc principio nixi, multa elegantissima Theoremata quadraturas spatiorum concernentia demonstarunt, quin et hoc ipso fundamento, tota illa Veterum methodus Exhaustionum dicta, nititur.[15] (Hermann, 1700, § 4, 5).

Hermann then transforms this principle by using the notion "incomparabilis", which he defines with regard to the infinitely small quantities in the following way: "Incomparabiles dicuntur duae quantitates, quarum una, infinities altera minor est, vel quod idem est, quae numero infinito multiplicata alteram demum vel restituere, vel superare valet."[16] (Hermann, 1700, §6, 8) After having defined the notion "incomparable," the definition of equality is presented in the following way: "Aequalitas consistit in incomparabilitate differentiae duarum pluriumve quantitatum cum ipsis quantitatibus."[17] (Hermann, 1700, 2) From this Hermann concludes that the proposition criticised by Nieuwentijt, namely that "infinitely small quantities combined with ordinary quantities can be regarded as zero and neglected," is now understandable and correct.

In Chapter 4, Hermann examines Nieuwentijt's proposition that the infinitesimals of the second or higher order are nothing, and the proof of this proposition. Hermann argues, again with reference to Leibniz, that we should assume different orders of infinity: "[...] infiniti sunt gradus tam infinitorum quam infinite parvorum." (Hermann, 1700, 40). Then he reworks the "calculation" that Nieuwentijt had performed incorrectly. Instead of multiplying b/mm by m, Nieuwentijt ought to have multiplied b/mm by mm. Then he would get b, which is a finite quantity. He would thus show that the assumption of b/mm as an infinitely small quantity does not lead to a contradiction. Differentials of second and higher order therefore are possible and do exist.

---

15 "One can see this proposition in the writings of Archimedes, especially in those concerning the dimension of the circle and the parabola. Most geometers who have written about the quadrature of spaces relying on this principle have demonstrated the most beautiful theorems concerning quadratures, thus the whole so called method of exhaustion of the ancients is based on this principle." A similar concept of equality can be found by the way in the unpublished paper De circuli quadratura of Nicolaus Cusanus written in 1450. Cf. (Nagel, 2007).

16 "Two quantities are called incomparable if one of them is infinitely smaller than the other, or what is the same, if the one multiplied by an infinite number can restore the other or exceed it."

17 "Equality consists in the incomparability of the difference of two [...] or more quantities with those quantities themselves."

Hermann illustrates his assumption of different orders of infinity by examining a set of hyperbolas all having the same axes as common asymptotes (Hermann, 1700, 42). If one looks at the space between one of these hyperbolas, the ordinate at a given x and the axis, it is evident that this space is infinite, but also that it is larger or smaller than the similar space produced by any other hyperbola of the set. Thus different orders of infinite spaces are possible.

In Chapter 5, Hermann again shows that it is possible to find the differential quotient of an exponential function like $x^v=y$. He repeats Leibniz's demonstration, but explicitly indicates that the text of the publication of 1695 in the *Acta Eruditorum* was corrupted by a misprint and that therefore Nieuwentijt's objections in his *Considerationes secundae* are baseless (Hermann, 1700, 4). With regard to the theory of exponential functions and their differentiation, Hermann refers to Johann Bernoulli's treatise in the *Acta Eruditorum* from March 1697, in which Bernoulli took himself to have already answered Nieuwentijt's questions concerning the exponential functions in a completely satisfactory way (Hermann, 1700, 53).[18] We cannot address here all of the other arguments Hermann raises in order to secure the foundations of the infinitesimal calculus. In the last chapter of his booklet, he summarizes the basic laws of the calculus formulating a definition, an axiom (or postulate) with two lemmas, and four propositions by which he hopes to satisfy Nieuwentijt's often pronounced desire to see the principles of the calculus together with their demonstrations.

In the second definition Hermann repeats what he has already said with regard to the infinitely small: "Quantitas vero infinite parva est, quae omni assignabili minor est et talis Infinitesima vel Differentiale vocatur."[19] (Hermann, 1700, 56) From this he concludes his first proposition, which states that a finite quantity to which an infinitely small quantity is added, will not be augmented by this addition. At the same time, this quantity will not be diminished if an infinitely small quantity is subtracted from it (Hermann, 1700, 56).[20] Hermann then returns to the main point, namely to the defini-

---

18 "Haec omnia solidius explicata inveniet B. Lector, in Cel. Viri Joh. Bernoulli schediasmate Act. Lips. An. 1697 mens. Mart. Inserto; et quae inibi Cel. Bernoulli dixit, ex asse forsitan Cl. Nieuventiit jam satisfecerunt." – The mentioned article of Johann Bernoulli is entitled "Principia calculi exponentialium seu percurrentium," in: *Acta Eruditorum*, March 1697, 125–133.

19 "A quantity is really infinitely small if it is smaller than any quantity which could be given, and such a quantity is called an infinitesimal one or a differential."

20 "Quantitas Finita, infinite parva sibi addita non augetur vel eadem demta non minuitur." Hermann repeats here a postulate of Johann Bernoulli's lectures to l'Hôpital. Cf. footnote 1.

tion of equality proposed by Leibniz. Referring to his own proposition, Hermann formulates the following corollary: "Hinc veritas illius aequalitatis definitionis quam supra in fine §. I. Cap. I. dedi, luculenter patet."[21] (Hermann, 1700, 56)

I do not have space to refer to the rest of Hermann's considerations. I would like to remark though that Hermann is convinced that, in addition to all the proofs which he and Leibniz had brought forward, the results achieved through the application of the calculus in mathematics and physics provide strong arguments for the solidity of its foundation. Without the calculus, for example, all the properties of special curves like the catenaria, the velaria, the elastica, the paracentric isochrone or the brachistochrone would have remained concealed. Likewise, deeper insight into the laws of nature would not be possible. Hermann therefore recommends the study of Newton's *Principia*, which he praises as a work more precious than gold, but also the papers of his master Jacob Bernoulli. The brilliant achievements due to the calculus in mathematics could be found in the articles of the *Acta Eruditorum* and further in L'Hôpital's *Analyse des infiniment petits*. It seems to me as if Hermann has somehow realized that L'Hôpital – as I said at the beginning of my paper – had formulated the laws of the differential calculus in his *Analyse* in such a sophisticated way that they could not serve as objects for attacks like those of Nieuwentijt.

Hermann thus shares with Leibniz not only the definitions of the basic notions of the calculus but also his pragmatic arguments. It is therefore not astonishing that Hermann's *Responsio* was very welcomed not only by his master Jacob Bernoulli, who reviewed the *Responsio* in the *Acta Eruditorum* (Bernoulli, 1701),[22] but also by Leibniz himself. Even Johann Bernoulli, whom Hermann had visited at Groningen,[23] was very pleased by the fervent support which the Leibnizian party received from Hermann. He recommended Hermann again to Leibniz, in spite of the fact that Hermann had been the personal assistant of his not-so-beloved brother.[24] Leibniz then let

---

21 "From this the truth of that definition which I have given above in Chapter 1 is obvious."

22 That Jacob Bernoulli is the author of this review is confirmed by Laeven (Laeven, 1986, 309).

23 Hermann visited Johann Bernoulli at Groningen in April 1701. There he showed him then his Responsio. Cf. Johann Bernoulli to Leibniz, May 7, 1701 (Leibniz, 1856, 668).

24 He also had recommended Hermann to the chair of Utrecht and Leiden, a recommendation to which Leibniz also supported. Cf. ex gr. Johann Bernoulli to Leibniz, January 15, 1704 (Leibniz, 1856, 737), and Leibniz to Johann Bernoulli, January 20, 1704 (Leibniz, 1856, 743).

Hermann be elected a member of the Brandenburg Society[25] and succeeded in promoting him to the chair of mathematics at Padua (cf. Robinet, 1991). So Hermann's *Responsio* became his entrance ticket to an academic career. Hermann himself was convinced that he had achieved his aim of defending the Leibnizian conception of the infinitesimal. Looking back on his criticisms of Nieuwentijt's objections, he writes at the very end of his *Responsio*:

> Animo ergo Cl. huic Viro satisfacienti bina duntaxat fundamentalia modo laudati infinitesimorum calculi theoremata cum demonstratione attulisse sufficere existimo, quanquam multa supersint quae eodem quoque modo demonstrari possent; intacta tamen relinquere consultius duxi, neque enim praesens negotium id efflagitare videbatur, neque chartae angustia illud permittere. Num vero hac in parte scopum attigerim, judicent alii; neque enim omnibus me satisfacturum promittere ausim, neque dicta theoremata omni accuratione me demonstrasse affirmavero, quamquam pro legitimis easdem venditare me posse, nullus dubitem.[26] (Hermann, 1700, 62)

3. The Discussion of Infinitesimals between Johann Bernoulli and Christian Wolff

It would of course be interesting to find out in what way Jacob Hermann shared Johann Bernoulli's opinion regarding infinitesimals. Unfortunately, Johann Bernoulli did not publish an explicit theory of infinitesimals. His considerations about this subject are rare and mostly hidden in his unpublished correspondence. I cannot give a detailed account of this highly interesting subject here, but I shall refer to a short passage in a letter Johann Bernoulli wrote to his long-standing correspondence partner, the mathematician and philosopher Christian Wolff.[27]

---

25 Hermann was elected, together with Michel Angelo Fardella, on September 9, 1707. Cf. Hartkopf, 1992, 146 and 94.

26 "To satisfy the desire of this man, I regard it as sufficient, to have presented at least two fundamental theorems of the infinitesimal calculus together with their proofs, though there are still many things left, which could be demonstrated in the same way. But I think it is a good advice not to go into detail, because my task did not seem to require it and the size of this booklet does not allow it. Whether or not I have now reached my aim is up to others to judge. I do not dare to promise that I have satisfied everybody nor can I affirm that I have demonstrated those theorems with the utmost accuracy, though I have no doubt that I have presented them in a pertinent and relevant way."

27 The correspondence of Johann Bernoulli and Christian Wolff comprises 96 letters written in the years between 1706 and 1745. The autograph letters of Wolff and the drafts of Johann

In 1730 Johann Bernoulli studied Christian Wolff's *Ontologia* (Wolff, 1730), which he had received as a present from the author. In a letter to Wolff dated August 31, 1730,[28] Bernoulli informs his partner that he had read the book with great pleasure. He admires the facility and the clarity of Wolff's style and he praises Wolff for his effort to make philosophy fit for the public. He then criticises Wolff's statements concerning the finite and the infinite (Wolff, 1730, 597). Wolff defined the infinite in mathematics as something to which we cannot assign limits beyond which it could not be augmented.[29] According to this definition, the finite is then defined as something to which we can assign limits, from where it starts and where it ends, i.e. beyond which it could be augmented.[30] Finally, Wolff defines the infinitely small as something to which we cannot assign limits beyond which it could not be further diminished.[31] That means that for any limit which we can assign it must be always the case, that the infinitely small is beneath of this limit. And he concludes: "Infinite parva impossibilia sunt" (Wolff, 1730, 601, § 803), and: "Quantitates infinitae & infinite parvae Mathematicorum non sunt verae quantitates, sed saltem imaginariae" (Wolff, 1730, 602, § 804).[32]

Bernoulli then points out to Wolff that he would disapprove this chapter if he had to understand it literally in the way it is written. Then he explains his own conceptions of "finite", "infinite" and "infinitesimal" to Wolff. The notions of finite or infinite, be they real or mathematical, are merely relative notions for him. We cannot find in them anything large or small and anything finite or infinite.[33] Johann Bernoulli thus believes that each quantity can be called finite, infinite and even infinitely small only if we compare it with other quantities of the same kind. To illustrate this idea he refers to the

---

Bernoulli are preserved in volume Ms L I a 671 of the University Library of Basel. I am currently preparing an edition of this correspondence.

28 Johann I Bernoulli to Christian Wolff, August 31, 1730, Ms UB Basel, L I a 671, Nr. 34.

29 "Infinitum in Mathesi dicimus, in quo nulli assignari possunt limites, ultra quos augeri amplius nequeat" (Wolff, 1730, 597).

30 "Finitum dicimus in Mathesi, cui assignari possunt termini, unde incipit & ubi definit, seu ultra quos augeri potest" (Wolff, 1730, 598, § 798).

31 "Inifinite parvum in Mathesi dicitur, cui nullus assignari potest limes, ultra quem imminui amplius nequit." (Wolff, 1730, 601 (by misprint "901"), § 802)

32 "Infinitely small quantities are impossible and thus the infinite or the infinitely small quantities of the mathematicians are not real quantities but only imaginary ones."

33 "[...] ego quidem statuo, notionem finiti et infiniti sive realis sive Mathematici esse mere relativam, et nihil in se spectatum esse magnum vel parvum, nihil quoque infinitum vel finitum [...]". (Johann Bernoulli to Christian Wolff, August 31, 1730, Ms UB Basel, L I a 671, Nr. 34)

common observation that objects close to us appear larger than more remote objects, so that we cannot say anything about their real size. A microscope makes small animals appear very large to us. Thus if we had eyes like microscopes we could see a little piece of dust like a gigantic mountain and not in its real size. And if our eyes would function like two different microscopes we could see the same object as at once smaller and larger, depending on which of our eyes is closed. Therefore, Bernoulli argues, we have no criterion by which we could decide which natural and real size an object has. By analogy to this, each quantity can be regarded as finite, infinite and infinitely small depending on the comparisons we are drawing.

Johann Bernoulli then turns to the general properties of a quantity. In his opinion each quantity is infinitely divisible or, more precisely, divisible without coming to an end, even if we are not able to divide it in reality (Bernoulli, 1993).[34] Then he states that a quantity consists of infinitely small parts, which in reality are not separated and cannot be divided by natural forces. He is convinced that these infinitely small parts do exist even if we are not able to bring them to mind. But these infinitely small parts are only infinitely small with regard to the whole of which they are parts. On the other hand they are infinite themselves with regard to their proper parts, of which they possess an infinite number, a fact which also is not imaginable for us (Bernoulli, 1993).[35]

After these statements Bernoulli abruptly concludes the discussion by writing:

> Habes hic, Vir amplissime, paucis quam habeam notionem de infinito et infinitatis gradibus, credo eam cum tua facile conciliari posse; qua alii de hac materia disputant, Geometriae imperiti, nituntur conceptibus confusis et abeunt plerumque in Logomachias.[36] (Ibid.)

---

34 "Quantum est divisibile in infinitum, seu potius divisibile sine fine, etsi actu ipso ita divisum non sit [...]". (Johann Bernoulli to Christian Wolff, August 31, 1730) This statement apparently refers to Aristotle's definition of a quantity which he gave e.g. in *Metaphysics* V, 13, 1020a, where he states that a quantity is something what can be divided into parts.

35 "[...] habet eas partes infinitesimas, etsi a se invicem non separatas et forte naturae viribus non separabiles, existunt tamen, de hoc dubitari non potest et existunt nobis non cogitantibus. Dantur revera quantitates infinite parvae, sed quae tales sunt tantum respectu totorum quorum sunt partes, alias enim et ipsae sunt infinitae respectu suarum partium, quas etiam habent numero infinitas, nobis pariter non cogitantibus [...]". (Johann Bernoulli to Christian Wolff, August 31, 1730)

36 "You receive here, Sir, some few [remarks] of what I think of the notions of the infinite and of the degrees of infinity. I presume that it will be easy to bring it in accord with your opinion. What other people, who are not trained in mathematics, debate on this subject, is based on confused concepts and ends almost with logomachies."

Of course this isolated statement of Johann Bernoulli in his letter to Christian Wolff, concerning infinitesimals, is not sufficient to serve as a basis for characterising his conception in general. But even taking into account how limited our only source here is, we can perhaps try to sketch some characteristic aspects of his meta-mathematical conception. For Johann Bernoulli, infinitesimals do exist. They are parts of a quantity and because they are infinitely divisible themselves they are quantities too. Their characteristic property of being infinitesimally small is only relational. Therefore Bernoulli calls them "imaginary." In his calculations he can treat them as finite quantities until he decides whether or not he can neglect them. Thus differentials are sometimes regarded as constant and finite quantities and sometimes as being zero. Another consequence of this pragmatic concept is that Johann Bernoulli can regard higher order differentials as infinitesimals of the differentials of the next lower order without any conceptual problems. Utility and success in calculation are decisive for him. And as long as he gets convincing results he does not care about conceptual deficiencies.

It is obvious that Johann Bernoulli's pragmatic conception is closer to Wolff's. At the same time, we find in his conception some basic similarities with that of Leibniz, but also some characteristic differences. After the discovery of the calculus during his Paris period, Leibniz developed a new conception of infinitesimals. Although he originally identified the "infinitesimalia" as fixed quantities and as "inassignabilia," in his later period he came to regard them as variables. In addition to this, by the time of the later period he introduced – as we have seen – a new definition of equality. With this definition, he could regard x and $x + dx$ as equal for calculation without any logical contradiction. Thus, for Leibniz, infinitesimals are on the one hand fictions, but on the other hand they are fictions with which he can calculate. Bernoulli similarly regards infinitesimals as imaginary quantities. But in contrast to Leibniz he holds differentials to be constants. Thus he had to adopt that x and $x+dx$ would become equal only under the condition that dx can be regarded as zero in comparison to x. In this regard, Bernoulli's conception, which he explains to Christian Wolff in 1730, is almost the same as the one he had in mind when he gave his lectures to L'Hôpital in 1691/92.

4. Conclusion

In the hands of masters like Leibniz and Bernoulli both conceptions could lead of course to new and unexpected results independent of their differing foundations. Jacob and Johann Bernoulli as well as the other members of their circle demonstrated the success of their pragmatic attitude in a brilliant manner even when their philosophical conceptions of infinitesimals were inferior to those of Leibniz. Only infrequently did the weakness of their conceptions lead them to paradoxical results. But even a genius like Leonhard Euler, the most prominent member of the Bernoulli circle, sometimes fell victim to the hidden traps caused by insufficient conceptual foundation of infinitesimals.

Jacob Hermann has shown us how one could deal with the more sophisticated Leibnizian concept. But he was a Bernoulli disciple, too, and so it took a century until a rigorous foundation for the basic notion of the calculus was constructed. It seems to me that, for example, Cauchy's definition of a limit depends more on the ideas of Leibniz than on the pragmatic conceptions of the Bernoulli circle. At least in the background, Leibniz's new concept of equality seems to be present there.

A detailed story of the development of the notion "infinitesimal" is not yet written. Leibniz certainly was a turning point in this history. Likewise, disciples like Jacob and Johann Bernoulli as well as Jacob Hermann made contributions that propelled thinking about this central notion of the calculus. For this reason, it is worthwhile to have at least a short look into some of their considerations concerning infinitesimals as helpful contributions for clarifying this notion.

Douglas Jesseph

Truth in Fiction: Origins and Consequences of Leibniz's Doctrine of Infinitesimal Magnitudes

The status of infinitesimals in Leibniz's philosophy of mathematics is an issue whose resolution is not without difficulty. In many contexts Leibniz's account of his *calculus differentialis* is phrased in terms that are most readily interpreted as implying the real existence of infinitely small magnitudes. In other places, he claims that there are, in actual fact, no infinitely small magnitudes and the device of infinitesimals is simply a convenient fiction, useful for stating and deriving results, but without any serious ontological import. One can therefore sensibly ask whether Leibniz truly believed in the reality of infinitesimal magnitudes, but thought that the central results of his *calculus differentialis* might be formulated and derived by means that did not presuppose the reality of the infinitesimal. Pursuing this sort of interpretive strategy would obviously require that Leibniz's frequent claims about the fictionality of the infinitesimal be taken a something less than face value. That is not a decisive problem for an interpretation of Leibniz as a realist about infinitesimals, but it does suggest that one ought at least to consider the prospects for seeing Leibniz as committed to the view that the infinitesimal has the status of a "well founded fiction." I take Leibniz's claims about the fictionality of the infinitesimal to be his considered view on the subject, although I am not convinced that he held consistently to a "fictionalist" position from his earliest writings on the calculus.[1]

My purpose here is to trace what I take to be the origins of Leibniz's notion of the fictional infinitesimal, which I believe can be found in Thomas

[1] Richard Arthur (Arthur, 2008c) makes the case that Leibniz's fictionalism about infinitesimals can be dated as early as 1676. The evidence he cites certainly favors this approach to Leibniz's correspondence and various unpublished sources. As we will see, however, Leibniz often made public pronouncements that are most readily interpreted as endorsing the reality of infinitesimal magnitudes. This ambivalence would lead to some considerable perplexity and controversy, as we will see in section 3.

Hobbes's doctrine of *conatus*, and particularly the application that Hobbes made of this concept in the solution of geometric problems of tangency, quadrature and arclength determination – precisely the sorts of problems that the Leibnizian calculus was designed to solve. Having shown the role that the *conatus* concept plays in Hobbes's approach to mathematics, I will argue that some salient features of it appear in Leibniz's formulation of the calculus. In particular, the notion that *conatus* is a finite, but negligibly small, quantity is significant. Ultimately, I think that Hobbes's notion of *conatus*, or at least a near descendent of it, appears in Leibniz's claim that infinitesimal magnitudes are "well founded fictions" that can, in principle, be replaced by the consideration of finite quantities. In the context of his theory of the fictional infinitesimal, Leibniz's notion of "incomparably small" quantities plays a central role, and I think it can be shown that the Leibnizian theory of the incomparably small (yet finite) magnitude has its roots in the Hobbesian of *conatus*.

My first task is therefore to outline the role of *conatus* in the Hobbesian approach to geometry; with this material in hand, I will investigate some of Leibniz's pronouncements on the foundations of his calculus with the aim of showing that these owe a significant debt to Hobbes's proposals. In the end, Leibniz's account of infinitesimals emerges as a relatively stable fictionalism, but there are some interpretive difficulties that remain. In particular, it is not altogether clear how we are to understand Leibniz's frequent claims that infinitesimal magnitudes are, at least in principle, eliminable from a demonstration. Further, it is difficult to reconcile Leibniz's fictionalist account of the infinitesimal with traditional notions of rigorous demonstration.

1. Hobbes, *Conatus*, and the Mathematics of Motion

Hobbes first introduced the concept of *conatus* in his 1655 treatise *De Corpore* – a work presented as the first part of the elements of philosophy and containing Hobbes's doctrines on the nature of body as well as his exposition of a thoroughly materialistic philosophy of mathematics.[2] The fundamental idea behind the doctrine of *conatus* can be seen in the etymological

[2] The original of Hobbes' *De Corpore* (1655) was followed by an English translation entitled *Of Body* (1656) which altered (but did not improve) some of the failed mathematics in the original. Both items are universally known by the Latin title.

fact that the word derives from the Latin deponent verb *conor*, meaning to strive or attempt. Hobbes's preferred English term for *conatus* is 'endeavour', although current scholarship is practically unanimous in retaining the Latin term. In any case, the *conatus* of a body is its tendency, striving, or endeavor to move in a certain direction. Thus stated, the doctrine might seem to involve the attribution of some kind of mental states to otherwise inanimate bodies, but Hobbes had no such agenda.[3] As Hobbes defines it, *conatus* is essentially a point motion, or motion through an indefinitely small space: "Conatum esse motum per spatium et tempus minus quam quod datur, id est, determinatur sive expositione vel numero assignatur, id est per punctum et in instanti."[4] This definition employs Hobbes's idiosyncratic conception of points, in which a point is an extended body, but one sufficiently small that its magnitude is not considered in a demonstration.[5] It is useful to think of points in this sense as corresponding to the notion of a particle in physics – a body sufficiently small that the distance between any to parts of it can be ignored. Likewise, a line or curve could be taken as the trace of a particle. As Hobbes stated, "eo sensu, quo Terra *punctum* et via eus annua *linea ecliptica* vocari solet."[6] In explicating the definition of *conatus* he remarks that

> meminisse oportet per punctum non intelligi id quod quantitatem nullam habet sive quod nulla ratione potest dividi (nihil enim est ejusmodi in rerum natura), sed id cujus quantitas non consideratur, hoc est, cujus neque quantitas neque pars

---

[3] Indeed, Hobbes's account of the mind (and particularly volition) reduces all mental states to motions in the body. A mental state such as desire is simply the *conatus* of an animate body towards something, while aversion is the *conatus* away from something. The reult is that, far from attributing mental states to inanimate bodies, Hobbes took animate bodies to be those that are highly organized, self-preserving, and move in characteristice ways by communicating motion to their parts. This is made explicit in *Leviathan*: "seeing life is but a motion of Limbs, the beginning whereof is in some principall part within; why may we not say, that all *Automata* (Engines that move themselves by springs and wheeles as doth a watch) have an artificial life?" (Hobbes, 1651, preface).

[4] "*Conatus* is motion through a space and a time less than any given, that is, less than any determined whether by exposition or assigned by number, that is, through a point" (*De corpore* 3.15.2; OL I, 177).

[5] This definition follows from Hobbes' thoroghgoing materialism. He denies the existence of any abstract or immaterial substances and must consequently treat the objects of mathematics as bodies (since his ontology recognizes only bodies as real). A point therefore is a body whose magnitude is not considered, a line or curve is the path traced by a moving point, etc. For more on Hobbes's approach to mathematics see Jesseph, 1999, chapter 3.

[6] "This is the sense in which the Earth is usually called a point and the path of its annual revolution the ecliptic line" (*De corpore* 2.8.12; OL 1: 99).

> ulla inter demonstrandum computatur; ita ut punctum non habeatur pro indivisibili, sed pro indiviso, sicut etiam *instans* sumendum est pro tempore indiviso, non pro indivisibili.[7] (*De corpore* 3.15.2; OL 1, 177–8)

The result, as I have mentioned, is that *conatus* is a kind of "tendency toward motion" or a striving to move in a particular direction.

This definition allows for a further concept of impetus, or the instantaneous velocity of a moving point; the velocity of the point at an instant can be understood as the ratio of the distance moved to the time elapsed in a *conatus*. In Hobbes's terms "Impetum esse ipsam velocitatem, sed consideratam in puncto quolibet temporis, in quo fit transitus Adeo ut impetus nihil aliud sit quam quantitas sive velocitas ipsius conatus."[8] (*De corpore* 3.15.2; OL I, 178) The concepts of *conatus* and impetus are basic to Hobbes's analysis of motion, and it is no great exaggeration to say that his whole program for natural philosophy, which he deemed the true science of motion, is drawn from his account of *conatus* and impetus.[9]

The concepts of impetus and *conatus* can be applied to the case of geometric magnitudes as well as to moving bodies. Because Hobbes held that geometric magnitudes are generated by the motion of points, lines, or surfaces, he also held that one could inquire into the velocities with which these magnitudes are generated, and this inquiry can be extended to the ratios between magnitudes and their generating motions. For example, take a curve to be traced by the motion of a point, and at any given stage in the generation of the curve, this generating point will have a (directed) instantaneous velocity. This, in turn, can be regarded as the ratio between the indefinitely small distance covered in an indefinitely small time; this ratio will be a finite magnitude which can be expressed as the inclination of the tangent to the curve at the point.

---

[7] "[...] it must be re remembered, that by a point is not to be understood that which has no quantity, or which cannot by any means be divided; for there is no such thing in nature; but that, whose quantity is not at all considered, that is, whereof neither quantity nor any part is computed in demonstration, so that a point is not to be taken for an indivisible, but for an undivided thing; as also an instant is to be taken for an undivided time, and not for an indivisible time." (EW I, 206)

[8] "Impetus, *or quickness of motion, to be the swiftness or velocity of the body moved, but considered in the several points of that time in which it is moved. In which sense* impetus *is nothing else but the quantity or velocity of endeavour [conatus].*" (EW I, 207)

[9] See Jesseph, 2006, for an account of how these concepts work in the foundations of Hobbes's natural philosophy.

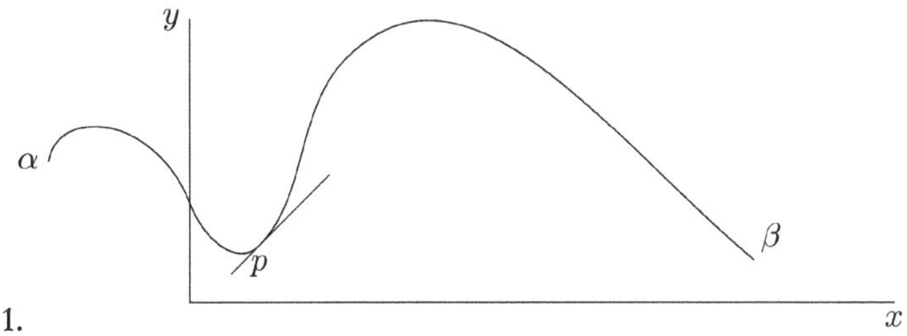

Figure 1.

Consider, for instance the curve *αb* as in Figure 1. The *conatus* of its generating point at any instant will be the "point motion" with which an indefinitely small part of the curve is generated; the impetus at any stage in the curve's production will be expressed as the ratio of the distance covered to the time elapsed in the *conatus*. Representing the time by the *x*-axis and the distance moved by the *y*-axis, then (assuming time to flow uniformly) the instantaneous impetus will be the ratio between the instantaneous increment along to the *y*-axis to the increment along the *x*-axis. The tangent to the curve at the point *p* is the right line that continues or extends the *conatus* at *p*; or, equivalently, the tangent is the dilation or expansion of the point motion into a right line.

It is important to observe here that the tangent is constructed as a finite ratio between two quantities that, in themselves, are small enough to be disregarded. That is to say, the ratio between two "inconsiderable" quantities may itself be a considerable quantity. Hobbes emphasized this feature of his system when he stressed that points may be larger or smaller than one another, although in themselves they are quantities too small to be considered in a geometric demonstration. Thus, in discussing the comparisons that may be made between one conatus and another, Hobbes declared:

> sicut punctum cum puncto, ita conatus cum conatu comparari potest et unus altero major vel minor reperiri. Nam si duorum angulorum puncta verticalia comparentur inter se, erunt in ratione ipsorum angulorum aequalia vel inaequalia; vel si linea recta secet plures circulorum concentricorum circumferentias, puncta sectionum erunt inaequalia in ratione ipsarum perimetrorum.[10] (*De corpore* 3.15.2; OL I, 178)

[10] "And yet, as a point may be compared with a point, so one endeavour [*conatus*] may be compared with another endeavour [*conatus*], and one may be found to be greater or less than another. For if the vertical points of two angles be compared, they will be equal or unequal in the same proportion which the angles themselves have to one another. Or if a strait line cut many circumferences of concentric circles, the inequality of the points of intersection will be in the same proportion which the perimeters have to one another." (EW I, 206–7)

Hobbes's concepts of *conatus* and impetus can also be applied to the general problem of quadrature by analyzing the area of a plane figure as the product of a moving line and time. Hobbes himself was eager to solve problems of quadrature (most notably the quadrature of the circle), and it is here that his concept of conatus is put most fully to work. Indeed, the third part of *De Corpore* (which bears the title "On the Ratios of Motions and Magnitudes") is Hobbes's attempt to furnish a general method for finding quadratures. In the very simplest case, the whole impetus imparted to a body throughout a uniform motion is representable as a rectangle, one side of which is the line representing the instantaneous impetus while the other represents the time during which the body is moved. More complex cases can then be developed by considering non-uniform motions produced by variable impetus. In chapters 16 and 17 of *De Corpore* Hobbes approached a variety of different quadrature and tangency problems, and in so doing he presented a number of important results that belong to the "pre-history" of the calculus. Of special interest in this context is Hobbes's appropriation of important results from Cavalieri's 1646 treatise *Exercitationes Geometricae Sex*, which were incorporated into chapter 17 of *De Corpore* as an investigation into the area of curvilinear figures.

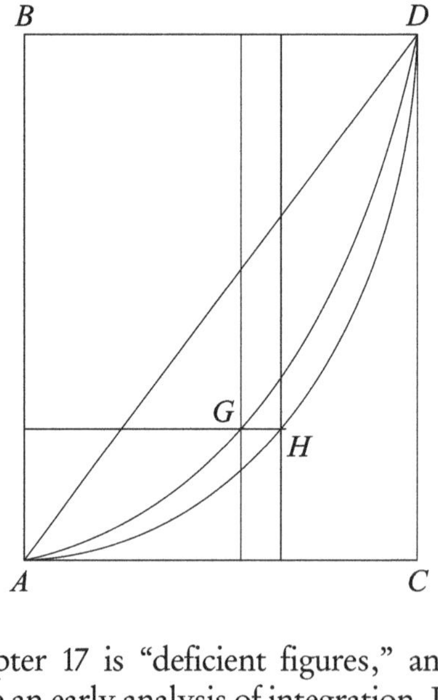

Figure 2.

The subject of chapter 17 is "deficient figures," and it presents something very much like an early analysis of integration. In Hobbes's parlance the deficient figure *ABDGA* in Figure 2 is produced by the motion of the right line *BD* through *BA*, while *BD* diminishes to a point at *A*. The "complete figure" corresponding to the deficient figure is the rectangle *ABDC*, produced by the motion of *BC* through *AB* without diminishing. The complement of the deficient figure is *DGAC*, the figure that, when added to the deficient figure, makes the complete figure. Hobbes proposed to determine the ratio of the area of the deficient figure to its complement, given a specified rate of decrease of the quantity *BD*. He concluded that the ratio of the deficient figure to its complement is the same as the ratio between corresponding lines in the deficient figure and their counterparts in the complement. As he stated the theorem in article 2 of chapter 17:

> Figura deficiens facta a quantitate continuo decrescente donec evanescat, secundum rationes ubique proportionales et commensurabiles est ad complementum suum ut ratio totius altitudinis ad altitudinem quolibet tempore diminutam, ad rationem quantitatis integrae, quae figuram describit, ad eandem tempore eodem diminutam.[11] (*De corpore* 3.17.2; OL I, 209)

Thus, if the rate of diminution of *BD* is uniform the line *AD* will be a right line (the diagonal of the rectangle), and the deficient figure will be to its complement in the ratio of one to one. In more complex cases, as when *BD* decreases as the square of the diminished altitude, the area of the deficient figure will be twice that of its complement. And, in general, if the line *BD* decreases as the power n, the ratio of the deficient figure to its complement will be n:1.

In the fourth of his six *Exercationes Geometricae* Cavalieri pursued a result that historians of mathematics generally characterize as the attempt to prove the geometric equivalent of the theorem that $\int_0^a x^n dx = \frac{a^{(n+1)}}{n+1}$. Except for differences in diagrams and terminology, Cavalieri's fourth *Exercitatio* delivers the same results as Hobbes's account of deficient figures. The central theorem, which is the analogue of the result we just saw stated by Hobbes, reads:

[11] "A deficient figure, which is made by a quantity continually decreasing to nothing by proportions everywhere proportional and commensurable, is to its complement as the proportion of the whole altitude to an altitude diminished in any time is to the proportion of the whole quantity which describes the figure, to the same quantity diminished in the same time." (EW I, 247)

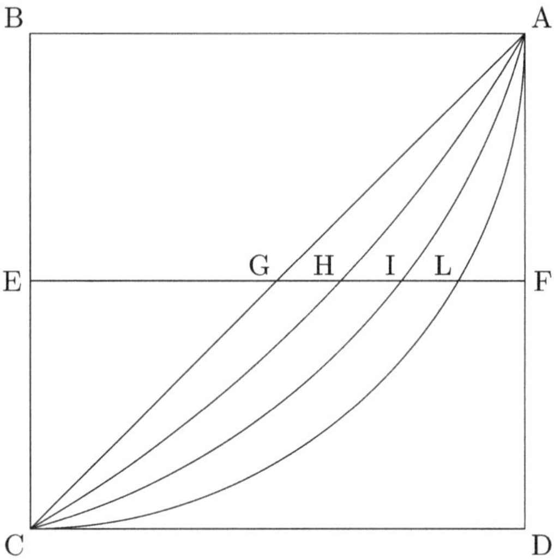

Figure 3.

In quocunque parallelogrammo, ut *BD*, regula basi *CD*, si agatur ipsi, *CD* quaecunque parallela, *EF*, & ducatur diameter, *AC*, quam illa secet in *G*; erit ut, *DA*, ad *AF*, ita, *CE*, vel, *EF*, ad, *FG*. Dicatur autem, *AC*, diagonalis prima. Rursus ut *q.DA*, ad *q.AF*, ita fiat, *EF*, ad *FH*, & ita ubique fieri intelligatur in parallelis ipsi, *CD*, ita ut omnes homologae ipsi, *HF*, terminentur ad curvam, *AHC*. Pariter ut *c.DA*, ad *c.AF*, ita quoque fiat, *EF*, ad, *FJI*, & sic in caeteris, descripta curva *CIA*. Et ut *qq.AD*, ad *qq.AF*, ita sit *EF*, ad *FL*, & sic in caeteris, descripta curva, *CLA*. Quod et in reliquis potest fieri supponi potest. Dicatur autem, *CHA* diagonalis 2. *CIA*, diagnoalis 3. *CLA*, diagonalis 4. &c. Similiter triangulum, *AGCD*, vocatur 1. spatium diagonalium parallelogrammi, *BD*, trilineum, *AHCD*, 2. spatium, *AICD*, tertium, *ALDC*, 4., &c. Dico ergo parallelogrammum, *BD*, duplum esse primi spatij, triplum secundi, quadruplum 3. quintuplum quarti, &c.[12] (Cavalieri, 1647, 279)

12 "In any parallelogram such as *BD* with the base *CD* as *regula*, if any parallel to *CD* such as *EF* is taken, and if the diameter *AC* is drawn, which cuts the line *EF* in *G*, then as *DA* is to *AF*, so *CD* or *EF* will be to *FG*. And let *AC* be called the first diagonal. And again as DA^2 is to AF^2, let *EF* be to *FH*, and let this be understood in all the parallels to *CD*, so that all of these homologous lines *HF* terminate in the curve *AHC*. Similarly, as DA^3 is to AF^3, let also *EF* be to *FI*, and likewise in the remaining parallels, to describe the curve *CIA*. And as AD^4 is to AF^4, let *EF* be to *FL*, and likewise in the remaining parallels to describe the curve *CLA*. Which procedure can be supposed continued in other cases. Then let *CHA* be called the second diagonal, *CIA* the third diagonal, *CLA* the fourth diagonal, and so forth. Similarly let the triangle *AGCD* be called the first diagonal space of the parallelogram, the trilinear figure *AHCD* the second diagonal space of the parallelogram, *AICD* the third, *ALCD* the fourth, and so on. I say therefore that the parallelogram *BD* is twice the first space, triple the second space, quadruple the third space, quintuple the fourth space, and so forth."

Hobbes and Cavalieri employed different proof procedures in attempting to establish this result. Although I cannot go into the details, it is worth observing that Hobbes's procedure (at least in some of its guises) employs the idea of a *conatus* or the "aggregate of the velocities" whereby lines in a figure are generated.[13] In relying on a concept of the aggregate of velocities or *conatus* Hobbes differs, at least superficially, from Cavalieri, who had reasoned from a ratio between "all the lines" in two figures to the same ratio in their areas.[14]

Nevertheless, there is more than enough similarity between Hobbes and Cavalieri on this point to warrant the conclusion that Hobbes borrowed quite heavily from the Italian mathematician.[15] Indeed, Hobbes's nemesis John Wallis remarked that "those propositions which Mr. Hobs had concerning the measure of the [higher-order parabolas] were not his own, but borrowed from somebody else without acknowledging his author," and they "were to be found demonstrated in an exercitation of Cavalierius *De Usus Indivisibilium in Potestatibus Cossicis*" (Wallis, 1656, 7). In fact, Hobbes seems to have taken Cavalieri's demonstrations and re-cast them in a way that emphasized the consideration of point motion or *conatus*, and he evidently saw himself as reforming Cavalieri's doctrines to bring them within the purview of what he termed his "method of motions."[16]

It is well known that Leibniz was profoundly influenced by his reading of Hobbes, and he seems to have been particularly enamored of the Hobbesian concept of *conatus*. In a famous 1670 letter to Hobbes, Leibniz announced his great admiration for "abstractis motuum rationibus, in qvibus jacta a Te fundamenta."[17] (22 July, 1670, A II, 1 (2006), 92) To the extent that the concept of *conatus* is the basis for Hobbes's analysis of motion, this endorsement suggests that Leibniz was ready to follow Hobbes in using the concept for the analysis of all phenomena produced by motion. Indeed,

---

13 In the 1668 version of Chapter 17, Article 2, of *De Corpore* Hobbes argued "as the aggregate of all the velocities by which the right lines [...] generated in the same manner is to the aggregate of the times designated by the right lines, [...], so the plane surface *DCFEB* is to the plane surface *ABEFC*." See Jesseph, 1999, 365–7, for an account of this demonstration.

14 For an overview of Cavalieri's methods see Andersen, 1985.

15 Cavalieri was not Hobbes' only source for his analysis of parabolic curves. Much of chapter 16 of *De Corpore* (which treats "motion by concourse") owes a significant debt to Galileo's analysis of the parabola in the *Two New Sciences*. For more on the Galilean background to Hobbes's mathematics, see Jesseph, 2004.

16 On Hobbes' "method of motions," which he thought capable of solving any geometric problem, see Jesseph, 1999, 234–7.

17 "The foundations you have laid concerning the abstract principles of motion".

scholars today generally accept that "Leibniz's early writings on natural philosophy are virtually steeped in *De Corpore*" (Bernstein, 1980, 29). In particular, Leibniz's reading of Hobbes appears to have been the source for much of his (admittedly limited) mathematical knowledge before his stay in Paris in the 1670s (Hofmann, 1974, 6–8), and other aspects of Leibniz's philosophy bear the signs of a distinctly Hobbesian influence (Ross, 2007). Significantly, Ursula Goldenbaum's contribution to this volume draws attention to Leibniz's marginalia in his copy of Hobbes's *De Corpore*, and these show beyond any serious doubt that Leibniz was very much taken with the Hobbesian system.

Some of the clearest evidence of Hobbes's influence can be found in Leibniz's 1671 *Theoria Motus Abstracti*, which employs the concept of *conatus* to investigate the nature of motion. In a 1671 letter to Henry Oldenburg, Leibniz announced that his theory of abstract motion provides the basis for the solution of any number of mathematical and philosophical puzzles. The theory, he claimed,

> invictas propemodum Compositionis continui difficultates explicat, Geometriam indivisibilium, et Arithmeticam infinitorum confirmat; ostendit nihil esse sine partibus in rerum natura; infinitus actu cujuscunque continui partes esse; doctrinam de angulis esse de quantitatibus inextensorum; Motum esse Motu fortiorem, ergo et conatum conatu: conatum autem esse motum per punctum in instanti; punctum ergo puncto majus esse.[18] (A II, 1 (2006), 166 = Oldenburg, 1965–77, VIII, 22)

The "geometry of indivisibles" and the "arithmetic of infinities" to which Leibniz refers are the contributions of Cavalieri and John Wallis. Cavalieri's method of indivisibles is mentioned explicitly among the "Fundamenta praedemonstrabilia" of the *Theoria motus abstracti*, as a theory "ejus veritas evidenter demonstratur, ut cogitentur quaedam, ut sic dicam, rudimenta seu initia linearum figurarumque qualibet dabili minora."[19] (Leibniz, 1671, 9 = A II, 2, 265) Wallis's 1655 treatise *Arithmetica Infinitorum*, although not

[18] "[...] explains the hitherto unresolved difficulties of continuous composition, confirms the geometry of indivisibles and arithmetic of infinities; it shows that there is nothing in the realm of nature without parts; that the parts of any continuum are in fact infinite; that the theory of angles is that of the quantities of unextended bodies; that motion is stronger than motion, and *conatus* stronger than *conatus* – however, *conatus* is instantaneous motion through a point, and so a point may be greater than a point" (Oldenburg, 1965–77, VIII, 26, slighly changed by me, D.J.).

[19] "Whose truth is obviously demonstrated so that we must think of certain rudiments, so to speak, or beginnings of lines and figures, as smaller than any given magnitude whatever".

mentioned explicitly in the text, is evidently referred to in the letter to Oldenburg when Leibniz refers to the "arithmetic of infinities." In light of this, it is no great interpretive leap to see Leibniz connecting the doctrine of *conatus* with the classic problem of quadrature, just as Hobbes had done, and thus to find part of the origin of the calculus in Leibniz's close reading of *De Corpore*.

It would be a vast overstatement to claim that Leibniz's calculus is simply the application of Hobbes's ideas. It is well known that Leibniz's mathematical thought was also strongly influenced by Galileo's approach to the geometry of indivisibles, for example, and the influence of Huygens cannot be overlooked, nor can Pascal's investigations into infinite sums and differences.[20] All of these are, without question, part of the background to Leibniz's calculus. Nevertheless, we can agree that Hobbes was one among many whose writings stimulated the development of the Leibnizian approach to the calculus.[21] However, there is one important difference between the Leibnizian and Hobbesian conceptions of conatus that is significant: Leibniz's language (at least in the *Theoria motus abstracti*) strongly implies that *conatus* be a literally infinitesimal quantity, while Hobbes regards it as having finite magnitude, but one so small as to be disregarded. In the end, however, Leibniz adopted a doctrine not far removed from Hobbes's.

2. Incomparable Magnitudes and the Question of Rigor

Traditional criteria of rigorous demonstration forbid the use of infinitary methods, and the standard formulation of Leibniz's calculus certainly seems to run afoul of such restrictions. The problem is particularly appar-

---

20 See Knobloch's commentary to Leibniz, in DQA, for an account of Leibniz's debt to Galileo; Herbert Breger's contribution to this volume indicates some of the significant debts that Leibniz's approach to the calculus owes to Pascal and Huygens.

21 This is not to say that Leibniz found Hobbes' philosophy completely acceptable, even in its account of the foundations of mathematics. Leibniz lists Hobbes's metaphysical and mathematical errors in the *Theoria Motus Abstracti* thus: "*Hobbius* tollit mentes incorporeas, tollit indivisibilia vera, atque ex eo principio in dubium revocat inventum Pythagorae hecatomba dignum, 47. I^mi^ Euclidis, fundamentum Geometriae: negat radicem quadrati, seu ut ego vocare soleo, numerum quadratillorum, de quo alibi, conincidere numero partium lateris, fundamentum non Algebrae tantum, sed et Geodesiae, multaque alia de motu tradit parum demonstrata: Quanquam caeteroquin nihil laudi ejus Viri, cujus profunditatem maximi facio, detractum velim." (Leibniz, 1671, 29 = A, VI, 2, 275). On Hobbes's eventual rejection of fundamental principles of geometry, see Jesseph, 1999, chapter 6.

ent in Leibniz's early publications of his method: mysterious terms *dx* and *dy* appear in equations for curves and increments, only to vanish when their work is done, seeming to hover between something and nothing. It is therefore no great surprise that "traditionalist" opponents would make a case against the *calculus differentialis*, charging Leibniz and his associates with violating standards of rigor that guarantee the security and demonstrative status of mathematics. In replying to these critics, Leibniz employed something very much like Hobbes's notion of points and *conatus* as finite but negligible quantities, although he phrased his defense in terms of "incomparably small" magnitudes.

In reply to the criticisms voiced by Nieuwentijt, who had held that the infinitesimal quantities *dx* and *dy* were illegitimately dismissed from calculations, Leibniz declared such quantities "incomparably small" and legitimate objects of mathematical study. To Nieuwentijt's requirement that only those quantities are equal whose difference is zero, Leibniz insisted

> Caeterum aequalia esse puto, non tantum quorum differentia est omnino nulla, sed et quorum differentia est incomparabiliter parva; et licet ea Nihil omnino dici non debeat, non tamen est quantitas comparabilis cum ipsis, quorum est differentia. Quemadmodum si lineae punctum alterius lineae addas, vel superficiei lineam, quantitatem non auges. Idem est, si lineam quidem lineae addas, sed incomparabiliter minorem. Nec ulla constructione tale augmentum exhiberi potest.[22] (AE 1695 = GM V, 322)

There is an obvious parallel between such "incomparably small" elements of lines or surfaces and Hobbes's conception of points, for it is exactly the hallmark of Hobbes's points that – though finite – they are too small to be considered in any demonstration. Leibniz's preference here for the language of the incomparable rather than the infinitesimal raises the question of whether such incomparable magnitudes are to be thought of as literally infinitesimal or whether they should be treated as finite but negligible quantities in the manner of Hobbes's points.

At first sight, one might interpret Leibniz's reply to Nieuwentijt as defending the reality of infinitesimals, seeing the term "incomparably small" as a kind of euphemism for "infinitesimal." But I think such an interpre-

[22] "I think that those things are equal not only whose difference is absolutely nothing, but also whose difference is incomparably small; and although this difference need not be called absolutely nothing, neither is it a quantity comparable with those whose difference it is. Just as when you add a point of one line to another line or a line to a surface you do not increase the magnitude, it is the same thing if you add to a line a certain line, but one incomparably smaller. Nor can any increase be shown by any such construction."

tation ultimately fails. Leibniz declared that it is enough to show that incomparably small quantities can be justly neglected in a calculation, and his reply to Nieuwentijt (AE 1695 = GM V, 322) refers to certain lemmas published in February of 1689 (AE 1689 = GM, VI, 144–160) for the full justification of this procedure. These lemmas of 1689 are contained in Leibniz's *Tentamen de Motuum Coelestium Causis*, which appeared in the *Acta Eruditorum* of Leipzig. But when we turn to them for enlightenment, two points become tolerably clear. First, such "incomparable" quantities were intended explicitly to avoid references to infinitesimals and instead to replace infinitesimal magnitudes with finite differences sufficiently small to be ignored in practice. Second, the doctrine of the incomparable has a very strong analogy with Hobbes's treatment of points, *conatus*, and impetus. The paragraph expounding these lemmas opens with the declaration that

> Assumi inter demonstrandum *quantitates incomparabiliter* parvas, verbi gratia differentiam duarum quantitatum communium ipsis quantitatibus incomparabilem. Sic enim talia, ni fallor, lucidissime exponi possunt. Itaque si quis nolit adhibere *infinite parvas*, potest assumere tam parvas quam sufficere judicat, ut sint incomparabiles et errorem nullius momenti, imo dato minorem, producant. Quemadmodum terra pro puncto, seu diameter terrae pro linea infinite parva habetur respectu coeli, sic demonstrari potest, si anguli latera habeant basin ipsis incomparabiliter minorem, angulum comprehensum fore recto incomparabiliter minorem, et differentiam laterum fore ipsis differentibus incomparabilem; item differentiam sinus totius, sinus complementi et secantis fore differentibus incomparabilem; item diffferentiam chordae, arcus et tangentis.[23] (AE 1689 = GM VI, 150–1)

The use intended for such incomparably small magnitudes is to avoid disputes about the nature or existence of infinitesimal quantities, and Leibniz holds that it would always be possible to use ordinary quantities similar to

---

[23] "I have assumed in the demonstrations *incomparably small* quantities, for example the difference between two common quantities which is incomparable with the quantities themselves. Such matters as these, if I am not mistaken, can be set forth most lucidly in what follows. And then if someone does not want to employ *infinitely small* quantities, he can take them to be as small as he judges sufficient to be incomparable, so that they produce an error of no importance and even smaller than any given [error]. Just as the Earth is taken for a point, or the diameter of the Earth for a line infinitely small with respect to the heavens, so it can be demonstrated that if the sides of an angle have a base incomparably less than them, the comprehended angle will be incomparably less than a rectilinear angle, and the difference between the sides will be incomparable with the sides themselves; also, the difference between the whole sine, the sine of the complement, and the secant will be incomparable to these differences."

the unassignable ones, and that such methods are of great use in finding tangents, determining maxima and minima, and investigating the curvature of lines (GM VI, 150). In other words, the lemmas on incomparable magnitudes are to serve as a foundation for the calculus which permits the talk of infinitesimals to be reinterpreted in terms of incomparable (but apparently finite) differences. These lemmas figure prominently in Leibniz's writings on the foundations of the calculus, since he almost invariably refers back to them in later discussions on the nature of the infinitesimal. It is also significant that Leibniz's "incomparably small" not only satisfies Hobbes's definition of a geometric point (a quantity sufficiently small that its magnitude cannot be regarded in a demonstration) but the doctrine is illustrated by the example of taking the Earth as a point with respect to the heavens, precisely as Hobbes had done.

3. Fictional Infinitesimals and Incomparable Magnitudes.

When we turn to Leibniz's treatment of the foundations of the calculus after 1700, the theme of the fictionality of the infinitesimal becomes much more clearly defined. There were two controversies in the Parisian *Academie des Sciences* that drew Leibniz into a discussion of the nature of infinitesimals, and in both cases he elaborated a theory in which the infinitesimal turns out to be a fictional entity, albeit a fiction that is sufficiently well-grounded that it cannot lead from true premises to a false conclusion. The first of these controversies was initiated by Michel Rolle, who argued that the notion of an infinitesimal was not only inconsistent, but that the calculus that employed it could lead to error.[24] The second controversy concerned the logarithms of negative numbers and pitted Leibniz against Johann Bernoulli. I cannot treat either of these in detail, but Leibniz's pronouncements in both cases offer a chance to see the ultimate status of his theory of the infinitesimal.

In a much-discussed letter to M. Pinsson in August of 1701, parts of which were published in the *Journal de Sçavans*, Leibniz offered his opinion on the controversy sparked by Rolle's criticisms. The letter responds to an anonymous criticism of the infinitesimal which Abbé Gouye had published in an earlier issue of the *Journal*. Leibniz argued in reply that

24 See Mancosu, 1996, chapter 6, and Blay, 1986, for accounts of this controversy.

> on n'a pas besoin de prendre l'infini icy à la rigeur, mais seulement comme lors qu'on dit dans l'optique que les rayons du soleil viennent d'un point infiniment eloigné et ainsi sont estimés paralleles. Et quand il y a plusieurs degrés d'infini ou infiniment petit, c'est comme le globe de la terre est estimé un point à l'egard de la distance des fixes, et une boule que nous manions est encor un point en comparaison du semidiametre du globe de la terre, de sorte que la distance des fixes est comme un infini de l'infini par rapport au diametre de la boule. Car au lieu de l'infini ou de l'infiniment petit, on prend des quantités aussi grandes et aussi petites qu'il faut pour que l'erreur soit moindre que l'erreur donnée. De sorte qu'on ne differe du style d'Archimede que dans les expressions qui sont plus directes dans nostre Methode, et plus conformes à l'art d'inventer.[25] (29 August, 1701, A I, 20, 493–4)

These remarks are of a piece with Leibniz's earlier claims about the eliminability of infinitesimal magnitudes: he denies that the calculus really needs to rely upon considerations of the infinite and again insists that it can be based on a procedure of taking finite but "negligible" errors that can be made as small as desired. And again, it is worth observing that Hobbes used essentially the same language, comparing the earth to a point in comparison to the heavens.

The more ardent partisans of the infinitesimal (notably Johann Bernoulli, Varignon, and the L'Hôpital) were deeply concerned by Leibniz's apparent concession to the critics of the calculus. Varignon wrote to Leibniz in November of 1701 requesting a clarification of Leibniz's views on the reality of infinitesimals and expressing the fear that the publication of the letter to M. Pinsson had done harm to the cause because some had taken him to mean that the calculus was inexact and capable only of providing approximations. Varignon therefore asked Leibniz "de vouloir bien nous envoyer au plustost cette déclaration nette et précise de vôtre sentiment sur cela"[26] (28 November, 1701, GM IV, 90).

---

[25] "There is no need to take the infinite here rigorously, but only as when we say in optics that the rays of the sun come from a point infinitely distant, and thus are regarded as parallel. And when there are more degrees of infinity, or infinitely small, it is as the sphere of the earth is regarded as a point in respect to the distance [of the sphere] of the fixed stars, and a ball which we hold in the hand is also a point in comparison with the semidiameter of the sphere of the earth. And then the distance to the fixed stars is infinitely infinite or an infinity of infinities in relation to the diameter of the ball. For in place of the infinite or the infinitely small we can take quantities as great or as small as is necessary in order that the error will be less than any given error. In this way we only differ from the style of Archimedes in the expressions, which are more direct in our method and better adapted to the art of discovery."

[26] "That you send us as soon as possible a clear and precise declaration of your thoughts about this."

In his reply to Varignon Leibniz issued a summary statement of his views on the infinite and its role in the calculus. This statement brings together themes we have already seen: the fictional nature of infinitesimals, the possibility of basing the calculus upon a science of incomparably small (but still finite) differences, and the equivalence of the new methods and the Archimedean techniques of exhaustion. After assuring Varignon that "mon dessein a esté de marquer, qu'on n'a point besoin de faire dependre l'analyse Mathematique des controverses metaphysiques, ny d'asseurer qu'il y a dans la nature des lignes infiniment petites à la rigueur"[27] (2 February, 1702, GM IV, 91). Leibniz once again suggests that incomparably small magnitudes be taken in place of the genuine infinite. These incomparables would provide as many degrees of incomparability as needed for the purposes of the calculus, and although these are really finite quantities they may still be neglected, in accordance with the notorious "lemmas on incomparables" in the 1689 *Tentamen de Motuum Coelestium Causis* (Ibid., 91–2). Leibniz's account of the nature of infinitesimals thus brings us again to his account of incomparably small (yet finite) magnitudes. But, as I have noted, this doctrine seems very much of a piece with Hobbes's notion of *conatus*.

The full scope of the "fictionalist" reading of the infinite was not made widely known, largely because Leibniz and his associates had reason to fear that any public retreat from a full commitment to the reality of the infinitesimal would complicate the already difficult battle for the acceptance of the calculus. As Leibniz explained in a late letter to the French mathematician Pierre Dangicourt:

> Quand [nos amis] disputérent en France avec l'Abbé Gallois, le Père Gouge [sic!] & d'autres, je leur témoignai, que je ne croyois point qu'il y eût des grandeurs véritablement infinies ni véritablement infinitésimales, que ce n'étoient que des fictions, mais de fictions utiles pour abréger & pour parler universellement, comme les racines imiginaires dans l'Algébre, telles que $\sqrt[2]{(-1)}$ [...] Mais comme Mr. le Marquis de l'Hospital croyoit que par là je trahissois la cause, ils me priérent de n'en rien dire, outre ce que j'en avois dit dans un endroit des Actes de Leipsic, & il me fut aisé de déférer à leur priére.[28] (11 September, 1716, Dutens III, 500–501)

---

27 "My intention was to point out that it is unnecessary to make mathematical analysis depend on metaphysical controversies or to make sure that there are lines in nature which are infinitely small in a rigorous sense."

28 "When [our friends] were disputing in France with the Abbé Gallois, father Gouye and others, I told them that I did not believe that there were actually infinite or actually infinitesimal quantities; the latter, like the imaginary roots of algebra ($\sqrt{-1}$), were only fictions,

The final piece in the puzzle of Leibniz's theory of the infinitesimal, and one that leads us back to Hobbes, is the late note *Observatio quod rationes sive proportiones non habeant locum circa quantitates nihilo minores, et de vero senso methodi infinitesimalis*, which appeared in the *Acta Eruditorum* in April of 1712. It was sparked by a controversy over the nature of ratios between positive and negative quantities, which grew to include the cases of logarithms and roots of negative numbers.[29] Johann Bernoulli (who was also a firm believer in the reality of infinitesimals and a chief partisan in favor of the Leibnizian calculus in the Académie) held that logarithms of negative numbers were the same as those of positive numbers, so that the logarithm of $-a$ is the same a the logarithm of a. Leibniz treated the issue of negative quantities in ratios, logarithms, and roots as fictions that could be harmlessly employed in calculation, but which did not correspond to anything mathematically real. In Leibniz's view, there is no ratio of +1 to − 1 (as Bernoulli required), since otherwise this would be the same as the ratio of −1 to +1, a result which he took to be inconsistent with the very notion of a ratio. Further, Leibniz characterized the fictionality of the infinitesimal in language that seems to have been almost borrowed from Hobbes.

In objecting to the notion that there could be a proper ratio between positive and negative quantities, Leibniz remarked:

> Porro, ut nego rationem, cujus terminus sit quantitas nihilo minor, esse realem, ita etiam nego, proprie dari numerum infinitum vel infinite parvum, etsi Euclides saepe, sed sano sensu, de linea infinita loquatur. *Infinitum continuum* vel *discretum* proprie nec unum, nec totum, nec quantum est, et si analogia quaedam pro tali a nobis adhibeatur, ut verbo dicam, est modus loquendi; cum scilicet plura adsunt, quam ullo numero comprehendi possunt, numerum tamen illis rebus attribuemus analogice, quem infinitum appellamus. Itaque jam olim judicavi, cum infinite parvum esse errorem dicimus, intelligi dato quovis minorem, revera nullum; et cum ordinarium, et infinitum, et infinities infinitum conferimus, perinde esse ac si conferremus ascendendo diametrum pulvisculi, diametrum terrae, et diametrum orbis fixarum[30]. (AE 1712 = GM V, 389)

---

which however could be used for the sake of brevity or in order to speak universally [...] But as the Marquis de l'Hôpital thought that by this I should betray the cause, they asked me to say nothing about it, except what I already had said in the Leipzig *Acta*, and it was easy for me to comply with their request."

29 See Marchi, 1974, for an overview of the controversy.

30 "Just as I have denied of the reality of a ratio, one of whose terms is less than zero, I equally deny that there is properly speaking an infinite number, or an infinitely small number, or any infinite line or a line infinitely small [...]. The infinite, whether continuous or discrete, is not properly a unity, nor a whole, nor a quantity, and when by analogy we use it in this

This note repeats themes we have already seen, but one striking feature of the entire piece is Leibniz's vivid reminiscences about his mathematical work during his years in Paris. Leibniz recalls his encounters with the work of Arnauld, John Wallis, and Joachim Jungius in the 1670s, and it is precisely during this period that Leibniz was working on the *Theoria motus abstracti* and still very much under the influence of Hobbes. As Marc Parmentier has phrased the issue in commenting on this Leibnizian work:

> les premières lignes de son article laissent penser que cette récente polémique a révillé un souvenir personnel qu'une quarante d'années d'intense activité diplomatique, historique et scientifique, n'avaient pas réussi à effacer, et qui surgit soudain à sa mémoire dans son originelle clarité.[31] (Leibniz, 1995, 423)

We thus return to where we started, with Leibniz's notion of the fictional infinitesimal being connected with his early mathematical work which, as we have seen, is closely connected with his reading of Hobbes.

4. Conclusions.

This very brief account raises perhaps more questions than it answers. I would like to close by considering two important consequences of Leibniz's doctrine of the fictional infinitesimal. The first is the question of how Leibniz might guarantee that the infinitesimal is, indeed, a well-founded fiction. The second, and related, issue is whether there is a stable conception of mathematical rigor that underlies Leibniz's writings on the calculus.

A fiction is well-founded in the Leibnizian sense when it does not lead us astray, so that indulgence in the fiction is harmless. The basic idea here seems to be something to the effect that we can "speak with the vulgar" when we employ the language of the infinitesimal, but "think with the learned" when we recognize that there really are no such things. Yet we still

sense, it is a certain *facon de parler*; I should say that when a multiplicity of objects exceeds any number, we nevertheless attribute to them by analogy a number, and we call it infinite. And thus I once established that when we call an error infinitely small, we wish only to say an error less than any given, and thus nothing in reality. And when we compare an ordinary term, an infinite term, and one infinitely infinite, it is exactly as if we were to compare, in increasing order, the diameter of a grain of dust, the diameter of the earth, and that of the sphere of the fixed stars."

31 "The first lines of his article incline one to think that the recent polemic [over the nature of ratios] had revived a personal recollection that forty years of intense diplomatic, scientific, and historical activity had not been able to erase, and which suddenly came into his memory with its original clarity."

stand in need of some sort of guarantee that we will not, in fact, be led astray. In the mathematical context, this means that we need some kind of proof to the effect that infinitesimals can always, at least in principle, be eliminated, so that proofs that depend on them can be replaced by others that require only finite differences between finite quantities. Leibniz often made grand programmatic statements to the effect that derivations which presuppose infinitesimals can always be re-cast as exhaustion proofs in the style of Archimedes. But he never, so far as I know, attempted anything like a general proof of the eliminability of the infinitesimal, or offered anything approaching a universal scheme for re-writing the procedures of the calculus in terms of exhaustion proofs. The closest thing we have are the notorious "lemmas on incomparable magnitudes" from 1689, but these are really more promissory notes than rigorous proofs. What, then, are we to make of Leibniz's confidence that the infinitesimal is a well-founded fiction? He was certainly aware that some infinitesimal arguments could lead to paradox and contradiction, but it is unclear whether he had a surefire way of avoiding error.

A related issue is what the Leibnizian conception of mathematical rigor comes to once we have accepted the notion of a fictional infinitesimal. As classically understood, a rigorous argument is one that begins with transparently true first principles, proceeds by valid inference procedures, and deals only with objects that are clearly conceived. It is far from clear whether Leibniz would allow that the proof procedures of the calculus are, in fact, rigorous in this sense. After all, the infinitesimal is decidedly not the sort of thing we can conceive clearly, and it seems a bit odd to think that there might be transparently true first principles that deal with merely fictional objects. In the end, then, we might ask whether classifying infinitesimals as "useful fictions" can really deflect the criticism of the calculus which characterizes it as unrigorous. This is not an issue I'm in any position to resolve at the moment, and will leave it for another day.

François Duchesneau

Rule of Continuity and Infinitesimals in Leibniz's Physics

A constant paradox due to an apparently dual treatment of infinitesimals seems to afflict Leibniz's contributions to natural philosophy. Expressions involving differentials and integrals are often introduced in Leibniz's analyses as mere modes of representation for transformation processes that call for causal explanations grounded in meta-empirical reasons. However, it may also seem that Leibniz interprets these meta-empirical reasons by allowing an analogy between the entities his physical theories identify as conceptual referents and the models or representation schemes geometric infinitesimals afford. Leibniz was eager to generalize his findings and express their overarching heuristic value by setting up a harmonious network of conceptual connections that would prolong or expand the lines of analysis he had initially sketched. Indeed, Leibnizian scientific statements are not free of ambiguities, but for Leibniz, analogies, when duly controlled, could become crucial means for promoting the art of discovery (*ars inveniendi*). Many of his mathematical papers witness to this type of audacity (Leibniz, 1989, 11–52). On the other hand, his scientific methodology itself favors hypothetical constructions (Bertoloni Meli, 1993; Duchesneau, 1993; Duchesneau, 2006). With hypotheses, truths of reason may be applied to the analysis of contingent truths expressing the connection of natural phenomena: along this line, a condition of valid hypothesizing consists in framing relevant mathematical models. Taking this methodological pattern as background, Leibniz's frequent appeals to differential and integral modeling in physics may be interpreted in a manner that resolves some of the more or less apparent ambiguities associated with the epistemological status of Leibnizian infinitesimals.

1. Symbolic constructs and analogues of causal reasons

We should first recall the methodological stance underpinning the reference to infinitesimals in Leibniz's natural philosophy. Among many texts, I shall quote from the the *Nouveaux Essais sur l'entendement humain* (1704) where Leibniz counters Locke's negative assessment of the capacity of experimental philosophy based on observation and experience of natural phenomena to be transformed into demonstrative knowledge:

> Je crois bien que nous n'irons jamais aussi loin, qu'il seroit à souhaiter; cependant il me semble qu'on fera quelques progrés considerables avec le temps dans l'explication de quelques phenomenes, parce que le grand nombre des experiences, que nous sommes à portée de faire, nous peut fournir de *data* plus que suffisans, de sorte qu'il manque seulement l'art de les employer, dont je ne desespere point qu'on poussera les petits commencemens, depuis que l'analyse infinitesimale nous a donné le moyen d'allier la Geometrie avec la physique et que la dynamique nous a fourni les loix generales de la nature.[1] (*Nouveaux Essais*, 4.3.26, A VI, 6, 389)

According to Leibniz in this instance, the advancement of scientific knowledge will result from a *mise en forme* of experimental data: this *mise en forme* aims at providing explanations through theoretical representation of the relevant causes. (Duchesneau 1993, 203–204) To this end we need, on one hand, models based on algorithms for the transposition and analysis of data; and on the other principles for theory building. These should express the architectonic requirements for establishing a system of nature: such a system will be ordered according to determining reasons, and these reasons will incline without necessitating (*incliner sans nécessiter*) and represent *ex hypothesi* an optimal contingent configuration of natural entities. (Rutherford 1995, 238–244) Leibniz states that he has offered exemplary contributions under both headings. First, he developed the infinitesimal calculus which affords a methodological tool for the analysis of phenomena; second, he established the dynamics as an integrative theoretical corpus of general laws supposed to rule over nature.

[1] "I do believe that we shall never advance as far as one might wish; yet it seems to me that considerable progress will eventually be made in explaining various phenomena. That is because the great number of experiments which are within our reach can supply us with more than sufficient data, so that all we lack is the art of employing them; and I am not without hope that the small beginnings of that will be extended, now that the infinitesimal calculus has given us means for allying geometry with natural science and now that dynamics has supplied us with the general laws of nature." (NE, 389)

Leibniz's innovative contributions to dioptrics and catoptrics offer an appropriate instance of this dual aspect of theoretical structuring which involves both analytic models inspired by the calculus as well as architectonic principles like the rule or law of continuity. Summarizing his previous demonstration of the fundamental laws of refraction and reflection to be found in the *Unicum opticæ, catoptricæ et dioptricæ principium* (1682) (Dutens III, 145–150), Leibniz argues:

> Pourveu qu'on se figure, que la nature a pour but de conduire les rayons d'un point donné à un autre point donné par le chemin le plus facile, on trouve admirablement bien toutes ces loix, en employant seulement quelques lignes d'Analyse, comme j'ay fait dans les Actes de Leipzig.[2] (GP IV, 340)

The hypothesis in the *Unicum principium*, formulated at the very time Leibniz was beginning to reveal the virtues of his calculus, comprised three stages. The first employed the constant connection of observed empirical facts. The second afforded a conjecture about the resistance of physical media to the passage of light rays. The third presented a geometrical expression for the path of light, calibrated according to a calculus *de maximis et minimis*. The third stage subsumed both the empirical connections and their hypothesized physical causation. In a way, this theoretical scheme accommodated the explanatory hypothesis to the norm of an analysis that would be both architectonic and geometrical.

If we now turn to the later *Tentamen anagogicum*, subtitled *Essay anagogique dans la recherche des causes* (c. 1696), we may argue that Leibniz aimed then at implementing this architectonic norm by an analytic representation based on integrating infinitesimal metric relations. This analytic representation would express the manner in which determining causal factors operate in the generic instances of light refraction and reflection. In the model used to determine how the Snell-Descartes refraction law applies to variously curved separation surfaces, analysis, according to the *Tentamen*, is supposed to be founded "sur l'evanouissement de la différence ou sur l'unicité des jumeaux reunis, et nullement sur la comparaison avec toutes les autres grandeurs".[3] (GP VII, 275) This entails that the various sets of possible differences involved in the angular derivation of light rays are presumed

---

[2] "Provided we figure out that nature aims at directing the rays [of light] from a given point to another given point by the easiest path, we discover astonishingly well all these laws by using only a few lines of analysis, as I did in the *Leipzig Proceedings* [*Acta eruditorum*]." (In the following, it will be my translation if no other source is given, F.D.)

[3] "[…] On the disappearance of a difference or on the unique result of reuniting twins, and not at all on a comparison of the greatest and smallest with all other magnitudes." (L 480)

to be reabsorbed at the limit in a unique optimal determination. Seeking the limit of variation in infinitesimal differences is a means to establish the harmonious system resulting from the intervention of the hypothesized physical causes. Thus, the most determined motion is shown to be unique by exhausting the least differences expressed in the measures of quantities. In sum, the method of optimal forms (*de formis optimis*), characterized by Leibniz as achieving a maximum or a minimum beyond given maxima and minima, promotes in physics a method of analysis that determines in the various cases the tendency inscribed in the form of curves from the serial progression of tangents. This tendency itself then leads to a limit beyond any assignable magnitude. The resulting determination is that of a limit which ceases being quantitative and becomes qualitative, thus indicating the result of a choice for the best. Infinitesimal calculus provides a methodological devise – a kind of *artifice méthodologique* – to guide us to a causal explanation through architectonic schematization. For the sake of analogy, this stance is likened to that of projective geometry, in which elements considered as forming homologous series orient the eye towards the axis of symmetry that provides them with a single overarching determination. In Leibnizian physics: "c'est par une technique de l'idéal – à savoir l'analyse – que va être prouvée la réalité de la cause finale."[4] (Belaval, 1960, 409)

In fact, Leibniz was interested in a method of discovering physical sufficient reasons by using mathematical determinations to symbolize a deeper order of real, but contingent, determinations. Leibniz thus distinguished between mathematical determinations which imply absolute, but abstract, necessity due to the rational makeup of the corresponding algorithms, and architectonic determinations which imply only "une nécessité de choix, dont le contraire importe imperfection"[5] (GP VII, 278). From this perspective, geometrical determinations may be viewed as "des demy-determinations" (half-determinations) (GP VII, 279). These do not suffice *per se* to determine the unique relevant causal pattern behind phenomenal connections, but serve for figuring out, or symbolizing, the integrative combination of mechanical means by which the architectonic plan of nature might be achieved. This subordinate role of infinitesimal modeling may be found ambiguous if judged in strict epistemological terms. The lines of analysis to which Leibniz alludes can represent either a mere technique of metric determination for dynamic factors or a method for constructing the

[4] "It is by an idealizing technique, namely analysis, that the reality of final causes will be proved."

[5] "[…] A necessity of choice whose contrary means imperfection." (L 484)

representation of relevant theoretical entities. Of course, Leibniz takes infinitesimal modeling as a way of devising symbolic analogues that enable us to discursively reach the level of theoretical entities and general laws of nature. A later generation of enlightened scientists, influenced by Newton's methodological approach, will find in such a use of analytic models a mere way of mathematically expressing physical parameters which reveal themselves *a posteriori* from the collected data of experience (Blay, 1992). But, from Leibniz's viewpoint, physics is concerned with a world in which the relations of order and causality are nested in each other to infinity, thus reflecting the fact that they result from an optimal architectonic design.

Naturally, an appropriate method of analysis should therefore take advantage of mathematical models involving gradual transitions to the limit beyond any assignable difference. Applied to such complex objects as phenomenal reality, Leibniz's general analytic method would afford specific merits. As he puts the matter in his 1694 *Considérations sur la différence qu'il y a entre l'analyse ordinaire et le nouveau calcul des transcendantes*:

> Notre méthode étant proprement cette partie de la Mathématique générale, qui traite de l'infini, c'est ce qui fait qu'on en a fort besoin, en appliquant les Mathématiques à la Physique, parce que le caractère de l'Auteur infini entre ordinairement dans les opérations de la Nature.[6] (GM V, 308)

The algorithm of the calculus can free analysis from those incommensurable and transcendent quantities which impede the continuous progression of ordinary analysis in resolving equations. By providing continuous analogical constructions, the calculus authorizes transitions to the limit and thus finite resolutions of infinite quantitative relations: by differentiating or summing up along various degrees, it can set *quodam modo* equivalences between infinite and finite ratios. This way of proceeding is considered by Leibniz as providing constructions which are deemed in practice sufficient:

> Et j'ai donné par là une voye générale, selon laquelle tous les problêmes, non seulement des différences ou sommes, mais encore des différentio-différentielles ou sommes des sommes et au delà, se peuvent construire suffisamment pour la practique: comme j'ai donné aussi une construction générale des quadratures par un mouvement continu et réglé.[7] (GM V, 308)

---

6 "Our method being properly that part of general mathematics which deals with the infinite, this fact explains how much we need it for applying mathematics to physics, since the character of the Infinite Creator penetrates ordinarily the operations of Nature."

7 "And I gave this way a general method according to which all problems, not only of differences or sums, but also of second-order differentials or sums of sums, and beyond, can be

Continuity as framed in the calculus results from symbolic patterns devised to enable transposing metric relations beyond any assignable difference. Indeed, this analogical operation (which Leibniz associates with ordinary analysis) extends the order of generation of finite relations in equations. But the artificial device of infinitesimal differences offers a new mode of intelligibility concerning the generation of those finite ratios. Everything occurs as if infinitesimal magnitudes could be used to represent the continuous generation of discrete quantities more properly than any other geometrical model; and yet these differences only possess ideal and symbolic existence, figuring only as determining transitions to the limit. However, the increased objective value of such a model depends on our matching the various continuous analytic series into which physical factors are transposed with hypothesized architectonic sufficient reasons exemplified in the corresponding theoretical entities. Borrowing a formula of the *Nouveaux Essais* concerning the presumed correspondence of infinite series with incommensurable entities, we could argue in Leibnizian fashion that: "une certaine progression de Synthèse devroit étre melée avec nostre analyse pour y mieux reussir" (*Nouveaux Essais,* 4.3.6, A VI, 6, 377).[8] And this is why physical theory is called upon to provide analogical referential grounding for the abstract symbolic constructs of mathematical analysis. At least, this is my proposed reading of the crucial statement made by Leibniz to Varignon:

> On peut dire de même que les infinis et infiniment petits sont tellement fondés que tout se fait dans la Geométrie, et même dans la nature, comme si c'estoient des parfaites realités, temoins non seulement nostre Analyse Geometrique des Transcendentes, mais encor ma loy de la continuité, en vertu de laquelle il est permis de considerer le repos comme un mouvement infiniment petit (c'est à dire comme equivalent à une espece de son contradictoire), et la coincidence comme une distance infiniment petite, et l'égalité comme la derniere des inegalités etc. [...][9] (Letter to Varignon of February 2, 1702, GM IV, 93)

sufficiently synthesized for the practice; as I also gave a general construction of quadratures by a continuous and regular progression."

8 "If we are to have better success with that, our analysis should be mingled with some measure of synthesis." (NE, 377)

9 "So it can also be said that infinites and infinitesimals are grounded in such a way that everything in geometry, and even in nature, takes place as if they were perfect realities. Witness not only our geometrical analysis of transcendent curves but also my law of continuity, by virtue of which we may consider rest as infinitely small motion (that is, as equivalent to a particular instance of its own contrary), coincidence as infinitely small distance, equality as the limit of inequalities, etc." (L 544)

In this passage, attention should be drawn to the tension between the statement that infinite quantities might be taken for perfect realities, and the conditional clauses modulating the meaning of it, especially the "as if" and "we may consider," as well as the reminder that we are only involved in a strategy of comparison, analogy, and equivalence between contradictory terms when their difference is drawn to the limit. The letter to Varignon underlines that homogenous continuity is of the nature of an abstract and ideal determination; but everything works out as if there were such a determination as would allow us to project these homogenous transformation schemes on the structure of reality. If infinitesimal determinations seem to match real determinations, it is not because they correspond with the ultimate analytic structure of natural entities, since the latter is unreachable, but rather because the rules governing those determinations afford the best symbolic representation of the infinite series of sufficient reasons that causally determine the order of phenomena as objects of distinct representation.

The principle of continuity in a physics structured according to the analytic method will reveal an architectonic ordering of constituent parameters as these form continuous series involving transitions to the limit and analogous to those to which the calculus applies. Thus the principle of continuity entails a high level of generality: it sets the type of order that prevails in generating continuous series of states for given factors. This order is characterized by law-like progression in infinitely gradual sequences of states, and these sequences imply the integration of apparently contrary states when a transition to such a limit entails lesser than any determinable difference. By this modeling of connected states we can symbolically apprehend the type of rationality governing change, a rationality expressed by the immanent order of transitions in nature. This point is rightly expressed in Leibniz's *Initia rerum mathematicarum metaphysica*:

> Hinc etiam sequitur *Lex Continuitatis* a me primum prolata, qua fit ut lex quiescentium sit quasi species legis in motu existentium, ut lex curvilineorum est quasi species legis rectilineorum, quod semper locum habet, quoties genus in quasi-speciem oppositam desinit. Et hic pertinet illa ratiocinatio quam Geometræ dudum admirati sunt, qua ex eo quod quid ponitur esse, directe probatur id non esse, vel contra, vel qua quod velut species assumitur, oppositum seu disparatum reperitur. Idque continui privilegium est; *Continuitas* autem in tempore, extensione, qualitatibus, motibus, omnique *naturæ transitu* reperitur, qui nunquam fit per saltum.[10] (GM VII, 24–25)

[10] "Hence follows the Law of Continuity that I initially formulated, from which it results that the law of entities at rest is a quasi-species of the law of entities in motion, as the law of

One of Leibniz's constant lines of argument in his more technical writings involves discarding the false notion of infinitesimals as discrete entities, or even symbols for discrete entities, since the analogous reference to infinitesimals in natural philosophy has to do with extending relational properties pertaining to infinite continuous series and thereby subverting the apparent discreteness of perceptually isolated states. In terms of experiential awareness of physical properties, differentials, integrals, and their various degrees do not represent any strictly determined features of the real, but they symbolize the capacity of change that is presumed to underpin any transition between any apparently discrete states that have distinct determinations. As stated in the *Justification du calcul des infinitésimales par celui de l'algèbre ordinaire* (1702), this boils down to a way of expressing things that takes some liberty – an acceptable degree of tolerance – with strict truth concerning natural entities, at least as these are known from experience or represented through geometrical schematization:

> Cependant quoyqu'il ne soit point vray à la rigueur que le repos est une espece de mouvement, ou que l'égalité est une espece d'inégalité, comme il n'est point vray non plus que le Cercle est une espece de polygone régulier: neantmoins on peut dire, que le repos, l'égalité, et le cercle terminent les mouvements, les [in]égalités et les polygones reguliers, qui par un changement continuel y arrivent en evanouissant. Et quoyque ces terminaisons soyent exclusives, c'est à dire non comprises à la rigueur dans les varietés qu'elles bornent, neantmoins elles en ont les proprietés, comme si elles y estoient comprises, suivant le langage des infinies ou infinitesimales, qui prend le cercle, par exemple, pour un polygone regulier dont le nombre de costés est infini.[11] (GM IV, 106)

curved lines is a quasi-species of the law of straight lines, which always happens whenever the genus terminates in a quasi-species. And here applies this argument which geometers have admired, according to which from the supposition that something is, it is directly proved that it is not, or contrariwise, from the supposition of a sort of species, the opposite or the disparate is found. And this is the privilege of what is continuous: thus continuity is found in time, extension, qualities, motion, and all transitions in Nature, which never happen *per saltum*." – This piece was written in or after 1714 mentioning a publication of Leibniz in the *Acta eruditorum* for 1714 (GM VII, 17).

11 "Although it is not at all rigorously true that rest is a kind of motion or that equality is a kind of inequality, any more than it is true that a circle is a kind of regular polygon, it can be said, nevertheless, that rest, equality, and the circle terminate the motions, the inequalities, and the regular polygons which arrive at them by a continuous change and vanish in them. And though these terminations are excluded, that is, are not included in any rigorous sense in the variables which they limit, they nevertheless have the same properties as if they were included in the series, in accordance with the language of infinites and infinitesimals, which takes the circle, for example, as a regular polygon with an infinite number of sides." (L 546)

One can partially conclude from such statements that the mathematical models at work in infinite analysis afford symbolic analogues for continuously developing metric relations in boundary cases representing an essential difference, though one less than any assignable quantity. There is a protracted analogy here that can be granted some objective value in the representation of natural phenomena because there seems to be a parallel analogy involved in the formulation of the general laws of nature presumed to obey an overarching rule of continuity. When Leibniz formally introduced his principle of continuity in 1687, he presented it as a general principle of order grounded in mathematical necessity and yet applicable to physics because of the essentially geometric makeup of the order of nature:

> Il a son origine de l'infini il est absolument necessaire dans la Geometrie, mais il reussit encore dans la physique, par ce que la souveraine sagesse, qui est la source de toutes choses, agit en parfait geometre, et suivant une harmonie à laquelle rien de se peut adjouter.[12] (*Lettre de M. L. sur un principe general utile à l'explication des loix de la nature par la consideration de la sagesse divine*, GP III, 52)

This means that physical changes could be analogically accounted for by appealing to resolution processes that apply to isomorphic infinite relations. There are, however, limitations on the kinds of relations to which the law of continuity applies; negative and positive numbers are fundamentally different kinds of magnitudes, but the negative is not a limiting case of the positive in the way that zero is a limiting case of ever-smaller positive magnitudes. Further, the epistemic possibility of subsuming limit cases under the law of continuity does not imply that the infinite and infinitesimal magnitudes involved in such schematic representations must represent any real properties of physical agents. However, Leibniz also grounded the principle's application to the realm of nature on the notion that the architectonic plan according to which natural realities unfold, results from a transcendent calculus, itself analogically reflected in the formulation and use of infinitesimal algorithms by our finite understandings. Thus the use of equations based on the analytic reduction of infinite relations for representing the interconnectedness of physical variables is metaphysically founded, even if infinitesimals cannot be granted ontological status as determined physical magnitudes.

---

12 "The principle has its origin in the *infinite* and is absolutely necessary in geometry, but it is effective in physics as well, because the sovereign wisdom, the source of all things, acts as a perfect geometrician, observing a harmony to which nothing can be added." (L 351)

2. The Significance of Infinitesimals in the Dynamics

A crucial issue concerning the physical meaning of infinitesimals in such a system of analogical representations nested in one another involves the epistemological status of the theoretical concepts structuring Leibniz's dynamics. The issue is all the more important since Leibniz himself repeatedly justified the conceptual comprehensiveness of the laws of the dynamics by the possibility of representing them by means of models borrowed from differential and integral calculus. This is particularly clear in his correspondence with Burcher De Volder, Johann I Bernoulli, and Jakob Hermann. And it is clearly the case that some authoritative commentators, such as Martial Gueroult, have thus, following Leibniz's suggestions, attempted to translate the main relations underpinning the Leibnizian mechanical analysis by using the language of infinitesimals (Gueroult, 1967, 28–49). Such reconstitutions are principally grounded in the arguments of the first part of the *Specimen dynamicum* (1695). Though it would be appropriate to take into consideration a wider set of texts, let us turn to this one in particular.

One of the main differences between the *Specimen dynamicum* and an earlier presentation of the dynamics like that of the *Dynamica de potentia* (1690), is that the former exposition tended to be more axiomatic, resting at the level of geometrical expressions and of *a priori* model framing. Leibniz in the *Specimen* was more preoccupied with establishing the demonstrable connection between models and theoretical entities and with identifying the causal order subjacent to the series of mechanical states which geometrical constructions and equations would express. An elliptic masterpiece, the *Specimen dynamicum* suggests the complex ordering of ingredients in a nascent Leibnizian physical theory beyond the provisional harmony of geometry-inspired models. We may consider the *Specimen* as an attempt to link the basic arguments of the dynamics to their analytic presuppositions in terms of architectonic principles *and/or* theoretical concepts. The "and/or" clause is not innocent in this instance. Indeed, Leibniz resorts to concepts and arguments derived from the metaphysics to provide rational grounding to his dynamics, but his main strategy to this purpose consists in analogically stretching the meaning of the physical concepts involved by appealing to architectonic principles and infinitesimal models (Duchesneau, 1994).

Leibniz conceived the power to act at the foundation of physical reality as a disposition spontaneously tending towards its fulfillment. He called this disposition *nisus* or *conatus*. These terms, especially *conatus*, had been historically linked to the development of Leibniz's thought. They derived from the influence of Hobbes and Cavalieri at the time Leibniz was writing the *Hypothesis physica nova* and *Theoria motus abstracti* (1671) (Bernstein, 1980; Beeley, 1996). They initially signified indivisible elements of motion that, combining their determinations, would cause extended masses to move. If the term thereafter lost the connotations it owed to inadequate mathematical and physical models, it reemerged in the context of the dynamics to signify the embryonic causal ingredients of force in any physical entity that begins to act. This model expressed the inherence of dynamic dispositions in the well-founded phenomenal bodies that compose physical reality. At the same time the model implied that extensive properties themselves would result from the diffusion or continuation of those elementary tendencies to act, counterbalancing in time the mutual resistance of interacting bodies.

Leibniz presents the theoretical framework of his dynamics by distinguishing between active force and passive force and by further dividing each term, following a criterion of level, into primitive and derivative force. We thus get the well-known four notions: *vis activa primitiva*, *vis passiva primitiva*, *vis activa derivata*, and *vis passiva derivata*. Indeed, Leibniz identifies the primitive active and primitive passive force with entelechy and prime matter, which are metaphysical concepts, but, following the suggestion contained in the letter to De Volder of April 3, 1699, we may equate these primitive terms with fundamental laws governing series of mutations that are reflected at the phenomenal level in the changing active and passive dispositions of bodies. As Leibniz notes:

> Itaque Materiæ resistentia duo continet, impenetrabilitatem seu antitypiam et resistentiam seu inertiam et in his, cum ubique in corpore sint æqualia vel extensioni ejus proportionalia, principii passivi seu materiæ naturam colloco, ut in vi Activa varie sese per motus exercente Entelechiam primitivam et ut verbo dicam aliquid Animæ analogum agnosco, cujus natura in perpetua quadam ejusdem seriei mutationum lege consistit, quam inoffenso pede decurrit.[13] (GP II, 171)

The result is a collection of theoretical constructions grounded in abstractive analogy and justified by the application of regulative architectonic principles. The primitive active and passive forces afford a theoretical scheme

[13] "The resistance of Matter comprises both impenetrability or antitypy, and resistance or inertia, and in these, since they are everywhere in body equal or proportional to its extension, I locate the nature of the passive principle or matter; likewise in Active force exerting itself diversely through motion, I recognize a primitive entelechy, or to spell it out, something analogous to a Soul whose nature consists in a certain perpetual law for the consistent series of mutations it undergoes continuously."

representing the determining reason required to account for the analogical projection of serial laws on mechanical properties and effects. Leibniz's approach is remarkably analytical: it aims at reaching reasons sufficient to justify, at a non-phenomenal level, the harmony and interrelatedness of phenomena. The required analytic means can be sought for at either a more metaphysical or a more epistemological level without jeopardizing the rationality of the intended theoretical justifications. This is because these levels, predicating respectively non-phenomenal entities (such as the primitive active and passive forces) and laws based on meta-mathematical analogies (such as those governing mechanical effects) are presumed to afford corresponding expressions. But may a Leibnizian philosophy of nature merely consist in a system of regulative representations sufficient to account for the regular and well-founded sequences of phenomena?

Empirically derived concepts and principles of rational origin conjoin to account for derivative forces. The postulated primitive forces are placed beyond any extensive empirical reference. In fact, Leibniz acknowledges that the notion of *nisus* provides a transition from the empirical to the rational level through analogical progression. To these *nisus* underpinning the interplay of derivative forces correspond "[...] leges actionum, quæ non ratione tantum intelliguntur, sed et sensu ipso per phænomena comprobantur".[14] (GM VI, 237) The empirical reference for these *nisus* are the motions to which the analysis of phenomena can be reduced. It is by an analytic decomposition of motions into their generative factors that Leibniz expects to formulate a theory of *nisus* capable of enlightening the order of derivative forces. He presents this approach as an attempt to nominally define terms whose combination will elicit approximating representations of the forces at work in the phenomenal world. These representations embody a *more geometrico* abstract symbolization of the real relations that are presumed to originate from the deeper metaphysical framework. It is properly a model that Leibniz wants to build. Speaking of infinitely small *nisus* generating *impetus* through summation, Leibniz would not mean that: "hæc Entia Mathematica reapse sic reperiri in natura, sed tantum ad accuratas æstimationes abstractione animi faciendas prodesse".[15] (GM VI, 238)

---

14 "For to these derivative forces apply the laws of action, which are not only known by reason but also verified by sense itself through phenomena." (L 437)

15 "But I do not mean that these mathematical entities are really found in nature as such but merely that they are means of making accurate calculations of an abstract mental kind." (L 438)

From this somewhat nominal representation whose initial sketch was to be found in the *Phoranomus seu De potentia et legibus naturæ* (1689) (Leibniz, 1991), Leibniz infers a twofold integration system, that of *vis mortua* and *vis viva*. Dead force represents an integration of *conatus* in the instant without duration determining in any manner the outcome of that integration. This is the case of the ball in the tube uniformly rotating around a point on a horizontal plane, or that of the pebble in the sling, while the ball or the pebble are still restrained from moving. Living force only arises through motions actualized in time which entail a summation at another integrative level. If the force of gravitation or that of elastic tension for instance has already exercised itself for some time, a continuous succession of impressions issuing from dead force obtains which translates into the generation of living force:

> Hinc patet duplicem esse Nisum, nempe elementarem seu infinite parvum, quem et *solicitationem* appello, et formatum continuatione seu repetitione Nisuum elementarium, id est impetum ipsum [...]. Hinc vis quoque duplex: alia elementaris, quam et *mortuam* appello, quia in ea nondum existit motus, sed tantum solicitatio ad motum [...], alia vero vis ordinaria est, cum motu actuali conjuncta, quam voco *vivam*.[16] (GM VI, 238)

This would explain Galileo's apparently enigmatic statement that the force of percussion would be infinite compared to the mere tendency inherent in gravity (*Phoranomus*, I, § 18, Leibniz, 1991, 478; *Dynamica de potentia*, GM VI, 451).

A possible symbolic transposition of the various relations involved in Leibniz's presentation might elicit the following propositions. For the *nisus* or *conatus*, considered as a vector quantity:

$$dv = gdt \ .$$

For the *impetus* reduced to quantity of motion in the instant:

$$m\int_0^t gdt = mv \ .$$

[16] "Hence the nisus is obviously twofold, an elementary or infinitely small one which I also call a *solicitation* and one formed by the continuation or repetition of these elementary impulsions, that is, the impetus itself. [...] Hence force is also of two kinds: the one elementary, which I also call *dead* force, because motion does not yet exist in it but only a solicitation to motion [...]; the other is ordinary force combined with actual motion, which I call *living* force." (L 438)

For *impetus* following its temporal effect:

$$m\int_0^t gdt == m\int_0^t vdt \quad .$$

For *vis mortua*:

$$m\int_0^t gdt = mv \quad .$$

For *vis viva*:

$$m\int_0^t gdt = mm\int_0^t vdt = ms = mv^2$$

The expression of *impetus* integrated in time joins with that of *vis viva*, as the expression of *conatus* integrated in the form of *impetus* in the instant joined with that of *vis mortua*.

As Leibniz explained to De Volder, the whole system of definitions aims at conciliating such relations as originated from statics with the representation of the dynamic effects generated by forces in their causal activity. The algorithmic model of infinitesimal calculus makes it possible to represent the generating process of an effective *nisus* as a two-stage integration: the *impetus* represents an intermediary level between *conatus* as an elementary solicitation and *vis viva* as a summation of *impetus* effects in time; and thus a representation is found for the mediating role of a vectorial quantity of motion as contrasted with a scalar *mv* according to Descartes. Hence the analogies Leibniz borrows from geometry and infinitesimal calculus to figure out the ratios involved: *conatus* are like differentials of speed *dx*, speeds being noted *x*; and *vires vivæ* are like integrals of speeds times speed differentials (Letter to De Volder in reply to the latter's of November 21, 1698, GP II, 156): $\int xdx$. Or otherwise expressed following geometrical symbolization, *vires vivæ* are proportional to x^2, meaning that they develop as the squares of the speeds.

In the summation of *conatus* and *impetus*, homogeneity in contrast to disparity is gained by means of these symbolic expressions: they allow reconstituting continuous transitions to the limit between the various mathematical analogues of *conatus* and *impetus*. And in a way those symbolic transpositions we have just recast seem more determined than most of Leibniz's actual statements, which follow a more flexible analytic line. An additional manuscript piece to the *Brevis demonstratio erroris memorabilis Cartesii* (1686)[17] for

[17] The full title of this article published in the *Acta eruditorum* reads: *Brevis demonstratio erroris memorabilis Cartesii et aliorum circa legem naturalem, secundum quam volunt a Deo eandem semper quantitatem motus conservari, qua et in re mechanica abutuntur.* The manuscript

instance is content with superposing the various levels of integration. In some limit cases in which we deal only with dead forces, the two principles of conservation, that of quantity of motion and that of quantity of *vis viva*, are made to coincide, but we have to admit of a "divorce" (*divortium*) between the two hypotheses when the development of a new infinite series of determinations requires transiting to a superior level of integration. We are then faced with incommensurable analytic figures: "Est autem potentia viva ad mortuam vel impetus ad conatum ut linea ad punctum vel ut planum ad lineam."[18] (GM VI, 121) In this type of argument, everything boils down to layering up integrative functions to represent geometrical relations that would result from infinite summations. In the same way, the *Specimen dynamicum* seems to insist on one hand on the infinite summation of elementary *nisus* or *conatus* which results in generating the instantaneous *impetus*, and on the other hand on the infinite summation of *impetus* or dead or elementary forces forming the *vis viva* which results in effective motion. This strategy is that of a superposition of analogous instances. If the symbolic operation linking these various infinitesimal models according to an analytic demonstrative scheme were perfectly self-sufficient, would resorting to a system of primitive forces as metaphysical entities be necessary in Leibniz's physical system? Our symbolic reconstructions of Leibniz's arguments hinge on substituting in the equations of the dynamics values of speed, mass, momentum, etc., which are deemed analogically equivalent from the pragmatic point of view to infinite progressions or regressions to limits involving less than assignable differences. From a Leibnizian perspective, a theoretical foundation is required to ground these symbolic equivalences, since they should be made to match higher level architectonic reasons. The analytic and analogical process linking mechanical effects at various levels of integration requires a justification in terms of sufficient reasons that may extend further than simply affording a symbolic recasting of nominal definitions.

In the *Specimen dynamicum*, the metaphysical level of explanation involves two types of features: the primitive active and passive forces as basic theoretical entities, and the architectonic principles, such as the law of continuity, implementing their methodological function in constructing the physical theory. These two types of features are combined so as to provide a synthetic framework that can integrate the various analytic schemes de-

---

piece related to the arguments of the *Brevis demonstratio* was published by Gerhardt (GP VI, 119–123).

[18] "The living force is to the dead force, and the impetus is to the conatus, as the line to the point, or the plane to the line."

ployed to account for the system of nature. We conceive Leibniz's approach as an attempt to harmonize these various analytic pathways and the analogical models involved by interpreting them according to "metaphysic laws." These laws would constitute the *a priori* structure for the dynamics as a science of power and action. This *a priori* structure formed the heart of the more systematic *Dynamica de potentia* and influenced Leibniz's arguments in major pieces of the later correspondence with Denis Papin (Ranea, 1989), De Volder, Johann I Bernoulli, Christian Wolff and Hermann. The architectonic principles are part of the *a priori* structure, but they also offer a formal means for integrating the various analytic models and analogical schemes which apply the resources of infinitesimal representations to account for dynamic processes in the realm of phenomena.

To sum up, in the dynamics, differentials and integrals applied to the processes of generation and exertion of force can only be conceived as symbolic analogues entering equations which are themselves equivalent to nominal definitions. The pattern for these nominal definitions expressing dynamic processes is afforded by subsuming mechanical factors and their mathematical infinitesimal analogues under the aegis of architectonic principles, especially the principle of continuity. But Leibniz aimed at a higher level of theoretical justification in the form of an *a priori* derivation of the general laws of dynamics. One aspect of such an account is provided by the metaphysical scheme of the primitive active and passive forces ontologically grounding the relative and nominal relations symbolized by the various layers of mathematical summation models. The other and more technical aspect of this account consists in the attempt to ground the laws of dynamics, especially the *vis viva* conservation principle, on a more general law, that of conservation of formal or essential action, and to substitute an *a priori* style of demonstration for the *a posteriori* style initially followed in the *De corporum concursu* (1678) (Leibniz, 1994), and later on in the *Brevis demonstratio* and *Specimen dynamicum*. This attempt, which is only alluded to in the published *Specimen*, was the central objective of the unpublished *Dynamica de potentia* (1689–1690) (GM VI, 281–488). As reflected in some of his major scientific correspondences of the later period, the challenge for Leibniz rested in substituting real (or what I qualified as "quasi-real") definitions for the nominal definitions justifying the *a posteriori* statement of dynamical laws and combining the corresponding parameters in an analytically consistent way. I shall not presently enter the analysis of this higher system of explanatory concepts, which I have considered elsewhere (Duchesneau, 1994; Duchesneau, 1999).

But a question may be raised concerning the status of the infinitesimal analogues in this complementary scheme for Leibniz's dynamics. The demonstrative synthesis in the *Dynamica de potentia* implied that definitions, such as those of "formal effect" and "formal" or "essential action" were not presumed to depend on such analogues. Instead, these definitions would depend only on distinct conceptual elements explicating their own possibility and on the ever present architectonic principles which fostered theoretical constructions capable of systematically accounting for the empirical laws. In such constructions, the formalizing analogies and geometrical analytic models were still subjacent, and they afforded certain means for translating the order of the so-called abstract concepts into equations that might also fit the requirements of phenomenal analysis. For instance, this position is made manifest when Leibniz discusses with Hermann the possibility of an *a priori* way of demonstrating the main tenets of his dynamics. In this circumstance, Leibniz clearly discards the epistemic supposition that any infinitesimal symbolization might own a *prima facie* ontological meaning, not to say, a true reference to the structure of reality. Challenged by Leibniz, Hermann had in fact produced an abstract formal representation of the *causa agens* behind non-constrained motions: this representation was based on a combinatorial summation of various differential elements (Letter to Leibniz of December 22, 1712, GM IV, 384). In his critique of Hermann's model, Leibniz points out that his correspondent has displayed significant technical ability in resorting to infinitesimals; but these infinitesimal magnitudes would bear only a nominal and symbolic connection to such phenomena as gravity or percussion. Moreover, they would obviously fall short of reaching to the real notion of a force restoring itself continuously and to the true ground of a universal principle of conservation of formal action. Leibniz suggests that the alternative set of symbolic analogues he himself proposed might be more justified because these can be derived from a system of abstract definitions representing the "metaphysical" intelligibility of such possible conceptual components as may characterize a constantly self-restoring force. He insists:

> Non admitto causam agentem quæ mobili m tempore dt dat celeritatem dc, esse ut $mdc : dt$; nec video quomodo hoc possit probari, nisi assumas ut definitionem. Sed tunc non capio, nec video, quomodo ex hac notione cum spatio conjuncta formes potentiam, et cur non alius pari jure diceret causam agentem esse ut $mdl : dt$, vel aliud quiddam? Deinde in simplicissimis Elementis, ut hic, non quæritur, quid causa agens in alio producat, sed quid in se ipsa, nempe causa.[19] (Letter to Hermann of February 1, 1713, GM IV, 388)

[19] "I do not admit that the acting cause which gives mobile m at time dt velocity dc is as $mdc : dt$; and I do not see how it can be proved if it is not assumed as a definition. But then I do

With the kind of infinitesimal models used by Hermann we could only symbolize the relative transformations involved in the interaction of phenomenal bodies. In contrast, a Leibnizian system of sufficiently real definitions permits us to reach to the causal and therefore essential reality of force and then, at a later stage, to combine the more appropriate differential and integral analogues to account for the manifold applications of dynamic laws to the phenomenal world. In such cases a summation of gravity differentials might for instance appear relevant where they would otherwise seem to have no basis in reality. Leibniz clearly states that the deeper demonstration of the law of conservation of formal action at the foundation of his dynamics is afforded by his *a priori* proof based on "metaphysical principles". Thus he declares to Hermann:

> Sed probationem altiorem habeo ex principiis metaphysicis, quam nempe desideras, ubi non est necesse procedi per elementa infinite parva, nec opus est adhibere effectum violentum aut suppositionem, qualis est gravitatis.[20] (Letter to Hermann of September 9, 1712, GM VI, 378–379)

By synthesizing this proof from the primeval abstract and distinct notions of dynamics with the aid of architectonic rules, we should be able to avoid referring, among premises, to notions of infinitesimals. These are only symbolic analogical devices applicable, at a subordinate rational stage, to the analysis of derivative forces and phenomenal interactions.

3. Conclusion

Our focus in this paper has been exclusively on the epistemological status of infinitesimals in Leibniz's physics. The best expression to qualify Leibnizian infinitesimals in this capacity might be that of "symbolic analogues." They intervene in situations that call for symbolizing continuously developing metric relations: while signifying essential differences in the order of phenomenal transformations, these relations point to limits involving less than assignable quantities. Their objective value is nevertheless pre-

---

not understand nor see how from that notion jointly with space you can form force, and why someone else may not with equal right say that the acting cause is as *mdl* : *dt*, or anything else. Then in the most simple Elements, like here, one does not ask what the acting cause produces in some other thing, but what it produces in itself, therefore in the cause."

20 "I possess the higher proof from metaphysical principles which you wish for, wherein it is not required to proceed through infinitely small elements, nor is it needed to involve a violent effect and a hypothesis like that of gravity."

sumed on the basis of the continuity schemes prevailing in the architectonic makeup of the laws of nature. This epistemological stance is further illustrated in the dynamics where differentials and integrals are used as symbolic analogues to signify modes of generation and exertion of phenomenal forces that correspond to the architectonic and constitutive makeup of the more essential metaphysically-grounded forces. In this instance, representations of infinitesimals should be viewed as analytic devices to be ideally dispensed with or superseded in the aimed at synthetic derivation of the fundamental laws of power and action: this derivation would start from concepts of dynamical properties, especially formal or essential action, for which abstract quasi-real definitions could be provided. For Leibniz, this meant that if we need to use models based on infinitesimal calculus to represent the production and transformation of dynamic effects in nature, these models can only be considered as symbolic devices for determining parametric limits in cases of lesser than any assignable magnitudes. But, at the same time, the infinitesimal algorithm specifies the rational relations on which such analytic processes of transition to a limit should be grounded: it would therefore belong to a system of demonstrated and/or demonstrable mathematical truths. Though infinitesimals are mere operative fictions, the rational framework for representing differentiation and integration processes conforms to the requirements of the principle of continuity. Because this architectonic principle rules jointly over the integral and differential transformation processes specified by the calculus and over the causal laws sovereign wisdom has imposed on the system of nature, the harmony of essential determinations at both levels justifies that we resort to analogies of infinitesimals in accounting for those transformation processes which occur in the physical world. Infinitesimal calculus is a valuable analytic tool in the analysis of mechanical phenomena for the obvious epistemological reason that it successfully fulfills this function in the various domains of physics. But, in Leibnizian fashion, a metaphysical reason could also be evoked to justify that pragmatic and epistemic conformity. This reason would consist in presuming that an overarching transcendent calculus prevailed in the original determination of the laws of nature, and that the same architectonic requirements expressed by the principle of continuity inform the geometry of the infinite as they do the representation schemes grounding our analyses and explanations in the philosophy of nature.

Donald Rutherford

Leibniz on Infinitesimals and the Reality of Force

Leibniz's efforts to apply his differential calculus to the analysis of physical phenomena constitute one of the most forward looking aspects of his natural philosophy. Concealed in these efforts, however, are significant problems about the interpretation of the calculus and of his new science of dynamics. These problems come together in Leibniz's conception of force as a momentary endeavor that is represented in the calculus as an infinitesimal quantity. The idea of force as an infinitesimal element of action that is responsible for continuous changes in a body's state of motion has an undeniable intuitive appeal. Nevertheless, Leibniz articulates other views that make it difficult to see how such a conception of force can be defended.

According to Leibniz's dynamics, which he develops in opposition to Descartes's geometrical physics, active and passive forces are the only real properties of matter. As he writes in the programmatic *Specimen dynamicum* (1695): "Nihilque adeo in ipso reale est, quam momentaneum illud quod in vi ad mutationem nitente constitui debet. Huc igitur redit quicquid est in natura corporea praeter Geometriae objectum seu extensionem."[1] (GM VI, 235) Since for Leibniz the object of geometry – spatial extension – is merely "ideal," the real properties of matter are limited to its active and passive forces: its tendencies to initiate and to resist change. In the case of active force, Leibniz further insists that this force is something "momentary," by which he appears to mean that it lacks any finite duration.

For this reason, it is tempting to think of the basic elements of force as infinitesimal quantities: either infinitesimal amounts of endeavor that are

[1] "[…] there is nothing real in motion but a momentary something which must consist in a force striving toward change. Whatever there is in corporeal nature over and above the object of geometry or extension reduces to this" (AG 118). See also the unpublished part II of the *Specimen Dynamicum* (GM VI, 247/AG 130). Editions of Leibniz's writings are cited according to the list of abbreviations. Where a translation is cited, I have made use of it, though I have sometimes taken the liberty of modifying it slightly; where none is cited, the translation is my own.

summed to produce finite forces, or force states of infinitesimal duration – or both. Yet although Leibniz himself gives currency to this conception of force, in his reflections on the calculus he also expresses strong reservations about the coherence of the idea of an infinitesimal magnitude. To think of the differential *dx* as referring to a quantity smaller than any finite quantity, he says, is to operate with a "fiction," which is useful for calculating but does not designate any real entity.[2] This is because, strictly speaking, there cannot be such an entity: an actual amount, length or duration that is smaller than any finite amount, length or duration.

Whether Leibniz is correct in this judgment is open to debate.[3] Clearly, however, it raises significant concerns for his theory of force. If, as he believes, infinitesimal quantities of distance or speed are "fictions," are we not obliged to say the same about the momentary forces to which he appeals in explaining continuous change? If we are, then we are faced with an apparent inconsistency at the heart of his dynamics. Far from being the only real properties of matter, momentary forces turn out to be mere fictions. While an instrumentalism that counted all physical properties as fictions invoked to save the phenomena has its attractions, it is at odds with Leibniz's convictions concerning the reality of force. On the face of it, then, there appears to be no way for him to combine his preferred interpretation of the calculus with his understanding of physical force: if force is real, it cannot be an infinitesimal quantity; if it is an infinitesimal quantity, it cannot be real.

In some passages, Leibniz suggests that the appeal to infinitesimal forces should be viewed merely as a heuristic and not as an attempt to describe the nature of physical reality. Nevertheless, he explicitly claims that force is something "momentary," and so the question remains of how to conceive of force in a way that is consistent both with the role assigned to it in the explanation of continuous change and with his fictionalism concerning infinitesimals. In what follows I offer an analysis of Leibniz's position that locates the central problem in the conception of force as an infinitesimal quantity that sums over time or space to produce finite changes in a body's state of motion, or finite quantities of force. Such a picture of infinitesimal forces as the underlying causes of physical processes is at best a heuristic, which can lead us astray if taken literally. I also argue, however, that a con-

---

[2] "The infinitesimal calculus is useful with respect to the application of mathematics to physics; however, that is not how I claim to account for the nature of things. For I consider infinitesimal quantities to be useful fictions" (GP VI, 629/AG 230).

[3] For a modern defense of the mathematical coherence of infinitesimals, see Bell, 1998.

ception of force as a determinate finite quantity, while adequate for the purposes of physics, gets us no closer to what force really is. In the end, I propose, it is Leibniz's view that, in and of itself, force is not a mathematically representable property. To understand the sense in which force is real, we must turn to a different theoretical framework altogether, that of metaphysics, wherein force is represented as a modification of a substantial power, or principle of change. From this perspective, force is something momentary for Leibniz, because at any assignable moment each substance has a determinate tendency to change, yet this tendency itself lacks any measurable duration.

1. Representing Continuous Change

An anonymous essay published in the inaugural volume of the proceedings of the Berlin Academy of Science (1710) offers the following description of Leibniz's differential calculus:

> Hic dx significat elementum, id est incrementum vel decrementum (momentaneum) ipsius quantitatis x (continue) crescentis. Vocatur et *differentia*, nempe inter duas proximas x elementariter (seu inassignabiliter) differentes, dum una fit ex altera (momentanea) crescente vel decrescente; [...] Porro ddx est elementum elementi seu *differentia differentiarum*, nam ipsa quantitas dx non semper constans est, sed plerumque rursus (continue) crescit aut decrescit. Et similiter procedi potest ad dddx seu d^3x, et ita porro [...].[4] (Leibniz, 1710, 159–60 = GM VII, 222–3)

The account of the calculus presented here lends itself to an interpretation in terms of infinitesimal quantities. A continuous change in a finite quantity *x* is conceived to occur through the addition or subtraction of an element symbolized by *dx*, which is the difference between two minimally different values of *x*. For the change in *x* to be continuous, *dx* must be smaller than any finite difference; hence it is an infinitesimal quantity, added to or sub-

[4] "Here *dx* signifies an element, that is, a (momentary) increment or decrement of the (continuously) increasing quantity *x*. It is also called a *difference*, namely that between two minimally (or inassignably) different proximal [values of] *x*, where one arises from the other that is (momentarily) increasing or decreasing [...]. Furthermore, *ddx* is an element of an element, or *a difference of differences*, for the quantity *dx* itself is not always constant, but often in turn (continuously) increases or decreases. And similarly, one can proceed to *dddx* or d^3x, and so on [...]." – For the ascription of this text to Leibniz see Cajori, 1928–29, 2, 195.

tracted from x in a minimal interval of time (a "moment"). The author of the text further raises the possibility of a series of higher-order infinitesimals, representing inassignably small differences in the values of dx, of ddx, and so on. Thus, we appear to have a completely general interpretation of the calculus as a symbolism for representing continuous changes in both finite magnitudes and infinitely small magnitudes, in terms of the addition or subtraction of infinitesimal elements.

In his writings on dynamics Leibniz draws frequently on this interpretation of the calculus. His most prominent application of infinitesimals to the analysis of motion is found in the *Specimen dynamicum*. There he begins by distinguishing two senses of motion: the continuous path traveled by a body in a finite interval of time (*motus*), and the motion of a body in an instant (*Motio*):

> Quin etiam quemadmodum (non incommode ad usum loquendi doctrinalem) ab accessu jam facto faciendove distinguere possumus accessionem quae nunc fit, tamquam incrementum accessus vel elementum; aut quemadmodum descensionem praesentem a facto jam descensu, quem auget, distinguere licet; ita possemus praesentaneum seu instantaneum motus elementum ab ipso motu per temporis tractum diffuse discernere et appellare Motionem.[5] (GM VI, 237)

Again, a continuous change in a finite quantity, a body's motion through space, is explained in terms of the addition of "instantaneous elements of motion." The summation of these elements (ds) over time gives the elapsed motion, or path (s). Leibniz's stated purpose in distinguishing these two senses of motion is to clarify the significance of Descartes's measure of force as "quantity of motion," defined as the product of a body's size and speed.[6] Since ds represents the distance traveled by a body in an instant, it is

---

5 "To speak in a way not inappropriate for scientific use, just as we can distinguish the progress we are now making from the progress we have made or will make, considering our present progress as an increment or element of progress, or just as we can distinguish the present descent from descent already made, descent which it augments, so too we can distinguish the present or instantaneous element of motion from that same motion extended through a period of time, and call the former *motio*" (AG 120).

6 Cf. Descartes, *Principia Philosophiae*, II, 36 and 43 (AT VIII-1, 61 and 66–7). Although Leibniz routinely uses the term "velocitas" to refer to a body's speed, he recognizes the difference between speed and "directional speed" ("celeritas respectiva"), or velocity, and draws on it in his reformulation of the laws of motion; see GM VI, 493–94, and Garber, 1995, 314–19. In the present context the distinction is unimportant. (In the *Specimen dynamicum* Leibniz further complicates matters by labeling the modern notion of velocity "*conatus*": "However, just as a mobile thing existing in motion has motion [motum] in time, so too at any moment it has *speed* [*velocitatem*], which is greater to the extent that

proportional to the body's instantaneous speed. Consequently, the Cartesian quantity of motion is properly understood as a body's "momentary quantity of motion," which Leibniz identifies as its *impetus*. Quantity of motion itself, he argues, is more accurately explained as the quantity that "ex aggregatu impetuum durante tempore in mobili existentium (aequalium inaequaliumve) in tempus ordinatim ductorum nascatur."[7] (GM VI, 237).

Leibniz thus proposes to explain the generation of finite continuous quantities – a body's extended motion or its quantity of motion – as sums of infinitesimal quantities. Motion through space is generated by the successive addition of instantaneous elements of motion, and a body's quantity of motion is explained as the sum over time of its momentary quantities of motion. In cases of uniform motion, Leibniz recognizes that neither of these analyses is strictly necessary. If a body of mass m moves with constant speed v, its path over the interval t is given as the product vt, and its quantity of motion as mvt. Thus, the mathematical analysis of uniform motion can proceed independently of any positing of infinitesimal quantities. In Leibniz's view, however, non-uniform motion, in which a body's speed increases or decreases with respect to time, must be treated differently. In his unpublished *Dynamica* (1690), he appeals to the differential calculus as a means of representing the relevant changes in a body's state of motion:

> Quamdiu [...] velocitates mobilis eaedem per quasvis temporis partes (motu existente uniformi), sufficit calculus praecedens per quantitates vulgo receptas. Sed si variet ubique [...] velocitas in loco aut tempore, ad quantitates numero infinitas et magnitudine infinite parvas veniendum est seu ad incrementa aut decrementa vel differentias duarum quantitatum ordinariarum proximarum inter se. Exempli gratia: Dum grave motum accelerat, duae proximae sibi velocitates v et (v) a me dicentur habere differentiam infinite parvam dv, quae est incrementum velocitatis momentaneum, quo transit mobile a velocitate v ad (v). Itaque in Geometriam introduxi novum circa analysin infinitorum calculi genus, suo quodam Algorithmo alibi a me explicato instructum, ubi notis differentiae et summae eodem fere modo utor, quo notis radicis et potestatis in Algebra uti solemus.[8] (GM VI, 426–27)

---

more space is traversed in less time. Speed taken together with direction is called *conatus*" (GM VI, 237/AG 120).)

7 "[...] arises from the sum of the impetuses (equal or unequal) existing in a moving thing during a time, multiplied by the corresponding time" (AG 120).

8 "So long as the [...] speeds of the moving thing (existing with uniform motion) are the same throughout any parts of time, the preceding calculus by means of commonly received quantities suffices. But if [...] the speed varies everywhere in place or time, we must turn to

Leibniz's insistence on the need for the calculus in the analysis of non-uniform motion may seem at first glance unmotivated. In free fall, for example, where a body's speed increases by a constant factor with respect to time, an analysis of its motion in terms of "quantitates vulgo receptas [commonly received quantities]" suffices. To be charitable to Leibniz, we might read him as anticipating a more general treatment of motion, in which rates of change of speed (or the infinitesimal increments dv) need not be constant. By drawing on the calculus, he can represent the path of any body with respect to time, provided only that changes in its state of motion are continuous, or occur through infinitely small increments or decrements of speed.

Although this obviously is one of the signal achievements of the differential calculus, Leibniz's preference for an explanation of non-uniform motion in terms of infinitesimals can be traced to specific assumptions of his dynamics. With respect to free fall, Leibniz denies that Galileo's rule that a body gains equal increments of speed in equal times should be admitted as a genuine law of nature.[9] This is because he believes that a body's natural motion – the motion proper to it – is uniform and rectilinear.[10] Consequently, any variation in a body's speed or direction of motion must be ascribed to the action of forces that effect a change in it. His account of these forces is far from transparent. In fact, we find two diametrically opposed

quantities infinite in number and infinitely small in magnitude, that is, to increments and decrements, or differences of two ordinary quantities proximate with respect to each other. For example, when a heavy object accelerates in motion, two speeds v and (v) proximate to each other are said by me to have an infinitely small difference dv, which is a momentary increase of speed, by which the moving thing passes from speed v to (v). And so in geometry I have introduced a new kind of calculus concerning the analysis of infinites, laid out with its algorithm explained by me elsewhere, where I use signs for differences and sums in almost the same way as we are accustomed to use signs for roots and powers in algebra."

9 In a letter to Varignon of October 10, 1706, he writes: "The simpler way is that which does not make acceleration foundational, when there is no need to do so. I have made use of this for more than 30 years" (GM IV, 151). See also GM VI, 453–54.

10 *Dynamica*, part I, sec. 2, ch. 5, props. 1–2 (GM VI, 342) and part II, sec. 3, prop. 17 (GM VI, 502). The argument offered in Part II of the *Specimen dynamicum* for the rectilinear character of motion is premised on a claim about the nature of force: "since only force and the nisus arising from it exist at any moment (for motion never really exists, as we discussed above), and since every nisus tends in a straight line, it follows that *all motion is either rectilinear or composed of rectilinear motions*. From this it [...] follows that what moves in a curved path always tries [conari] to proceed in a straight line tangent to it" (GM VI, 252/AG 135). Leibniz's reasoning parallels that of Descartes in *Principia Philosophiae*, II, 39, though he rejects Descartes's grounding of rectilinear motion in "the immutability and simplicity of the operation by which God preserves motion in matter" (AT VIII-1, 63). For God's constant activity, Leibniz substitutes a tendency grounded in a body's inherent force.

explanations of how corporeal forces effect a change in a body's state of motion. In line with ordinary ways of thinking, Leibniz often represents these forces as ones that act mechanically through the impact of bodies. This is the basis of his vortex theory of planetary motion, presented in the *Tentamen de Motuum Coelestium Causis* (1689), which he defends against the rival theory of Newton's *Principia*. According to Leibniz, the orbital motion of celestial bodies is to be ascribed not to the action of a *sui generis* gravitational force ("action at a distance"), but to the impact of the moving particles of an aetherial fluid.[11]

Yet this is not Leibniz's deepest account of the nature and action of physical forces. Although we commonly explain changes in a body's state of motion by appeal to the action of external forces, in the strictest sense, Leibniz claims, there is no real causal interaction among things – no case in which one thing acts directly on another by transferring motion or force to it. Therefore, any change in a body's state of motion must be ascribed to the action of *internal* forces.[12] In general, the forces that explain the changes that occur in bodies in collision are elastic forces proper to each: "*Corpora non agunt immediate in se invicem motibus suis, nec immediate moventur, nisi per sua Elastra.*"[13] (*Dynamica*, part II, sec. 3, prop. 6; GM VI, 492)

---

[11] "[...] it can first of all be demonstrated that according to the laws of nature *all bodies which describe a curved line in a fluid are driven by the motion of the fluid.* For all bodies describing a curve endeavor to recede from it along the tangent (because of the nature of motion), it is therefore necessary that something should constrain them. There is, however, nothing contiguous except for the fluid (by hypothesis), and no conatus is constrained except by something contiguous in motion (because of the nature of the body), therefore it is necessary that the fluid itself be in motion" (GM VI, 149; trans. in Bertoloni Meli, 1993, 128–29). See also *Specimen dynamicum*, Part II: "For if we assume something we call solid is rotating around its center, its parts will try to fly off on the tangent; indeed, they will actually begin to fly off. But since this mutual separation disturbs the motion of the surrounding bodies, they are repelled back, that is, thrust back together again, as if the center contained a magnetic force for attracting them, or as if the parts themselves contained a centripetal force. Thus, the rotation arises from the composition of the rectilinear nisus for receding on the tangent and centripetal conatus among the parts" (GM VI, 252/AG 135–36). For discussion of the details of Leibniz's theory, see Aiton, 1984, and Bertoloni Meli, 1993.

[12] "Rigorously speaking, no force is transferred from one body to another, but every body is moved by an innate force [insita vi]" (A VI, 4, 1630/DLC, 333). See also A VI, 4, 1620; GP II, 195; and *Specimen dynamicum*, Part II: "every passion of a body is spontaneous, that is, arises from an internal force, even if it is on the occasion of something external" (GM VI, 251 /AG 134).

[13] "*Bodies do not act immediately on one another through their motions, nor are they immediately moved except through their own elasticity.*" – On the role ascribed to elastic forces, see Breger, 1984; Garber, 1995, 321–25.

For our purposes, it is unnecessary to negotiate between these two accounts of force, since the salient point applies to both: On the assumption that all change is continuous change, Leibniz maintains that it must be explained in terms of the action of infinitesimal forces. He refers to these forces generically as "dead force," indicating that their presence in a body does not depend upon its already being in a state of motion. In the *Specimen dynamicum*, Leibniz locates dead force within a complex taxonomy of forces. Within the category of force in general, he distinguishes, on the one hand, active and passive force, and on the other, primitive and derivative force. For the moment, we are interested only in the class of active derivative forces: the physical forces by which bodies act, and to which the laws of motion apply. Among these forces, Leibniz is primarily concerned to emphasize the difference between "dead force" (vis mortua) and "living force" (vis viva): "Hinc *Vis* quoque duplex: alia elementaris, quam et *mortuam* appello, quia in ea nondum existit motus, sed tantum solicitatio ad motum [...]; alia vero vis ordinaria est, cum motu actuali conjuncta, quam voco *vivam*."[14] (GM VI, 238)

Dead force is a theoretical primitive for Leibniz. It is an elementary "endeavor" or "tendency" to motion that is present both in bodies at rest and in bodies in motion. The former are conceived as objects that would move were some impediment to motion removed. Examples include a lever balanced by a counterweight, a stretched spring, or a body suspended from a height. In an object at rest, dead force is the force by which motion is initiated; in an object already moving, dead force accounts for changes in its state of motion and for the accumulation of the living force by which it acts on other bodies.

Leibniz illustrates the operation of dead force with the example of a tube rotating with a constant speed about a fixed center. At the end of the tube nearest the center, a ball is suspended. When released, the ball tends to move outward toward the other end of the tube. On Leibniz's analysis, prior to its release, the ball has dead force, in the form of a "conatu[s] a centro recedendi," but this centrifugal conatus is "infinite parvum respectu impetus quem jam tum habet a rotatione" (GM VI, 238).[15] Upon its release, the ball acquires an outward motion, which increases through successive

---

14 "One force is elementary, which I also call *dead force*, since motion does not yet exist in it, but only a solicitation to motion [...]. The other force is ordinary force, joined with actual motion, which I call *living force*" (AG 121).

15 "It is obvious that, in the beginning, the conatus for receding from the center, [...] is infinitely small in comparison with the impetus it already has from rotation" (AG 121).

impressions of dead force until its centrifugal impetus is comparable to its rotational impetus.[16]

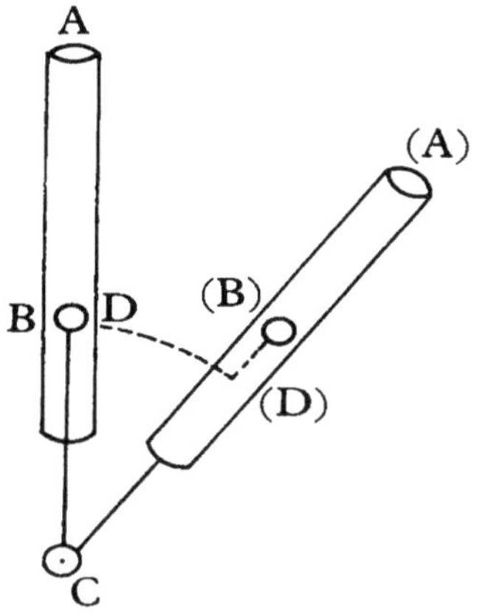

Figure 1.

In this example, Leibniz makes several assumptions about the nature of dead force and its relation to motion:

- Dead force is an infinitely small endeavor or tendency to motion.
- In the absence of impediments, the action of dead force produces an infinitely small change in a body's speed, and hence in its impetus.
- The accumulated effect of the action of dead force is a finite increase in a body's impetus.

So understood, the notion of dead force supports the analysis of motion in terms of infinitesimals. Assuming a continuous increase of speed, dead force is a body's power to pass from a resting speed v to a speed $v + dv$ in the interval dt. Because v itself is conceived as an infinitesimal change of distance, dead force can be construed as a second-order infinitesimal, which stands in the same relation of proportionality to speed and impetus as these first-order infinitesimals stand to motion and quantity of motion, respectively. Just as speed is conceived as the infinitesimal distance traveled by a body in an instant, so dead force is an infinitesimal endeavor that effects an infinitely small change in a body's speed or impetus.[17]

---

[16] Leibniz presents a fuller version of the argument in *Dynamica*, part II, sec. 1, props. 27–28 (GM VI, 451–52). The figure that follows in the text is reproduced from AG 121.

[17] "[…] just as the numerical value [*aestimatio*] of a motion extending through time derives from an infinite number of impetuses, so, in turn, impetus itself (even though it is something momentary) arises from an infinite number of increments successively impressed upon a given mobile thing […]. From this it is obvious that the *nisus* is twofold, that is,

Leibniz cites dead force as the cause not only of the generation of impetus but also of *vis viva*, or living force. The contribution that infinitesimal forces make to the production of these two types of dynamical property must be carefully distinguished. Solicitation, or dead force, impresses on a body an infinitesimal increment of speed dv, which can be greater or smaller depending upon the magnitude of the force. Impetus is generated through the addition of successive increments of speed; thus, assuming a constant force (as in free fall), impetus increases linearly with time. Descartes believed that a body's quantity of motion (or impetus) was a measure of its moving force, or its power to effect change in the state of another body through collision.[18] In his 1686 *Brevis demonstratio*, Leibniz showed that Descartes's measure of force cannot be correct, and that moving force is properly calculated as the product of a body's mass and the square of its speed.[19] It thus follows that in cases of uniform acceleration, where a body gains equal increments of speed in equal times, its living force increases in proportion to the square of time, or, equivalently, in proportion to the distance the body is moved. By the rules of Leibniz's calculus, we can infer that with each increment of speed dv, a body's living force increases by a factor of $2vdv$.[20] Therefore, while constant impressions of dead force over time produce a linear increase in the impetus of a moving body, they produce a geometrical increase in its living force. Leibniz highlights this difference in a 1699 letter to De Volder, in which he draws on the symbolism of his calculus:

> Eodem modo etiam fit, ut gravi descendente, si fingatur ei quovis momento nova aequalisque dari celeritatis accessio infinite parva, vis mortuae simul et vivae aestimatio observetur, nempe ut celeritas quidem aequabiliter crescat secundum tempora, sed vis ipsa absoluta secundum spatia seu temporum quadrata, id est secundum effectus. Ut ita secundum analogiam Geometriae seu analysis nostrae solicitationes sint ut dx, celeritates ut x, vires ut xx seu ut $\int xdx$.[21] (GP II, 156)

---

elementary or infinitely small, which I also call *solicitation*, and that which is formed from the continuation or repetition of elementary nisus, that is, *impetus* itself" (GM VI, 238/AG 121). See also GM V, 325; GM VI, 451–52; GP II, 154; and GM VI, 151, translated in Bertoloni Meli, 1993, 131.

18 Descartes, *Principia Philosophiae*, II, 40, 43 (AT VIII-1, 65–7).

19 For analyses of the argument, see Brown, 1984; Garber, 1995, 310–13.

20 An increment of speed dv produces a new force $F + dF$ that is proportional to $(v + dv)^2$, or $v^2 + 2vdv + dv^2$. Leibniz's rules for the calculus allow him to ignore the last infinitesimal product. Thus the newly acquired force is represented by the factor $2vdv$.

21 "It happens in the same way also with a falling weight that a measure of both dead and living forces is obtained, if it is imagined that at any moment it receives a new and equal infinitely small increase in speed. Namely, the speed increases in equal amounts according to

Leibniz's conclusion that there is a significant difference in the mathematical representations of a body's quantity of motion and of its moving force is of fundamental importance for physics. Against Descartes, he stresses that the property of force involves two distinct components: a body's power at a moment to move itself or another body through a given *distance* and the *speed* at which it is able to do so. Hence, a body possesses greater living force to the extent that it is able to move a greater mass a greater distance, and to do so more quickly (GP II, 220). On Leibniz's account, impetus measures the speed that a body acquires through successive impressions of dead force, but it does not measure the contribution that dead force makes to a body's capacity to effect change in its own state of motion or in the state of motion of another body. That is measured by a body's living force. Unfortunately, beyond telling us that living force "ex infinitis vis mortuae impressionibus continuatis nata [arises from infinite continual impressions of dead force]" (GM VI, 238/AG 122), and that dead force "vis vivae [...] non nisi infinitesimalis pars est [is only an infinitesimal part of living force]" (GM VI, 104/AG 255), Leibniz has little to say about how the two types of force are related.[22]

2. Infinitesimals as "Useful Fictions"

In the *Specimen dynamicum* Leibniz appeals to infinitesimal forces in explaining the initiation of motion, continuous increases or decreases of speed or impetus, and the accumulation of living force through motion. Yet despite the prominence he gives to these accounts, there is reason to doubt whether he intends them to be taken literally as descriptions of causal processes underlying physical change. In the *Specimen dynamicum* itself, he expresses reservations about interpreting infinitesimal quantities as real properties found in nature: "[...] non ideo velim haec Entia Mathematica reapse sic reperiri in natura, sed tantum ad accuratas aestimationes abstractione animi faciendas prodesse."[23] (GM VI, 238)

---

time, but the absolute force itself increases according to space or the square of the times, that is, in accordance with the effect. So, by analogy with geometry, or my analysis, solicitations are as dx, speeds are as x, and forces are as xx or $\int xdx$."

22 The interpretation of their relation remains a matter of controversy. For a brief survey of the literature, see Bertoloni Meli, 1993, 89–90.

23 "I would not want to claim on these grounds that these mathematical entities are really found in nature, but I only wish to advance them for making careful calculations through

Although this passage may seem to suggest a special problem about the physical instantiation of infinitesimal quantities, Leibniz's early reflections on the "labyrinth of the continuum," and his subsequent efforts to absolve his differential calculus of any commitment to the reality of infinitesimals, demonstrate that his concerns are broader than this. In texts from the 1670s, Leibniz returns repeatedly to the paradoxes posed by the notion of infinitely small magnitudes, particularly infinitesimals of space and time. He ultimately concludes that the idea of a determinate magnitude that is smaller than any finite magnitude is an incoherent one. Thus, infinitesimals cannot be appealed to as basic elements from which a spatial or temporal continuum is composed.[24]

Following the publication of his calculus, Leibniz is forced to return to this topic. When critics attack the calculus because of its perceived reliance on infinitesimals, Leibniz responds by rejecting the charge. The calculus, including higher-order differentials, can be defended as a mathematical tool without Leibniz needing to commit himself to the reality of infinitesimals. The calculus itself is identified with a set of rules for differentiation and integration (or methods for finding tangents, quadratures, etc.), and the difference *dx* is understood not as a determinate, infinitely small magnitude, but as an indeterminate "differentiam duarum quantitatum communium ipsis quantitatibus incomparabilem."[25] (GM VI, 151). In appealing to an "incomparable" difference, Leibniz makes no assumption about the absence of a strict proportionality between the finite quantities and their difference. Rather, he claims that, though finite, the difference always can be taken to be sufficiently small that no error results from it. Thus, the validity of reasoning using the calculus does not presuppose the existence of actual infinitesimals. In place of infinitely small quantities, one can take the differences "tam parvas quam sufficere judicat, ut sint incomparabiles et errorem nullius momenti, imo dato minorem, producant."[26] (GM VI, 151).

mental abstraction." (AG 121) – See also GM IV, 91; GP II, 305; GP VI, 629/AG 230 (cited in n. 2).

24 This commitment is in place by 1676. For discussion of the relevant arguments, see Arthur's Introduction to DLC, liv-lvii; Arthur, 2008c; Levey, 1998, and Levey, 2003.

25 "[...] difference of two ordinary quantities, incomparable with the quantities themselves."

26 "[...] to be as small as one judges sufficient, so that they are incomparable and produce an error of no importance, indeed one smaller than any given." – This account is advanced publicly in the "lemmata" to his 1689 *Tentamen de Motuum Coelestium Causis*. See Bertoloni Meli, 1993, 130–31. Leibniz elaborates on it in his letter to Varignon of February 2, 1702 (GM IV, 91–2/L 543). For further discussion, see Jesseph, 1998.

With this interpretation of the calculus in place, Leibniz can argue that, while talk of infinitesimals may have some heuristic value, in the strictest sense, he regards them "pro mentis fictionibus [as fictions of the mind]" (GP II, 305). Unlike the continuum, they have no reality even as ideal entities, but are merely imaginary: entities that are feigned to exist. For this reason, there is no case to be made for infinitesimals either as real physical entities or as mathematical ones. In a 1702 letter to Varignon, he writes:

> Pour dire le vray, je ne suis pas trop persuadé moy même, qu'il faut considerer nos infinis et infiniment petits autrement que comme des choses ideales ou comme des fictions bien fondées. Je croy qu'il n'y a point de creature au dessous de la quelle il n'y ait une infinité de creatures, cependant je ne crois point qu'il y en ait, ny même qu'il y en puisse avoir d'infiniment petites et c'est ce que je crois pouvoir demonstrer.[27] (GM IV, 110)

Leibniz categorically rejects the postulation of infinitely small parts of matter, and, we may assume, infinitely small quantities of force. The infinitely small does not exist as an element of the finite; rather, in the physical world there are only finite things, composed of smaller finite things, all the way down. He reiterates this view four years later to Des Bosses:

> Caeterum ut ab ideis Geometriae ad realia Physicae transeam, statuo materiam actu fractam esse in partes quavis data minores, seu nullam esse partem, quae non actu in alias sit subdivisa diversos motus exercentes. Id postulat natura materiae et motus et tota rerum compages, per physicas, mathematicas et metaphysicas rationes.[28] (GP II, 305)

If every part of matter is actually subdivided into smaller parts exercising different motions, then any quantity of force exerted by a part of matter will be a composite of the forces exerted by its parts. For Leibniz, however,

---

27 "To speak the truth, I am not at all persuaded myself that it is necessary to consider the infinite and infinitely small other than as ideal things or as well-founded fictions. I believe that there is no created thing beneath which there is not an infinity of created things; however, I do not believe that any of them are, or even that any of them could be, infinitely small – and this I believe can be demonstrated."

28 "To pass now from the ideas of geometry to the realities of physics, I hold that matter is actually fragmented into parts smaller than any given part; that is, there is no part of matter that is not actually subdivided into others exercising different motions. This is demonstrated by the nature of matter and motion and by the structure of the universe, for physical, mathematical, and metaphysical reasons." – See also his letter to Jacob Bernoulli of late August 1698 (GM III, 536).

this composition relation holds only among finite things. There is no ultimate account of the composition of finite things from infinitely small things, for infinitesimals of force or matter do not exist.

3. The Ideality of Finite Magnitudes

There seems, then, to be good reason to discount Leibniz's speculations in the *Specimen dynamicum* and elsewhere about infinitesimal forces, and to focus instead on an understanding of force as a physical property that always possesses a finite magnitude. Can we in this way arrive at an understanding of force as "aliquid reale et absolutum"[29] (GM VI, 248)? In Leibniz's view, we cannot. Although the idea of a finite quantity is – in contrast to that of an infinitesimal quantity – a mathematically coherent one, it identifies a property that is, according to Leibniz, merely ideal. Hence, a representation of force as a finite quantity cannot be a representation of force insofar as it is "real and absolute."[30]

The argument for this conclusion hinges on a crucial premise: Given the role played by force in the explanation of physical change (change measured in terms of time and distance), the magnitude of any such force is represented as the value of a continuous function. As a body accelerates, it gains force continuously in proportion to the square of time; when a body loses force through collision with another body, it does so continuously, or "through degrees" – change never occurring through a leap. A conception of physical properties as finite quantities that vary continuously with respect to time and space is integral to the modern conception of physical theory. Leibniz saw this more clearly than most. However, he also insists that continuous quantities as such are merely *ideal*:

> Sed continua Quantitas est aliquid ideale, quod ad possibilia et actualia, qua possibilia, pertinet. Continuum nempe involvit partes indeterminatas, cum tamen in actualibus nihil sit indefinitum, quippe in quibus quaecunque divisio fieri potest, facta est.[31] (GP II, 282)

---

[29] "[...] something real and absolute" (AG 131).

[30] Bracketed in this section is the problem of the duration of force. Even if a coherent conception of force as a finite quantity could be defended, there would still be the question of how such a force can be real, if only momentary. I return to this question in section 4.

[31] "Continuous quantity is something ideal, something that pertains to possibles and to actual things considered as possible. The continuum, of course, contains indeterminate parts. But in actual things nothing is indefinite, indeed, every division that can be made has been made in them" (AG 185).

The distinction between the actual or real (including the physically real) and the ideal is foundational to Leibniz's metaphysics; it is the escape route by which he extricates himself from the labyrinth of the continuum. His decisive observation concerns the structure of a mathematical continuum. A genuine continuum cannot be conceived as resolvable into, or composed from, basic elements. Infinitesimals are a candidate for such elements, but Leibniz rejects them as incoherent. This is why the account of continuous change in terms of the action of infinitesimal forces has only a heuristic value. In the end, Leibniz concludes that it is an error to think of a continuum in mereological terms. There are no actual, determinate divisions within it, and there is no way to generate it through the summation of parts. Instead, it is the nature of a continuum that it is a whole in which the possibility of indefinitely many arbitrary divisions can be conceived – more divisions than are measurable by any countable sequence.

This is a significant fact that Leibniz recognizes about the continuum. However, he is most struck by the metaphysical import of this fact. According to Leibniz, it is a mark of the real that it is composed of determinate parts, and that it is resolvable into a set of elements, or "true unities." Implicit in the latter claim is the assumption that whatever is real, is either a substance (an *unum per se*) or something whose existence can be explained in terms of the prior existence of substances (an *unum per aggregationem*). Since a continuum has no determinate parts and is not resolvable into elements, it cannot be real. Consequently, continua – including those of space and time – are only ideal, that is, the contents of ideas or concepts.[32]

Given this result, we can pose the following question about the property of force: Insofar as physical force is identified with a continuous quantity, or is represented as a function of continuous magnitudes, must it also be regarded as merely ideal? In one of his last letters to De Volder, Leibniz writes:

> [...] in extensione Mathematica, [...] nec prima Elementa, non magis quam inter numeros fractos minimus datur velut Elementum caeterorum. [Hinc Numerus, Hora, Linea, Motus seu gradus velocitatis, et alia hujusmodi Quanta idealia seu entia Mathematica revera non sunt aggregata ex partibus, cum plane indefinitum sit quo in illis modo quis partes assignari velit, quod vel ideo sic intelligi necesse est, cum nihil aliud significent quam illam ipsam meram possibilitatem partes quomodocunque assignandi.][33] (GP II, 276)

---

32 See GP II, 268/AG 178; GP II, 282/AG 185; GP IV, 568/L 583.

33 "[...] in mathematical extension [...] there are no basic elements, any more than a smallest number is found among the fractions, as the element of the rest. [Hence number, hour,

If continuous functions of time and distance, including a body's "degree of speed," are ideal quantities, then it would seem that the same should be said about force. An obvious objection to this inference is that, for Leibniz, force is not defined simply as a function of time and distance. His decisive objection to Descartes's physics is that force cannot be measured as the product of size and speed. The critical point here is usually taken to be Leibniz's claim that force is proportional not to speed, but to the square of speed (or velocity). No less significant, however, is his insistence that matter or mass cannot be identified with geometrical extension. If it could, then force would be exhaustively represented as a function of spatial and temporal variables, and we would have to conclude that force is merely ideal. Yet, Leibniz sees the "intimam corporum naturam [innermost nature of body]" (GM VI, 235/AG 118) not as extension but as *force*, and so there remains conceptual space for him to retain the idea that force itself is something real – and not merely, like space, time and degree of speed, ideal.

This clearly tracks the direction of Leibniz's thought. The problem, though, is how to conceive of, or to represent, force in a way that is consistent with our understanding of it as real. The charge, supported by Leibniz's analysis of the continuum, is that the resources of his science of dynamics do not allow us to do this. Any representation of a body's force as a value of a continuous function leaves us with a conception of force as ideal. For consider: for any body of fixed mass *m*, its force varies continuously as the square of its speed. Since *m* is constant, the only relevant factor in defining a body's acquisition of greater living force is its greater speed. Yet degree of speed is, according to Leibniz, an ideal quantity; hence, the increase in the body's force is represented in a way that we can only regard as ideal. Appeal to the magnitude of *m* is of no help, since it too is represented in the theory as the value of a continuous function, and in any case we have no access to *m* except via changes in a body's spatial and temporal parameters in response to the action of forces, themselves measured in terms of spatial and temporal parameters. In short, the science of dynamics offers us no way of representing a body's force – as a determinate finite quantity – that confirms its status for Leibniz as "something real and absolute."

line, motion or degree of speed, and other ideal quantities of this kind, that is, mathematical entities, are not in fact aggregated from parts, since the way in which someone may choose to assign parts in them is completely undetermined. Indeed, it is necessary that they be understood in this way, since they signify nothing other than the mere possibility of assigning parts in any way whatever.]" Leibniz indicates on his copy of the letter that the material enclosed in brackets was not included in the version sent to De Volder.

Another way to the same conclusion is offered by Leibniz's assertion that any part of matter is actually subdivided into smaller parts *ad infinitum*. Here he wishes to emphasize that in matter as it exists, there is not merely, as there is in mathematical extension, "the possibility of division in any way whatever." Any part of matter is actually infinitely divided in some determinate way. For Leibniz, this marks matter as having a different ontological status than a spatial continuum, and we may suppose that this difference applies also to matter's "innermost nature": force. Just as any part of matter is subdivided into infinite actual parts, so the force of that matter is divided into infinitely many smaller discrete forces. For two reasons, however, this fact is of limited value in helping us to understand the reality of force. First, while matter (and force) is posited to have the structure of an infinite envelopment of discrete parts, the science of dynamics depends upon the assumption that physical change is *continuous* change. Hence the basic tools we have for conceiving of physical force represent it in a way that fails to support the claim made for its reality. Second, the actual division of matter to infinity satisfies only a necessary condition for its reality. In this way we are able to conceive of any part of matter as an aggregate of prior things. However, because matter is represented by us as inherently spatial, such a resolution into prior parts is always incomplete: any division produces parts which themselves are further divided. At no point in the spatial resolution of matter do we reach "true unities," or substances – the only entities in terms of which the reality of matter and its force can be demonstrated.[34]

4. The Reality of Force

For Leibniz, I suggest, there is no fully adequate representation of force as a mathematical quantity. In saying this, I do not deny that Leibniz believes that there are correct answers to the question of how force ought to be measured within physical theory. He is confident that his dynamics offers a correct measure of the moving force of a body and that Descartes's physics does not. What I do deny is that, for Leibniz, any mathematical formula ex-

---

[34] In Leibniz's late (post-1700) writings, the "true unities" that ground the reality of matter and force are mind-like monads. To Varignon, he writes in 1702: "The fact is that simple substances (that is, those which are not beings by aggregation) are truly indivisible, but they are immaterial and only principles of action" (GM IV, 110). I examine this position in detail in Rutherford, 2004, and Rutherford, 2008.

pressible as a function of spatial and temporal variables is adequate to represent force as "something absolute and real." To represent force as an infinitesimal quantity, we have seen, is to represent it as something that is, strictly speaking, impossible. To represent it as a finite continuous quantity, is to represent it in a way that is mathematically coherent, but which involves a falsification of physical reality (it represents the real as ideal). These do not exhaust the possibilities of mathematical representation; we might suppose modeling the structure of matter/force through some form of discrete mathematics.[35] However, Leibniz does not envision this possibility and, in fact, makes it clear that he rejects any attempt to explicate force solely in mathematical terms. Force, instead, is something "metaphysical," which is not representable by the imagination, but can be grasped only by the intellect.[36] This is a point that many philosophers, particularly Cartesians, have failed to recognize, in Leibniz's view:

> Sed vulgo homines imaginationi satisfacere contenti rationes non curant, hinc tot monstra introducta contra veram philosophiam. Scilicet non nisi incompletas abstractasque adhibuere notiones sive mathematicas, quas cogitatio sustinet sed quas nudas non agnoscit natura, ut temporis, item spatii seu extensi pure mathematici, massae mere passivae, motus mathematice sumti etc.[37] (GP II, 249).

In order to understand force as *real*, we must set aside the imagination and mathematical modes of representation and rely instead on metaphysical concepts, known through the intellect. In the *Specimen dynamicum*, he writes:

> Hinc igitur, praeter pure mathematica et imaginationi subjecta, collegi quaedam metaphysica solaque mente perceptibilia esse admittenda, [...] Id principium

[35] This strategy is explored in Levey, 1998.

[36] In his *Lettre sur la Question si l'Essence du Corps Consiste dans l'Etendue*, published in the *Journal des Savans* in 1691, Leibniz writes: "there is in nature something other than what is purely geometrical, that is, extension and mere changes in it [...]. It is necessary to join to it some higher or metaphysical notion, namely, that of substance, action and force" (GP IV, 465). See also GP VI, 507/AG 192: "the laws of force depend upon some marvelous principles of metaphysics or upon intelligible notions, and cannot be explained by material notions or the notions of mathematics alone or by those falling under the jurisdiction of the imagination."

[37] "People are generally content to satisfy their imaginations and do not worry about reasons; hence so many monstrosities are introduced to the injury of the true philosophy. It is obvious that they use only incomplete and abstract notions, or mathematical ones, which thought supports but which nature does not know in their bare form; such notions as that of time, also of space or of what is extended only mathematically, of merely passive mass, of motion considered mathematically, etc." (L 529).

> Formam, an ἐντελέχειαν, an Vim appellemus, non refert, modo meminerimus per solam virium notionem intelligibiliter explicari.[38] (GM VI, 241–42)[39]

Leibniz's dynamics incorporates three metaphysical theses about the nature of force:

(1) The force that exists in matter is "quiddam prorsus reale" (GM VI, 247), and the "intimam corporum naturam" (GM VI, 235).[40]
(2) Derivative force, or "quod in actione momentaneum est," is something "accidentale seu mutabile" (GP II, 270).[41]
(3) Derivative force presupposes the existence of an active substance, or primitive active force, because "omne accidentale seu mutabile debet esse modificatio essentialis alicujus seu perpetui" (GP II, 270).[42]

We may see the third thesis as a way of reconciling the first two. As something accidental or changeable, derivative force is not a *per se* real being, or substance. The claim for its reality is justified, therefore, only if it is understood to exist as a modification of a prior substantial principle. In this way Leibniz moves the discussion of force squarely into the domain of metaphysics. The interpretative questions raised by this move are legion. Here I propose to address only two of them. The first concerns the nature of the substance, or active principle, in terms of which Leibniz explains the reality of physical force; the second, the support this account offers for his claim that physical force is something "momentary."

---

[38] "[...] we must admit something metaphysical, something perceptible by the mind alone over and above that which is purely mathematical and subject to the imagination [...]. Whether we call this principle form or entelechy or force does not matter, as long as we remember that it can be intelligibly explained only through the notion of forces" (AG 125).

[39] On Leibniz's association of mathematics and metaphysics with different modes of cognition, see his 1702 letter to Queen Sophie Charlotte, *On What Is Independent of Sense and Matter* (GP VI, 500–2/AG 187–88). His claim that metaphysical concepts in general are grasped through reflective self-knowledge throws considerable light on his belief that the paradigm of a substance is an immaterial soul. See GP II, 270/AG 180–81; GP II, 276/AG 182; and Rutherford, 1995, 83–5.

[40] "[...] something absolutely real" (AG 130); "innermost nature of body" (AG 118).

[41] "[...] what is momentary in action" is "accidental or changeable" (AG 180).

[42] "[...] everything accidental or changeable must be a modification of something essential or perpetual" (AG 180). Leibniz repeats this line of reasoning on many occasions: "we must consider derivative force (and action) as something modal, since it admits of change. But every mode consists of a certain modification of something that persists, that is, of something more absolute. [...] Therefore, derivative and accidental or changeable force will be a certain modification of the primitive power [virtutis] that is essential and that endures in each and every corporeal substance" (GM VI 102–3/AG 254).

To the extent that it intersects with his dynamics, Leibniz frames his theory of substance in a vocabulary inherited from Aristotle. The primary commitment of the theory is to a conception of substance as an original ground or principle of change.[43] The basis of a substance's fulfilling this function is its intrinsic power (*potentia*). However, against Aristotle, Leibniz insists that this power is not simply a potential or capacity for action, but a fully actual endeavor (*conatus*, *tendentia*). To mark its actuality, Leibniz labels this power, insofar as it is identified with a substance, *entelechy*, or *primitive active force*. And he contrasts this entelechy with *derivative* force, which includes all of the particular moments of "effort," by which a substance strives to attain new states.[44]

The idea of "effort" brings us back to the notion of dead force. We may recall that Leibniz defines dead force as an elementary endeavor or tendency to motion – a tendency evident in a compressed spring or a lever balanced by a counterweight. In elaborating the details of his dynamics, I stressed that dead force is explanatorily basic for Leibniz: it is the cause of any initiation of motion. Now, as we saw, in some of his writings, Leibniz attempts to subject the notion of dead force to the imagination, equating it with the force necessary to bring about an infinitely small change of speed. This move, I argued, cannot sustain a rigorous analysis, for the attempt to represent dead force as an infinitely small magnitude leads to the conclusion that dead force, strictly speaking, cannot exist. One response to this conclusion would be to say, so much the worse for dead force: like the infinitesimal itself, the notion may have some heuristic value, but it does not pick out any real entity in nature. In my view, this response is not supported by Leibniz's philosophy. Properly construed, as a moment of en-

---

[43] In the published essay *De ipsa natura* (1698), Leibniz poses the question "whether there is any *energeia* in created things." He responds by saying that he does not think "that it is in agreement with reason to deny all created, active force inherent in things"; and then continues: "Now let us examine a bit more directly [...] that nature which Aristotle not inappropriately called the *principle of motion and rest*; though, having taken the phrase rather broadly, that philosopher seems to me to understand not only local motion or rest in a place, but *change* in general and stasis or persistence" (GP IV, 504–5/AG 156).

[44] "Active force, which one usually calls force in the absolute sense, should not be thought of as the simple and common potential [*potentia*] or receptivity to action of the schools. Rather, active force involves an effort [*conatus*] or striving [*tendentia*] toward action, so that, unless something else impedes it, action results. And properly speaking, entelechy, which is insufficiently understood by the schools, consists in this" (GM VI 101/AG 252). See also *Nouveaux Essais*, II.xxi.1 (A VI, 6, 169/NE 169); *Nouveaux Essais*, II.xxii.11 (A VI, 6, 216/NE 216); *Theodicée*, §87 (GP VI, 149–50).

deavor realized in an entelechy, dead force is indeed a cause of change. It is not the principle of change – that role is assigned to substance itself; but dead force is a determination or modification of that principle, and hence something (derivatively) real. This is a point we miss if we attempt to reduce a metaphysical concept to a mathematical function.

The challenge of linking Leibniz's theory of substance to the details of his dynamics manifests itself in a variety of ways. One problem arises from the fact that Leibniz explicitly cites the mind or soul as the paradigm of an entelechy: "la plus claire idée de la puissance active nous vient de l'esprit. Aussi n'est elle que dans les choses qui ont de l'analogie avec l'esprit, c'est-à-dire dans les Entelechies" (A VI, 6, 172).[45] Not all entelechies are minds for Leibniz, but all entelechies are mind-like – what he calls in his late writings "monads." Yet if primitive active force belongs exclusively to mind-like substances, whose derivative forces are strivings for new perceptual states, can this account be of any use to Leibniz in grounding the reality of physical forces: conatus, impetus and *vis viva*? Elsewhere, I have argued that Leibniz does have a story to tell here, though it is one that takes us deep into the arcana of his idealism.[46]

A more pressing problem concerns Leibniz's characterization of derivative force as something "momentary [*momentaneum*]." This term strongly suggests an attempt to predicate temporal properties of derivative force: in itself, derivative force exists only for a moment. Yet this seems a disastrous route for Leibniz to take. We have seen that he rejects any attempt to quantify duration or distance in terms of infinitely small magnitudes. So, what precisely could he mean in claiming that derivative forces are "momentary"?

There are two ways in which Leibniz might respond to this question, neither of which requires the ascription of a temporal duration to derivative forces. The first involves the (admittedly counterintuitive) idea that, as used in this context, the term *momentaneum* carries no temporal connotation at all. Instead, the term is to be understood in a sense related to the technical notion of a "moment": a tendency to produce motion about a point or

45 "[...] the clearest idea of active power comes to us from the mind. So active power occurs only in things that are analogous to minds, that is, in entelechies" (*Nouveaux Essais*, II.xxi.4 (NE 172)). See also *Specimen dynamicum*, Part I: "primitive force (which is nothing but the first entelechy) corresponds to the *soul or substantial form*" (GM VI, 236/AG 119).

46 See Rutherford, 2004, 223–26. For Leibniz's defense of this strategy, see his letter to De Volder of June 30, 1704 (GP II, 270–71/L 537–38), and the discussion in Adams, 1994, 378–86.

axis.[47] To say that derivative force is *momentaneum*, then, would be to say merely that it is, or possesses, a tendency toward change, or the realization of a new state of a substance.

While this reading highlights an essential property of force – its nature as an inherent tendency – it arguably falls short of an adequate explanation of Leibniz's use of the term "momentary." One piece of evidence for this is that in his dynamical writings Leibniz establishes a mathematical relation between a body's momentary force or power and the temporal expression of that force – what he calls "action," defined as the product of power and time. In a 1713 letter to Hermann, he writes: "At potentia mihi per tempus extenditur, quia ipsa per se, meo sensu, tempus non involvit, sed est momentaneum quiddam, quod quovis momento replicatur, seu ducitur in tempus. Et ita prodit actio data."[48] (GM IV, 389).[49] At the very least, this passage affirms that momentary forces can be ascribed a temporal position – they exist before or after other momentary forces – and that in physics their summation over time is a measure of a body's action. Nevertheless, it is notable that Leibniz is careful not to assign a temporal dimension to power itself ("ipsa per se [...] tempus non involvit"). The tendency to change that is a body's power does not last for any length of time, finite or infinitesimal. Hence, power is not to be construed as an element, or temporally minimal part, of action.

In describing derivative force as "something momentary," I believe, Leibniz is best read as meaning not a tendency that exists *for* a moment, but a tendency that exists *at* a moment.[50] The temptation to think of derivative

---

47 The source of this concept is Archimedes, whose law of the lever states: Equal weights at equal distances are in equilibrium, and equal weights at unequal distances are not in equilibrium but incline towards the weight which is at the greater distance (Archimedes, *On the Equilibrium of Planes*, book I, postulate 1). The moment is defined as the product of the applied force and the distance from the point of its application to the rotational axis. Thus, for equal forces, a longer arm will produce a greater moment, or tendency to motion.

48 "But power for me is extended through time, since, in my sense, in and of itself it does not involve time, but is something momentary, which is replicated at any moment or is prolonged in time. And in this way it produces a given action."

49 See also his letter to the same correspondent of September 9, 1712: "the notion of *power* is such that, multiplied by the time in which it is exercised, it produces action; that is, power is that whose temporal exercise is action, for power cannot be known except from action" (GM IV, 379).

50 Though hardly decisive, Leibniz frames his view in this way in Part II of the *Specimen*: "since only force and the nisus arising from it exist at any moment [*quovis momento existat*] [...]" (GM VI, 252/AG 135).

forces as existing for a moment stems from a basic confusion about their nature. We think of derivative forces in this way, because we imagine them as discrete existences that can be conceived independently of the primitive active force of substance. We picture them in the way that Leibniz describes dead force in the *Specimen dynamicum*, as separate moments of endeavor, whose effects are infinitesimal increments of speed or *vis viva*. This, however, is an error. From the perspective of metaphysics, derivative forces are nothing more than the primitive force of substance, *conceived as determined in some particular way*. Recall that, on Leibniz's account, a substance's power is not merely a potential or capacity for acting; it is a fully determinate power, or entelechy, that is spontaneously exercised in action. Consequently, in designating a substance's derivative force, we are not referring to any entity over and above the substance itself; we are referring simply to some *way* or *mode* in which the substance exists – a mode in which it exhibits such-and-such tendency to change.[51]

The idea that derivative force is to be construed as a substance's tendency to change *at* a moment might suggest a general strategy for the interpretation of time-dependent physical properties. Contrary to Leibniz's assertion in the *Specimen dynamicum*, a body's speed should be conceived of not as an infinitesimal distance traveled in an instant, but as a tendency to change position *at* an instant. Likewise, a body's acceleration should be understood not as an infinitesimal increment of speed (or velocity), but as a tendency to change speed (or velocity) *at* an instant. Supporting this proposal is the fact that the modern interpretation of the calculus follows Leibniz in dispensing with infinitesimal differences, and replaces them with the

51 "*Primitive powers* [Les *Puissances primitives*] constitute the substances themselves, and *derivative* powers, or faculties, if you like, are only *ways of being* [*façons d'estres*], which must be derived from substances" (*Nouveaux Essais*, IV.iii.6 (A VI, 6, 379/NE 379)). "Derivative force is itself the present state when it tends toward or preinvolves a following state, as every present is pregnant with the future. But that which persists, insofar as it involves all cases, contains primitive force, so that primitive force is, as it were [velut], the law of the series, while derivative force is, as it were, a determination which designates some term in the series" (GP II, 262/L 533). Leibniz's terminology is not always consistent. In a later letter to the same correspondent (Burcher de Volder), he reserves the term "derivative force" for phenomenal physical forces, while uniting the substantial power and its tendencies under the heading of "primitive force": "I relegate derivative forces to the phenomena, but I think that it is obvious that primitive forces can be nothing but the internal tendencies [tendentias] of simple substances, tendencies by means of which they pass from perception to perception in accordance with a certain law of their nature" (GP II, 275/AG 181).

notion of a derivative: a continuous function that delivers the value of a time-dependent variable at a moment.[52]

Such an approach would have some affinity with Leibniz's account of derivative force; however, it would not succeed in establishing kinematic properties as real in Leibniz's sense. The reason for this, again, is his view of the ontology of space and time. If space and time are merely ideal, then so is any continuous function of spatial or temporal variables. Thus, while properties such as velocity and acceleration may be both mathematically and physically well defined, they do not meet the strictures of Leibniz's metaphysics; they do not pick out real properties of substance. According to Leibniz, derivative force *can* be understood in a way that is consistent with its status as a property of substance. Yet this is possible only insofar as derivative force is not represented mathematically in terms of spatial and temporal parameters, but is expressed in a properly metaphysical vocabulary.

In attempting to formulate such a vocabulary, Leibniz associates derivative force with the *states* of a substance. In studies from the 1680s, he defines "status [state]" as "rei praedicatum mutabile"[53] (A VI, 4, 633), and "mutatio [change]" as "complexum duorum statuum contradictoriorum sibi immediatorum"[54] (A VI 4, 869). Any change thus marks a transition from one state of a substance to another. Leibniz provides few details about how states are to be distinguished, but the clues he gives point to facts about their content as perceptual (or other mental) states.[55] Given this, we can surmise that there is a broad latitude in how we may pick out the states of a

52 For recent attempts to defend the reality of instantaneous velocity along these lines, see Smith, 2003; Lange, 2005. Lange, in particular, argues that instantaneous velocity is best understood as a tendency.

53 "[...] a changeable predicate of a thing."

54 "[...] an aggregate of two immediate, mutually contradictory states." Other variants of this definition: "Change is an aggregate of two contradictory states. But these states are necessarily understood to be immediate with respect to each other, since contradictories admit of no third thing" (A VI, 4, 556). "Change is an aggregate of two opposed states in one stretch of time, with no existing moment of change, as I demonstrated in a certain dialogue" (A VI, 4, 307). The dialogue to which Leibniz refers is the 1676 *Pacidius Philalethi*, which is one of the most important texts on this topic. For a detailed discussion, see Levey, 2003. See also A VI, 4, 563, 569, and 869, which expand the account to include a definition of temporal order, based on the causal connection of states.

55 See, e.g., the 1679 *De affectibus*, where definitions of "mutatio" and "determinatio" appear in close proximity to definitions of mental states: "*Cogitatio* est status mentis qui conscientiae causa proxima est [...]. *Sententia* est cogitatio ex qua sequitur conatus agendi ad externa" (A VI, 4, 1411).

substance; in the case of a monad, we may do so by referring to any change in its perceptual contents. The crucial point is the link Leibniz establishes between a substance's states and its derivative force. Having picked out any pair of proximal, mutually incompatible states, we have thereby designated a moment of change, which is explained in terms of one state's inherent tendency to give way to another. In Leibniz's technical vocabulary, the prior state is a "determinatio," which is defined as "status ex quo quid sequitur nisi quid aliud impediat"[56] (A VI, 4, 1426); and its ability to fulfill this causal role is ascribed to its inherent force or endeavor. Drawing together the strands of his account, Leibniz writes: "Porro ipsa Transitio, seu variatio, [...] nihil aliud est, quam complexus duorum statuum sibi oppositorum et immediatorum una cum vi seu transitus ratione, quae est ipsa qualitas"[57] (C, 9).[58]

With this account, according to which any state of a substance is endowed with an inherent tendency to change, we are again returned to the notion of dead force, now relocated to the internal dynamics of a substance. The primitive force of any substance is manifested in a continual succession of tendencies. This is evidenced both in a body's curvilinear

---

56 "[...] a state from which something follows unless something else impedes it."

57 "Moreover, the change, or variation, itself [...] is nothing but a complex of two states which are immediate and opposed to one another, together with a force or ground for the change, which itself is a quality."

58 As I read Leibniz, there is no canonical description of a substance's states or the changes that occur in them. Any state description involves the attribution of a determinate predicate to the substance, but states may be designated in ways that are more or less finely grained. A coarse-grained description may involve the attribution of successive conscious states to a substance, but underlying these states there are changes in unconscious states, which would be registered by a more finely grained analysis. The doctrine of *petites perceptions*, moreover, suggests that there is no limit to how fine-grained the analysis might become. At each stage in the analysis, we would pick out salient differences that mark a change in a substance's states, but at no point would we arrive at a smallest difference that revealed the change to be essentially discontinuous (cf. *Nouveaux Essais*, Préface (A VI, 6, 56–7)). Does this mean, then, that the activity of a substance is to be understood as continuous, in the way that we think of physical processes (e.g., increases of speed or temperature) as continuous? There are problems with saying this, since as we have seen Leibniz is adamant that spatial and temporal continua are (merely) ideal. Nevertheless, he seems committed to saying that the activity of the substance itself (primitive active force) is expressed continuously, and that the derivative forces that indicate a substance's tendency to change at a moment are abstractions from that activity. Such forces designate the substance's activity in a way that is parasitic on the ascription of discrete states to it, when in point of fact such states do not exist (or exist only relative to a certain mode of conceiving). Arthur reaches a similar conclusion about "the real continuity of substantial activity" in his Introduction to DLC, lxxxvii.

motion and in a soul's series of perceptual states.[59] However, it is Leibniz's view, expressed with growing confidence in his later writings, that the internal dynamics of the soul offer a better vantage point from which to grasp the reality of force than its external manifestations in bodily motion.[60] From this perspective we are able to recognize that derivative force cannot be separated, or even sharply distinguished, from the primitive force of substance. Primitive force is neither resolvable into, nor composed from, its successive tendencies. Rather, the tendencies are what it is to be that endeavor at some moment, defined in terms of proximal pairs of mutually exclusive states. Getting clear on the exact relationship between primitive and derivative force remains a task for Leibniz scholarship. I do not claim to have fully resolved that problem, but only to have drawn attention to the inadequate (from the point of view of metaphysics) conceptions of force and tendency that inform Leibniz's science of dynamics, particularly when those conceptions are tethered to the mathematics of the infinitesimal calculus.[61]

59 See Leibniz's reply to the second edition of Bayle's *Dictionary*: "The state of the soul, like that of the atom, is a state of change, a tendency. The atom tends to change its place, the soul to change its thoughts; each changes by itself in the simplest and most uniform way which its state permits" (GP IV, 562/L 579). Similar ideas are expressed at A VI, 4, 1426; GP II, 172/AG 172–73.

60 *Nouveaux Essais*, II.xxi.72 (A VI, 6, 210–11). From the perspective of the soul, a substance's endeavor is teleologically structured: it is a striving *for* the apparent good. See *Nouveaux Essais*, II.xxi.5: "volition is the effort or endeavor [conatus] to move towards what one finds good and away from what one finds bad, the endeavor arising immediately out of one's awareness of those things [...]. There are other efforts, arising from insensible perceptions, which we are not aware of; I prefer to call these 'appetitions' rather than volitions" (A VI, 6, 172–73/NE 172–73). I discuss this point in greater detail in Rutherford, 2005.

61 I am grateful to Dan Garber, John Whipple, and the participants in my 2006 Leibniz seminar at UCSD for discussion of some of the issues examined in this essay. Thanks are owed also to Ursula Goldenbaum and Doug Jesseph for helpful comments on the penultimate draft.

Daniel Garber

Dead Force, Infinitesimals, and the Mathematicization of Nature

The distinction between living and dead force is central to Leibniz's dynamics. As Leibniz understands the concept, dead force is the force that is associated with statics and gravitation, the kind of force exerted by a taut spring or an apple hanging on a tree. Leibniz contrasts dead force with *vis viva* or living force, which is associated with bodies that are actually in motion, the spring sprung or the falling apple. In nature, Leibniz asserts, living force and the real, finite motion it is associated with arises from an infinity of infinitesimal impressions of dead force. Now, dead force seems to be a very real part of nature for Leibniz. As such, dead force seems to be a real instantiation of the infinitesimal in nature. But, at the same time, Leibniz also seems to be quite skeptical indeed about the real existence of infinitesimals in mathematics. How can this be? How can Leibniz accept infinitesimal magnitudes in the physical world at the same time as he rejects infinitesimals in mathematics?

This is the tension that I would like to explore in this paper. I will begin by setting out the distinction between living and dead force as Leibniz gives it in one of his most careful expositions. This will allow us to see more clearly the tension that there is in Leibniz's views about force and infinitesimal magnitudes. But to resolve the tension, we must take something of a detour into Leibniz's views about mathematics, its relation to force in general and to the natural world, and the role that mathematics plays in the science of physics. This will allow us to see, in the end, just how dead force constitutes an infinitesimal magnitude for Leibniz, and, more importantly, how it doesn't.

1. Living and Dead Force: Some Key Texts

The notions of living and dead force appear in numerous of Leibniz's texts from 1673 on. But the distinction becomes particularly important in the late 1680s, when Leibniz is formulating his dynamics. A key text here is the *Specimen dynamicum* (SD) of 1695.

Leibniz's discussion of this distinction occurs in the context of a rather elaborate discussion of different varieties of force. At the most basic level, there are two related pairs of distinctions: primitive vs. derivative force, and active vs. passive force. So, in all, there are four principal varieties of force, primitive active and passive force, and derivative active and passive force.

Let us begin with the distinction between primitive and derivative forces. In the SD, Leibniz characterizes the primitive active force as corresponding to "the soul or substantial form;" the primitive passive force, on the other hand, is characterized as constituting "that which is called *primary matter* in the schools, if correctly interpreted." Form and matter are, of course, terms of art from the Aristotelian account of substance; form and matter join together to constitute a substance for Aristotle and his followers. And so for Leibniz as well. Leibniz writes in an essay he dated May 1702, written at the same time as he was attempting to explain his views to the Cartesian de Volder, and perhaps connected with that exchange:

> Vis activa primitiva quae Aristoteli dicitur ἐωτελέχεια ἡ πρώτη, vulgo forma substantiae, est alterum naturale principium quod cum materia seu vi passiva [primitiva] substantiam corpoream absolvit, quae scilicet unum per se est, non nudum aggregatum plurium substantiarum, multum enim interest verbi gratia inter animal et gregem.[1] (May 1702, GP IV, 395)

And so, it seems, the primitive forces, active and passive, come together to make up the corporeal substance, the genuine unity that, Leibniz claims, underlies the extended bodies of physics. In this way, the primitive forces are the constituents of substance that underly the derivative forces, those most of interest to the physicist. Leibniz writes in the SD:

> Vim ergo derivativam, qua scilicet corpora actu in se invicem agunt aut a se invicem patiuntur, [...] non aliam intelligimus, quam quae motui (locali scilicet)

---

1 "Primitive active force, which Aristotle calls first entelechy and one commonly calls the form of a substance, is another natural principle which, together with matter or [primitive] passive force, completes a corporeal substance. This substance, of course, is one *per se*, and not a mere aggregate of many substances, for there is a great difference between an animal, for example, and a flock." (On Body and Force, AG 252)

> cohaeret, et vicissim ad motum localem porro producendum tendit. Nam per motum localem caetera phaenomena materialia explicari posse agnoscimus.[2] (GM VI, 237)

Derivative force is, furthermore, that in terms of which we can frame the laws of physics. Leibniz writes, again in the SD: "[...] his [i.e. viribus derivativis] enim accommodantur leges actionum, quae non ratione tantum intelliguntur, sed et sensu ipso per phaenomena comprobantur."[3] (GM VI, 237) Leibniz uses a number of terms to describe the relation between primitive and derivative forces. In the SD he talks of derivative force as "primitivae velut limitatione, per corporum inter se conflictus resultans" (GM VI, 236).[4] Similarly, he writes to Johann Bernoulli on December 17, 1698:

> [...] si Animam vel Formam concipiamus, ut primam activitatem, cujus modificatione oriantur vires secundae [i.e. derivativae], ut extensionis modificatione oriuntur figurae, puto nos intellectui sic satis consulere. Nempe, ejus, quod essentia sua mere passivum est, nullae possunt esse modificationes activae, quoniam modificationes limitant magis, quam augent vel addunt [...].[5] (GM III, 552)

These passages suggest that derivative forces are to be understood as modes, accidents or the like, modifications of the primitive forces, which are understood as constituents of corporeal substances. Primitive active and passive forces, then, are the substantial ground of the derivative active and passive forces, which are their accidents or modes, as shape is an accident or mode of an extended thing. While these forces are connected with motion, it is very important not to confuse these forces with motions them-

[2] "Therefore, by derivative force, namely, that by which bodies actually act on one another or are acted upon by one another, I understand [...] only that which is connected to motion (local motion, of course), and which, in turn, tends further to produce local motion. For we acknowledge that all other material phenomena can be explained by local motion." (AG 120)

[3] "It is to these notions [i.e., the derivative forces] that the laws of action apply, laws which are understood not only through reason, but are also corroborated by sense itself through the phenomena." (AG 120)

[4] "[...] resulting from a limitation of primitive force through the collision of bodies with one another." (AG 119)

[5] "If we conceive of soul or form as the primary activity from whose modification secondary [i.e. derivative] forces arise as shapes arise from the modification of extension, then, I think, we take sufficient account of the intellect. Indeed there can be no active modifications of that which is merely passive in its essence, because modifications limit rather than increase or add."(AG 169)

selves. Leibniz puts it carefully in the SD: derivative force "motui [...] *cohaeret*, et vicissim ad motum localem porro *producendum* tendit."[6] (GM VI, 237 – My emphasis, DG) (On this, more later.)

Within the category of derivative force, Leibniz distinguishes between derivative active force and derivative passive force. Derivative passive forces are two: impenetrability, that by virtue of which one body cannot penetrate another, and resistance, that by virtue of which one body resists the acquisition of motion from another. (Again, we shall return to the derivative passive forces later in this essay.) And now, finally, within the category of active force, Leibniz distinguishes between living and dead forces. Since this is central to the theme of this essay, it is worth looking at Leibniz's account with some care.

Leibniz's discussion of this distinction in the SD is rather elaborate. (For ease of exposition I am dividing the passage and labeling its two parts 'A' and 'B.')

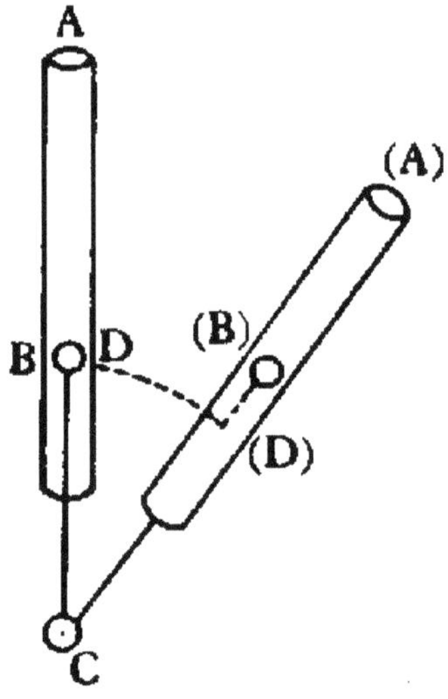

Figure 1.

He begins the exposition that will lead to the distinction between living and dead force with some distinctions one might make among various notions connected with motion:

> [A] Finge tubum AC [figure 1] in plano horizontali hujus paginae certa quadam uniformi celeritate rotari circa centrum C immotum, et globum B in tubi cavitate

6 "[...] is *connected with* motion [...] and [...] in turn, tends further to *produce* local motion."(AG 120)

existentem liberari vinculo vel impedimento atque incipere moveri vi centrifuga; manifestum est, initio conatum a centro recedendi, quo scilicet globus B in tubo tendet versus ejus extremitatem A, esse infinite parvum respectu impetus quem jam tum habet a rotatione seu quo cum tubo ipso globus B a loco D tendet versus (D) retenta a centro distantia. Sed continuata aliquamdiu impressione centrifuga a rotatione procedente, progressu ipso oportet nasci in globo impetum quendam centrifugum completum (D)(B) comparabilem cum impetu rotationis D(D). Hinc patet duplicem esse *Nisum*, nempe elementarem seu infinite parvum, quem et *solicitationem* appello, et formatum continuatione seu repetitione Nisuum elementarium, id est impetum ipsum [...].[7] (GM VI, 238)

For the moment I will skip over a sentence at the end of this paragraph, and continue on to the next paragraph:

[B] Hinc *Vis* quoque duplex: alia elementaris, quam et *mortuam* appello, quia in ea nondum existit motus, sed tantum solicitatio ad motum, qualis est globi in tubo, aut lapidis in funda, etiam dum adhuc vinculo tenetur; alia vero vis ordinaria est, cum motu actuali conjuncta, quam voco *vivam*. Et vis mortuae quidem exemplum est ipsa vis centrifuga, itemque vis gravitatis seu centripeta, vis etiam qua Elastrum tensum se restituere incipit. Sed in percussione, quae nascitur a gravi jam aliquamdiu cadente, aut ab arcu se aliquamdiu restituente, aut a simili causa vis est viva, ex infinitis vis mortuae impressionibus continuatis nata.[8] (GM VI, 238)

---

[7] "Consider tube AC rotating around the immobile center C on the horizontal plane of this page with a certain uniform speed, and consider ball B in the interior of the tube, just freed from a rope or some other hindrance, and beginning to move by virtue of centrifugal force. [See fig. 1] It is obvious that, in the beginning, the conatus for receding from the center, namely, that by virtue of which the ball B in the tube tends toward the end of the tube, A, is infinitely small in comparison with the impetus which it already has from rotation, that is, it is infinitely small in comparison with the impetus by virtue of which the ball B, together with the tube itself, tends to go from place D to (D), while maintaining the same distance from the center. But if the centrifugal impression deriving from the rotation were continued for some time, then by virtue of that very circumstance, a certain complete centrifugal impetus (D) (B), comparable to the rotational impetus D (D), must arise in the ball. From this it is obvious that the nisus is twofold, that is, elementary or infinitely small, which I also call solicitation, and that which is formed from the continuation or repetition of elementary nisus, that is, impetus itself." (AG 121)

[8] "From this it follows that force is also twofold. One force is elementary, which I also call dead force, since motion [*motus*] does not yet exist in it, but only a solicitation to motion [*motus*], as with the ball in the tube, or a stone in a sling while it is still being held in by a rope. The other force is ordinary force, joined with actual motion, which I call living force. An example of dead force is centrifugal force itself, and also the force of heaviness [*vis gravitatis*] or centripetal force, and the force by which a stretched elastic body begins to restore itself. But when we are dealing with impact, which arises from a heavy body which has already been falling for some time, or from a bow that has already been restoring its

In passage A Leibniz offers an analysis of various related notions of motion that go into understanding the way a ball behaves in a rotating tube. He distinguishes between the *impetus* a ball in motion around a center has from the *conatus for receding* it has by virtue of the centrifugal motion of the ball down the tube, which, in turn, results in a *centrifugal impetus* that the ball develops as it actually begins to move down the tube. Leibniz concludes passage A with a distinction between two kinds of *nisus*, an elementary *nisus*, and that which results from the repetition of the elementary *nisus*, which is impetus. In passage B, Leibniz then moves from a discussion of these different conceptions of *motion* to a discussion of the *forces* underlying the motion, presumably the forces that *cause* the motion. ("Hinc Vis quoque duplex [...].") Here is where the distinction between living and dead force enters. The dead force is the force that gives rise to the elementary *conatus* for receding. This is an infinitesimal whose infinite repetition gives rise to the living force that is associated with actual motion, the impetus itself: dead force is an infinitesimal quantity whose integral is living force; it is parallel to the infinitesimal *nisus* (an infinitesimal motion) whose integral is the actual motion of the body (Cf. GP II, 154; GM VI, 451–53).

The distinction between living and dead force is, in essence, the distinction between the force exerted by a dead weight and that exerted by a body in motion. Worrying about the relation between the two goes back long before Leibniz. Galileo, for example, raises the following problem in the *Discorsi*:

> Però, figurandoci, per esempio, uno di quei gran pesi che per ficcare grossi pali nel terreno si lasciano cadere da qualche altezza sopra uno de' detti pali (i quali pesi mi pare che gli addimandino berte), ponghiamo, v. g., il peso di una tal berta esser 100 libbre, l'altezza dalla quale cade essere quattro braccia, e la fitta del palo nel terreno duro, fatta per una sola percossa, importare 4 dita: e posto che la medisima pressura e fitta delle 4 dita, volendola noi far senza percossa, ricercasse che le fusse soprapposto un peso di mille libbre, il quale, operando colla sola gravità, senza moto precedente, chianleremo peso morto [...].[9] (Galilei, 1890–1909, VIII, 325)

shape for some time, or from a similar cause, the force in question is living force, which arises from an infinity of continual impressions of dead force." (AG 121–2)

9 "Imagine, for instance, one of those great weights (which I believe are called pile drivers) that are used to drive stout poles into the ground by allowing them to fall from some height onto such poles. Let us put the weight of such a pile driver at 100 pounds, and let the height from which this falls be four braccia, while the entrance of the pole into hard ground, when driven by a single such impact, shall be four inches. Next, suppose that we want to achieve the same pressure and entrance of four inches without using impact, and

It is this kind of distinction that Leibniz seems to have had in mind in the earliest text we have where Leibniz makes the distinction between living and dead force. The text in question is a letter Leibniz wrote to Mariotte in July 1673, discussing a problem that mixed static and dynamic elements. (Cf. A III, 1, 105–112) Like Galileo, Leibniz wants to compare the force exerted by a dead weight with that exerted by a moving body.

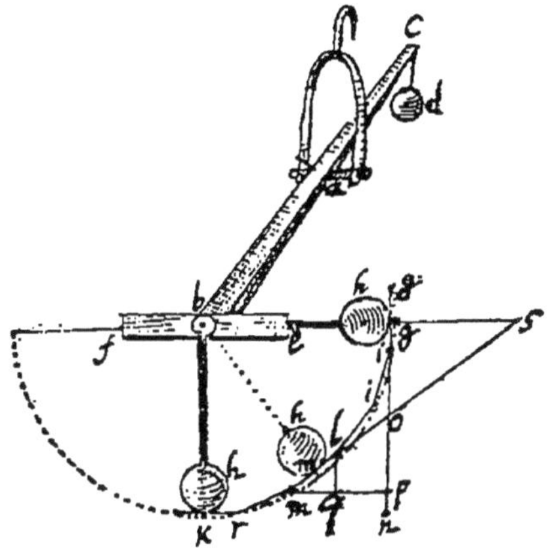

Figure 2.

In particular, Leibniz wants to compare the weight of a body *d* hanging off one end of a balance beam with the force exerted by a pendulum *h* swinging at the other end of the same beam (See fig. 2). In the course of the discussion, Leibniz distinguishes between "force morte" and "la force violente ou animée d'un choc."[10] (A III, 1, 107) Body *d* is always said to exert a "force morte."

There is no particular worry about the reality of these forces in Galileo or in this early text of Leibniz's: living force and dead force are just evident pieces of furniture in the physical world. Force later becomes central to Leibniz's metaphysical picture as well, as he begins to reconfigure the Cartesian world of material substances in terms of force. As he wrote in the important essay *De modo distinguendi phaenomena realia ab imaginariis* (1683–6?):

we find that this can be done by a weight of 1000 pounds, which, operating by its heaviness alone, without any preceding motion, we may call 'dead weight.'" (Galilei, 1974, 285–86)

10 "Dead force"; "violent force or force activated by a collision."

De corporibus demonstrare possum non tantum lucem, calorem, colorem et similes qualitates esse apparentes, sed et motum, et figuram, et extensionem. Et si quid est reale, id solum esse vim agendi et patiendi adeoque in hoc (tanquam materia et forma) substantiam corporis consistere [...].[11] (A VI, 4, 1504)

In this connection, there is a text in which Leibniz seems to give special prominence to dead force. In the essay *De primae philosophiae emendatione*, published in the *Acta eruditorum* in 1694, a year before the SD, Leibniz wrote:

Differt enim vis activa a potentia nuda vulgo scholis cognita, quod potentia activa Scholasticorum, seu facultas, nihil aliud est quam propinqua agendi possibilitas, quae tamen aliena excitatione et velut stimulo indiget, ut in actum transferatur. Sed vis activa actum quendam sive ἐντελέχειαν continet, atque inter facultatem agendi actionemque ipsam media est, et conatum involvit; atque ita per se ipsam in operationem fertur; nec auxiliis indiget, sed sola sublatione impedimenti. Quod exemplis gravis suspensi funem sustinentem intendentis, aut arcus tensi illustrari potest. Etsi enim gravitas aut vis elastica mechanice explicari possint debeantque ex aetheris motu, ultima tamen ratio motus in materia est vis in creatione impressa, quae in unoquoque corpore inest, sed ipso conflictu corporum varie in natura limitatur et coërcetur.[12] (GP IV, 469–70)

The essay (and this passage of the essay) are about the notion of force in general as important for the foundations of metaphysics and for our concept of substance. But the examples that Leibniz uses in this passage strongly suggest that dead force has an especially important role to play in this connection.

The reality of the *distinction* between living and dead force is underscored in the role that it plays in explaining the mistake that the Cartesians

---

11 "Concerning bodies I can demonstrate that not merely light, heat, color and similar qualities are apparent but also motion, figure, and extension. And that if anything is real, it is solely the force of acting and suffering, and hence that the substance of a body consists in this (as if in matter and form)." (L 365)

12 "Active force differs from the mere power familiar to the Schools, for the active power of faculty of the Scholastics is nothing but a close possibility of acting, which needs an external excitation or a stimulus, as it were, to be transferred into action. Active force, in contrast, contains a certain act or entelechy and is thus midway between the faculty of acting and the act itself and involves a conatus. It is thus carried into action by itself and needs no help but only the removal of an impediment. This can be illustrated by the example of a heavy hanging body which strains at the rope which holds it or by a bent bow. For though gravity and elasticity can and ought to be explained mechanically by the motion of the ether, the ultimate reason for motion in matter is nevertheless the force impressed upon it in creation, which inheres in every body but is variously limited and restrained in nature through the impact of bodies upon each other." (L 433)

make in holding to the conservation of quantity of motion. In the SD, for example, Leibniz traces the mistaken belief in the conservation of quantity of motion to the fact that until Leibniz came along and enlightened the world, people only knew about dead force. He writes:

> Veteres, quantum constat, solius vis mortuae scientiam habuerunt, eaque est, quae vulgo dicitur Mechanica, agens de vecte, trochlea, plano inlinato (quo cuneus et cochlea pertinent), aequilibrio liquorum, et similibus, ubi nonnisi de conatu primo corporum in se invicem tractatur, antequam impetum agendo conceperunt. Et licet leges vis mortuae ad vivam transferri aliquo modo possint, magna tamen cautione opus est, ut vel hinc decepti sint, qui vim in universum cum quantitate ex ductu molis in velocitatem facta confuderunt, quod vim mortuam in ratione horum composita esse deprehendissent. Nam ea res ibi speciali ratione contingit, ut jam olim admonuimus [...].[13] (GM VI, 239)

Insofar as dead and living force have very different causal effects, one would suppose that Leibniz thought them to be physically quite distinct from one another.

Now, the physical and metaphysical reality of force in general, and dead force in particular seems evident: these are important constituents of Leibniz's world. But Leibniz wants to build a mathematical physics. That is to say, he wants to subject these physical magnitudes to mathematics. It is not surprising that when he does so, he makes use of notions from his calculus. But, of course, this takes us directly to the central problem that I would like to address in this essay. As Leibniz presents it in the SD, dead force would appear to be a real instantiation of an infinitesimal quantity, an infinitesimal magnitude that really exists in nature. But, of course, Leibniz is not inclined to take a realistic view of infinitesimal magnitudes. Is the reality of dead force consistent with the very skeptical attitude that he takes to the reality of infinitesimal magnitudes? Now, one solution would be simply to deny the premise of this problem: one might hold that mathematics cannot rep-

---

13 "So far as one can establish, the ancients had knowledge only of dead force, and this is what is commonly called mechanics, which deals with the lever, the pulley, the inclined plane (where accounts of the wedge and the screw belong), the equilibrium of bodies, and the like. There we treat only the first conatus of bodies acting on one another, before those bodies have received impetus through acting. And although one might, in a certain way, be able to transpose the laws of dead force over into living force, great caution is needed; those who confused force in general with the product of bulk [*moles*] and velocity because they discovered that dead force is proportional to that product were misled in just such a way. For, as we once warned, this fact holds in this case for a special reason."(AG 122) Cf. also GM VI, 218–9, 267, 397.

resent force at all. But before accepting such an apparently radical solution to the question, I would like to ask, more generally, how Leibniz conceives of the relation between mathematics and physical reality.

2. Mathematics and the Physical World: The Case of Extension

I would like to begin not with the question of mathematics as it relates to living and dead force, but with extension and the physical world. As we shall see, force will enter here too, though it is passive force, not active force. Let me begin by returning to the passage quoted above from the *De modo distinguendi*:

> De corporibus demonstrare possum non tantum lucem, calorem, colorem et similes qualitates esse apparentes, sed et motum, et figuram, et extensionem. Et si quid est reale, id solum esse vim agendi et patiendi adeoque in hoc (tanquam materia et forma) substantiam corporis consistere [...].[14] (A VI, 4, 1504)

Leibniz's point here is that what is real in bodies is not their geometrical properties, but their forces, active and passive. As Leibniz conceives of it, the extension of bodies arises from the passive force bodies have. In an important passage appended to a letter to Arnauld from September 1687, Leibniz argues that insofar as matter is just passive force, "en ce sens la matiere ne seroit point étendue ny divisible, bien qu'elle seroit le principe de la divisibilité ou de ce qui en revient à la substance."[15] (GP II, 120) Similarly, in another passage dated at 1685 (?), Leibniz writes: "Materia est vis patiendi seu resistendi in quocunque corpore, ex qua sequitur extensio certa corporis, nisi Autor rerum secus velit."[16] (A VI, 4, 2326) The passage is primarily concerned with the problem of the Eucharist. In this context, the fact that God can create passive force without thereby creating something extended

[14] "Concerning bodies I can demonstrate that not merely light, heat, color and similar qualities are apparent but also motion, figure, and extension. And that if anything is real, it is solely the force of acting and suffering, and hence that the substance of a body consists in this (as if in matter and form)."(L 365)

[15] "[...] in this sense matter would not be extended or divisible, although it would be the principle of divisibility or of that which amounts to it in the substance." This is generally thought to be a later addition to the letter.

[16] "Matter is the force of being acted upon or of resisting in any body whatsoever, from which follows a certain extension in body, unless the Author of things desires otherwise."

is something that can only happen by miracle.[17] His larger point is simply that extension is not basic to body, but derives from the passive force in body, in particular, one would suppose, from impenetrability by virtue of which one body excludes other bodies from occupying the same place.

Leibniz offers a number of other arguments in which he distances extension from body. In a famous passage from the *Discours de métaphysique* (DM) he writes:

> *Que les notions qui consistent dans l'étendue enferment quelque chose d'imaginaire et ne sçauroient constituer la substance du corps* [...] je croy que celuy, qui meditera sur la nature de la substance, que j'ay expliquée cy dessus trouvera, ou que les corps ne sont pas des substances dans la rigueur metaphysique (ce qui estoit en effet le sentiment des Platoniciens) ou que toute la nature du corps ne consiste pas seulement dans l'étendue, c'est à dire dans la grandeur, figure et mouvement, mais qu'il faut necessairement y reconnoistre quelque chose, qui aye du rapport aux ames, et qu'on appelle communement forme substantielle, bien qu'elle ne change rien dans les phenomenes, non plus que l'ame des bestes, si elles en ont. On peut même demonstrer que la notion de la grandeur de la figure et du mouvement n'est pas si distincte qu'on s'imagine, et qu'elle enferme quelque chose d'imaginaire, et de relatif à nos perceptions, comme le font encor (quoyque bien d'avantage) la couleur, la chaleur, et autres qualités semblables dont on peut douter si elles se trouvent veritablement dans la nature des choses hors de nous.[18] (DM 12, A VI, 4, 1545)

What may underlie this claim is the observation that in reality, bodies do not and cannot have the geometrical shapes that we attribute to them: in reality, their boundaries are infinitely complex, and cannot be captured by

---

17 Adams, 1994, 349ff., discusses passages like these as part of an argument that Leibniz was an idealist in the period under discussion. However, the fact that extension can be separated from force only supernaturally makes these passages problematic for his case.

18 "*That the Notions Involved in Extension Contain Something Imaginary and Cannot Constitute the Substance of Body.* [...] I believe that anyone who will meditate about the nature of substance, as I have explained it above, will find that the nature of body does not consist merely in extension, that is, in size, shape, and motion, but that we must necessarily recognize in body something related to souls, something we commonly call substantial form, even though it makes no change in the phenomena, any more than do the souls of animals, if they have any. It is even possible to demonstrate that the notions of size, shape, and motion are not as distinct as is imagined and that they contain something imaginary and relative to our perception, as do (though to a greater extent) color, heat, and other similar qualities, qualities about which one can doubt whether they are truly found in the nature of things outside ourselves." (AG 44) There are many other passages in which Leibniz claims that our ideas of extension contain something imaginary. See, e.g., A VI, 4, 1622; DLC 315; A VI, 4, 1465; A VI, 4, 1612–13; etc.

geometry as it was known in Leibniz's day.[19] The ground of this view is a doctrine that the bodies of everyday experience are composed of corporeal substances, which, in turn, are composed of corporeal substances smaller still, bugs in bugs to infinity.[20] Because of that, the surfaces of ordinary objects are of infinite complexity, something not unlike a modern fractal, as Sam Levey has suggested. Thus Leibniz writes in the important *Specimen inventorum* of 1688 (?):

> Et vero quod paradoxum videri possit, sciendum est Extensionis non esse tam liquidam notionem quam vulgo creditur. Nam ex eo quod nullum corpus tam exiguum est, quin in partes diversis motibus incitatas actu sit divisum, sequitur nullam ulli corpori figuram determinatam assignari posse, neque exactam lineam rectam, aut circulum, aut aliam figuram assignabilem cujusquam corporis reperiri in natura rerum, tametsi in ipsa seriei infinitae deviatione regulae quaedam a natura serventur. Itaque figura involvit imaginarium aliquid, neque alio gladio secari possunt nodi quos nobis ex compositione continui male intellecta nectimus.[21] (A VI, 4, 1622)

Similarly, Leibniz writes in the *Primae veritates* of 1689(?):

> *Non datur ulla in rebus actualis figura determinata*, nulla enim infinitis impressionibus satisfacere potest. Itaque nec circulus, nec ellipsis, nec alia datur linea a nobis definibilis nisi intellectu, ut lineae antequam ducantur, aut partes antequam abscindantur.[22] (A VI, 4, 1648)

---

[19] On this question I am drawing on passages cited in Levey, 2003, and Levey, 2005, as well as Sam Levey's insightful analyses.

[20] Leibniz also offers a very strange and difficult argument to the unreality of shape from considerations drawn from the nature of time. This argument seems to be found in a single piece, a short fragment that the Akademie editors date at 1686 (A VI, 4, 1613–14; DLC, 297–99). It is discussed at length in Levey, 2003. I suspect that it was more of a philosophical experiment on Leibniz's part than a position that he seriously considered adopting.

[21] "Indeed, even though this may seem paradoxical, it must be realized that the notion of extension is not as transparent as is commonly believed. For from the fact that no body is so very small that it is not actually divided into parts excited by different motions, it follows that no determinate shape can be assigned to any body, nor is a precisely straight line, or circle, or any other assignable shape of any body found in the nature of things, although certain rules are observed by nature even in its deviation from an infinite series. Thus shape involves something imaginary, and no other sword can sever the knots we tie for ourselves by misunderstanding the composition of the continuum."(DLC 315)

[22] "There is no determinate shape in actual things, for none can be appropriate for an infinite number of impressions. And so neither a circle, nor an ellipse, nor any other line we can define exists except in the intellect, nor do lines exist before they are drawn, nor parts before they are separated off."(AG 34)

The suggestion here is that the geometrical shapes that we attribute to bodies when we perceive them as spheres or cubes or any other geometrical shape, bounded in lines and planes and curved surfaces is just an imposition of the imagination onto a reality that in itself is much more complex. This comes out most explicitly in a passage from a letter Leibniz wrote to the Princess Sophie in 1705:

> C'est nostre imperfection et le defaut de nos sens, qui nous fait concevoir les choses physiques comme des Estres Mathematiques [...]. Et l'on peut demonstrer qu'il n'y a point de ligne ou de figure dans la nature, qui donne exactement et garde uniformement par le moindre espace et temps les proprietés de la ligne droite ou circulaire, ou de quelque autre dont un esprit fini peut comprendre la definition [...]. Cependant les verités eternelles fondées sur les idées mathematiques bornées ne laissent pas de nous servir dans la practique, autant qu'il est permis de faire abstraction des inegalités trop petites pour pouvoir causer des erreurs considerables par rapport au but qu'on se propose [...].[23] (GP VII, 563–4)

In this way one may say that the extensionality of bodies is, in a way, phenomenal, the result of our imperfect senses which impose geometrical concepts onto bodies which are, in their real nature, quite something different and which don't fit them exactly.[24]

The question of the relation between geometry and the physical world is directly addressed in another important document. In response to Leib-

---

23 "It is our imperfection and the defects of our senses which makes us conceive of physical things as mathematical entities [...]. And one can demonstrate that there is no line or shape in nature that has the properties of a straight or circular line or of any other thing whose definition a finite mind can comprehend, or that retains it uniformly for the least time or space. [...] However, the eternal truths grounded on limited mathematical ideas don't fail to be of use to us in practice, to the extent to which it is permissible to abstract from inequalities too small to be able to cause errors that are large in relation to the end at hand [...]." There is a good discussion of this passage in Hartz and Cover, 1988, 501. Although I would claim that Leibniz's metaphysics of body and the ultimate make-up of substance is somewhat different when he wrote this letter than it was earlier in the 1680s and early 1690s, the view expressed in the passage quoted is very much continuous with the earlier period.

24 Some commentators have been tempted to read the no-exact-shape argument as an attempt to establish the claim that the world is made up of non-extended simple substances, and that the extension of bodies is an illusion in a strong sense. See, e.g., Adams, 1994, 229–32 and Sleigh, 1990, 112–14. But I think that it is more plausible to see Leibniz's intention here to point out the difference between what Sellars has called the manifest view of the world, the world as it appears to us, bodies with real geometrical shapes, and the scientific image of the world, bodies of infinite complexity, beyond our power to grasp in sense. See the excellent discussion of their views in Levey, 2005, 84–92.

niz's *Système nouveau* of 1695, Simon Foucher published a brief commentary. One of the issues that Foucher addressed was the question of the composition of the continuum. In response to these comments, Leibniz wrote one of the clearest accounts of the relation between the world of geometrical objects and the real world of bodies (Cf. GP IV, 491–2; AG 146–7). (Unfortunately, he chose not to publish the longer comments, making do with a short summary in the response he published.) In that text, Leibniz draws a clear distinction between the world of mathematical entities (lines, surfaces, numbers), and the world of concrete things. The problem of the composition of the continuum is concerned with the parts from which continua can be constructed. Leibniz's point is that the mathematical continuum does not have such parts, nor does it need them: its parts come from the division of the line, and these parts are not properly elements of that line. However, in real concreta, the whole is indeed composed of parts, though those parts don't make up a genuine mathematical continuum. The problem of the composition of the continuum is thus solved: the objects of geometry, which exist in the realm of the ideal, are continuous, but not composed of parts; the real objects that exist in the physical world are composed of parts, but they are not continuous.

It may look here as if Leibniz is denying that geometry truly represents bodies. But this cannot be right. It is important to remember that his metaphysics is intended to ground a fundamentally mechanistic conception of the physical world. From his earliest years, Leibniz consistently held that everything can be explained through size, shape, and motion, but that this mechanistic conception of the world requires a foundation in something that goes beyond extension and motion. In response to the Cartesians, who want to say that bodies *just are* the objects of geometry made real, Leibniz wants to emphasize the difference between real concrete bodies and the ideal world of geometrical objects. But this is not to say that Leibniz wants to deny extension to bodies altogether. In the course of his comments on Foucher, Leibniz makes the following observation: "Cependant le nombre et la ligne ne sont point des choses *chimeriques*, quoyqu'il n'y ait point de telle composition, car ce sont des rapports qui renferment des verités eternelles, sur lesquelles se reglent les phenomenes de la nature."[25] (GP IV, 491–2) Or, as he wrote to Sophie,

[25] "However, number and line are not *chimerical* things, even though there is no such composition, for they are relations that contain eternal truths, by which the phenomena of nature are ruled." (AG 146–7)

> Cependant les verités eternelles fondées sur les idées mathematiques bornées ne laissent pas de nous servir dans la practique, autant qu'il est permis de faire abstraction des inegalités trop petites pour pouvoir causer des erreurs considerables par rapport au but qu'on se propose [...].[26] (GP VII, 563–4)

The view seems to be that geometrical extension is something ideal that exists outside the world of concrete things. However, concrete things in the world *instantiate* geometrical relations, at least approximately, insofar as real extension is infinitely complex and not genuinely continuous. Real extension is thus both more and less than geometrical extension: more insofar as it is infinitely complex, and less insofar as it is not continuous. But yet geometry is applicable to the world of concreta, a world that in its nature is ultimately characterized in terms of force. That is, there are real forces in the world, which give rise to infinitely complex structures that instantiate geometrical relations, at least approximately. Bodies are extended insofar as geometry is (approximately) true of them. However, in a metaphysical sense, what is really there is force. In this way he says, again in the notes on Foucher: "L'etendue ou l'espace, et les surfaces, lignes et points qu'on y peut concevoir, ne sont que des rapports d'ordre, ou des ordres de coexistence [...]."[27] (GP IV, 491) Geometry in this way can be said to represent something that is really in body, even if it has properties that the concrete body it represents does not, such as continuity: mathematical representation is not identity. Indeed, this is one way of putting Leibniz's point, and this is exactly where Descartes erred, in confusing the mathematical representation of bodies in geometrical terms with their concrete reality.

3. Mathematics and the Physical World: The Case of Living Force

The question of the reality of extension in the physical world is, in essence, the question about the relation between mathematics (geometry) and the primitive passive force in bodies. Leibniz is very careful to acknowledge that even though we must carefully separate the physical, that is passive force, from its mathematical representation, that is, extension, there is a

[26] "[...] However, the eternal truths grounded on limited mathematical ideas don't fail to be of use to us in practice, to the extent to which it is permissible to abstract from inequalities too small to be able to cause errors that are large in relation to the end at hand [...]."

[27] "Extension or space and the surfaces, lines, and points one can conceive in it are only relations of order or orders of coexistence [...]." (AG 146)

sense – and an important sense, insofar as extension central to the mechanist conception of physics – that bodies are extended. Leibniz takes a similar view about the relation between (living) active force – vis viva – and its mathematical representation as mv^2.

This view is what is behind Leibniz's exposition in DM 17 and 18. In DM 17, Leibniz gives an exposition of his famous argument against Descartes' principle of the conservation of quantity of motion in a version very similar to the version given in the *Brevis demonstratio* published in the *Acta eruditorum* in 1686. Since the argument is familiar, and much discussed elsewhere, I will be brief. Descartes had argued that the total quantity of motion is conserved, where that physical magnitude is understood as the size times the (scalar) speed of all the bodies in a system, which can be represented as $m|v|$. What Leibniz wants to establish in this argument is that is quantity is not the same as what he calls "force." One thing that this means to Leibniz is that the mathematical expression '$m|v|$' is not an adequate representation of the amount of "force" a body has. But Leibniz also wants to articulate something deeper, the metaphysical point that "force" is not the same as motion.

Now, force in this context seems to be the ability to produce an effect that is associated with a body in motion. Thus he begins the argument with the assumption that "[...] un corps tombant d'une certaine hauteur acquiert la force d'y remonter, si sa direction le porte ainsi, à moins qu'ils ne se trouvent quelques empechemens".[28] (DM 17, A VI, 4, 1556–57) At the end of the same section he notes: "On voit [...] comment *la force* doit estre estimée par la quantité de l'effect qu'elle peut produire, par exemple par la hauteur, à la quelle un corps pesant d'une certaine grandeur et espece peut estre elevé [...]."[29] (DM 17, A VI, 4, 1558) Though it will be a few years before Leibniz establishes his full catalogue of forces, this force, associated with the actual motion of a body, is clearly what he calls living force in the SD.

In DM 17 Leibniz establishes that living force is different from quantity of motion by way of the following argument. Consider two bodies; let A be one unit in size, and B be four. (See figure 3.)

28 "[...] a body falling from a certain height acquires the force to rise up that height, if its direction carries it that way, at least, if there are no impediments."(AG 50).

29 "[...] we see that force must be calculated from the quantity of the effect it can produce, for example, by the height to which a heavy body of a certain size and kind can be raised [...]."; AG 50.

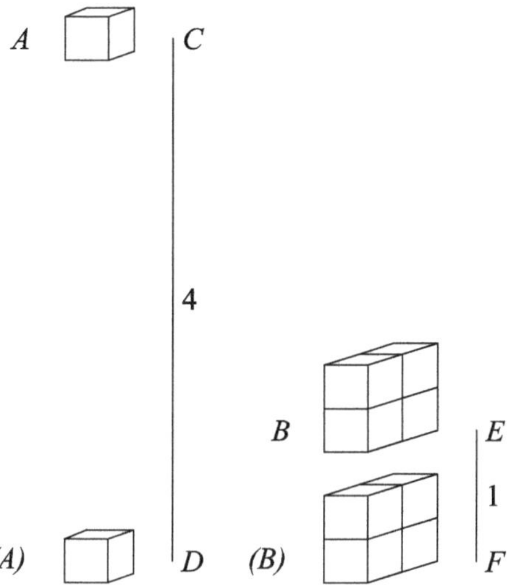

Figure 3.

Now, Leibniz reasons, it takes exactly as much work to raise *A* four feet (from *D* to *C*) as it does to raise *B* one foot (from *F* to *E*), since one can regard the larger body *B* as being made up of four smaller bodies, each identical to *A*, and each of which is being raised one foot. And so, when *A* and *B* fall through those respective distances, and their speeds converted to the horizontal, they should have exactly the same force, that is, ability to produce an effect. Now, Leibniz argues, when *A* falls, by the Galilean law of free fall it will acquire two degrees of speed, while *B* acquires one. But if that is the case, then after the fall, *A* will have two units of quantity of motion while *B* will have four. Leibniz concludes:

> Donc la quantité de mouvement du corps (*A*) au point *D* est la moitié de la quantité de mouvement du corps (*B*) au point *F*, et cependant leur forces sont egales; donc il y a bien de difference entre la quantité de mouvement et la force, ce qu'il falloit monstrer.[30] (DM 17, A VI, 4, 1557–58)

In the section following Leibniz draws some conclusions. Let me give the whole text before discussing the individual consequences:

> *XVIII. La distinction de la force et de la quantité de mouvement est importante entre autres pour juger qu'il faut recourir des considerations metaphysiques separées de l'etendue afin d'expliquer les phenomenes des corps.*

30 "Therefore the quantity of motion of body (A) at point D is half of the quantity of motion of body (B) at point F; yet their forces are equal. Hence, there is a great difference between quantity of motion and force – which is what needed to be proved."(AG 50)

> Cette consideration de la force distinguée de la quantité de mouvement est assez importante, non seulement en physique et en mechanique pour trouver les veritables loix de la nature et regles du mouvement, et pour corriger même plusieurs erreurs de practique qui se sont glissés dans les écrits de quelques habiles mathematiciens, mais encor dans la metaphysique pour mieux entendre les principes. Car le mouvement, si on n'y considere que ce qu'il comprend precisement et formellement, c'est à dire un changement de place, n'est pas une chose entierement reelle, et quand plusieurs corps changent de situation entre eux, il n'est pas possible de determiner par la seule consideration de ces changemens à qui entre eux le mouvement ou le repos doit estre attribué, comme je pourrois faire voir geometriquement, si je m'y voulois arrester maintenant. Mais la force ou cause prochaine de ces changemens est quelque chose de plus reel, et il y a assez de fondement pour l'attribuer à un corps plus qu'à l'autre; aussi n'est ce que par là qu'on peut connoistre à qui le mouvement appartient d'avantage. Or cette force est quelque chose de different de la grandeur, de la figure et du mouvement, et on peut juger par là que tout ce qui est conçû dans les corps ne consiste pas uniquement dans l'etendue, comme nos modernes se persuadent.[31] (DM 18, A VI, 4, 1558–59)

First of all, Leibniz concludes that (living) force and quantity of motion are distinct: this falls directly out of the argument.[32] But Leibniz goes further: he wants to use this argument as a step toward a more metaphysical point. Once we have distinguished quantity of motion from living force, Leibniz goes on to argue that living force is really in bodies in a way in which motion is not.

---

[31] "18. *The Distinction between Force and Quantity of Motion Is Important, among Other Reasons, for Judging That One Must Have Recourse to Metaphysical Considerations Distinct from Extension in Order to Explain the Phenomena of Bodies.*
This consideration, the distinction between force and quantity of motion, is rather important, not only in physics and mechanics, in order to find the true laws of nature and rules of motion and even to correct the several errors of practice which have slipped into the writings of some able mathematicians, but also in metaphysics, in order to understand the principles better. For if we consider only what motion contains precisely and formally, that is, change of place, motion is not something entirely real, and when several bodies change position among themselves, it is not possible to determine, merely from a consideration of these changes, to which body we should attribute motion or rest, as I could show geometrically, if I wished to stop and do this now. But the force or proximate cause of these changes is something more real, and there is sufficient basis to attribute it to one body more than to another. Also, it is only in this way that we can know to which body the motion belongs. Now, this force is something different from size, shape, and motion, and one can therefore judge that not everything conceived in body consists solely in extension and in its modifications, as our moderns have persuaded themselves [...]." (AG 51)

[32] On this see Brown, 1984. He emphasizes that the point of this argument is *not* to establish the conservation of the mathematical quantity mv^2.

The unreality of motion here is a consequence of the fact that when we consider motion in the strictest sense, as the change of position of bodies with respect to one another, "[...] il n'est pas possible de determiner par la seule consideration de ces changemens à qui entre eux le mouvement ou le repos doit estre attribué."[33] (DM 18, A VI, 4, 1559) And if it is not possible to determine which is the real subject of motion, which of two bodies in relative motion are *really* moving with respect to one another and which are at rest, then the motion in question isn't fully real. (The implicit assumption here is that for motion to be real, such a distinction must be at least in principle possible.) On the other hand, "la force ou cause prochaine de ces changemens est quelque chose de plus reel, et il y a assez de fondement pour l'attribuer à un corps plus qu'à l'autre."[34] (Ibid.) Therefore what is real in bodies in motion is just their (living) force, that which is the underlying cause of the motion, something which can, in principle, be attributed to one body rather than another. As he wrote in an essay from 1683, "Revera enim dici non potest cuinam subjecto insit motus, et proinde nihil in motu reale est, praeter vim et potentiam in rebus inditam [...]."[35] (A VI, 4, 1465)

In other texts Leibniz offers other considerations as to why living force is more real than the motion with which it is associated. A letter to Bayle from February 1687 suggests a somewhat different argument to what is substantially the same conclusion. Leibniz writes:

> J'adjouteray une remarque de consequence pour la Metaphysique. J'ay monstré que la force ne se doit pas estimer par la composition de la vistesse et de la grandeur, mais par l'effect futur. Cependant il semble que la force ou puissance est quelque chose de reel dès à present, et l'effect futur ne l'est pas. D'où il s'ensuit, *qu'il faudra admettre dans les corps quelque chose de different de la grandeur et de la vistesse, à moins qu'on veuille refuser aux corps toute la puissance d'agir.*[36] (GP III, 48)

---

[33] "[...] it is not possible to determine, merely from a consideration of these changes, to which body we should attribute motion or rest [...]."(AG 51)

[34] "[...] the force or proximate cause of these changes is something more real, and there is sufficient basis to attribute it to one body more than to another." (Ibid.)

[35] "It cannot really be said just which subject the motion is in. Consequently, nothing in motion is real besides the force and power vested in things [...]." (DLC 263)

[36] "I would like to add a remark of consequence for metaphysics. I have shown that force ought not to be estimated by the product of speed and size, but by the future effect. However, it seems that force or power is something real at present, while the future effect is not. From which it follows *that we must admit in bodies something different from size and speed, at least unless one wants to refuse bodies all power of acting.*" (quoted in Sleigh, 1990, 118) The dating is from Müller and Krönert, 1969, 80.

This is clarified by some things that he says in another text written not long after. In a letter to Paul Pellisson-Fontanier in July 1691:

> La notion de la force est aussi claire que celle de l'action et de la passion, car c'est ce dont l'action s'ensuit lors que rien ne l'empesche [...] et au lieu que le mouvement est une chose successive, laquelle par consequent n'existe jamais, non plus que le temps, parce que toutes ses parties n'existent jamais ensemble: au lieu de cela, dis-je, la force ou l'effort, existe tout entier à chaque moment, et doit estre quelque chose de véritable et de réel. Et comme la nature a plûtost égard au veritable, qu'à ce qui n'existe entierement que dans nostre esprit, il s'est trouvé (suivant ce que j'ai démontré) que c'est aussi la mesme quantité de la force, et non pas la mesme quantité du mouvement (comme Descartes avoit crû) qui sc conserve dans la nature. Et c'est de ce seul principe que je tire tout ce que l'experience a enseigné sur le mouvement, et sur le choc des corps contre les regles de Descartes, et que j'établis une nouvelle science que j'appelle la Dynamique dont j'ay projetté des Elemens.[37] (A I, 6, 226–7)

Here the point seems to be that insofar as motion never really fully exists at any given time, it isn't really real. But, Leibniz suggests, force *is* really there at any given moment.[38] As such, force must be distinct from the geometrical properties of body, the reality that undergirds motion, indeed, the reality that undergirds body itself.

But even though motion is distinct from living force, motion can be used as a *measure* of that force, as long as we don't imagine that force is directly proportional to motion. As Leibniz writes in the *Discours de métaphysique*, "on voit [...] comment *la force* doit estre estimée par la quantité de l'effect qu'elle peut produire, par exemple par la hauteur, à la quelle un corps pes-

---

37 "The notion of force is as clear as that of action and passion, because it is that from which action follows when nothing prevents it [...] On the other hand, motion is a successive thing, which, consequently, never exists, any more than time does, since all of its parts never exist together. Unlike that, I say, force or effort exists completely at each moment, and must be something true and real. And since nature takes account of that which is true rather than that which exists only entirely in our mind, one finds (as I have demonstrated) that it is also the same quantity of force, and not the same quantity of motion (as Descartes thought) that is conserved in nature. And it is from this principle alone that I draw everything that experience has taught about motion and about the impact of bodies, against Descartes' rules, and that I have established a new science which I call dynamics, whose elements I have set out." See also Costabel, 1973, 130–31; and SD, GM VI, 235; AG 118.

38 Cf. the discussion of Leibniz's notion of force in the 1694 "De primae philosophiae emendatione, et de notione substantiae," GP IV, 469–70; L 433.

ant d'une certaine grandeur et espece peut estre elevé [...]."[39] (DM 17, A VI, 4, 1558) Or, as Leibniz puts it more illuminatingly in a letter to the Marquis de L'Hôpital on 15/25 January 1696:

> Je demeure d'accord avec vous, qu'un corps agit par sa masse et par sa vistesse; aussi n'est ce que par ces choses que je determine la force mouvante. Mais il ne s'en suit point que les forces sont en raison composée des masses et des vistesses.[40] (A III, 6, 617)

Leibniz does agree – like any good mechanist – that the only physical magnitudes that enter into physical law are things like size and speed. In a number of texts Leibniz goes on to show that living force can be measured by mv^2, which is always proportional to the ability that a body in motion has to produce an effect. But, he would argue, though living force *gives rise* to motion, they are distinct notions, and there is not necessarily going to be a direct proportionality between the one and the other, as Descartes thought.

Again, Leibniz's point is against the Cartesians, parallel to the point that he made against the Cartesians on the extension of body. For Descartes, motion is a broadly geometrical notion, the ratio of distance and time. As a geometrical notion, it is a genuine feature of the world of bodies, and one of the central explanatory principles in physics, a quantity that as such characterizes body at its metaphysically most basic level. And just as Leibniz wants to get rid of the idea that extension is something that is really in bodies, in the full-blooded metaphysical sense, he wants to oppose the idea that motion is really in bodies, in the full-blooded metaphysical sense. Instead, Leibniz argues, what is really in bodies in motion is living force. But even so, although it doesn't characterize reality at its most metaphysically basic level, the notion of motion does apply to body, and the basic laws that govern force can be given in terms of the mathematical notion of motion. In place of Descartes' law of the conservation of $m|v|$, Leibniz proposes a law of the conservation of mv^2. But, we must remember, even though the term 'mv^2' is the mathematical representation of living force, it is not literally identical to living force itself, something that goes beyond size and motion. However, as long as we are aware of the differences between living

---

39 "[...] we see that force *must be calculated* from the quantity of the effect it can produce, for example, by the height to which a heavy body of a certain size and kind can be raised [...]." (AG 50)

40 "I remain in agreement with you that a body acts by its mass and by its speed, and also that it is only by these things that I determine the moving force. But it doesn't at all follow that forces are directly proportional to the product of the masses and the speeds."

force and its mathematical representation, we can use its mathematical representation in the systematic presentation of the laws of nature. Even though motion and force are different, 'mv^2' is the appropriate representation of living force in the realm of mathematics, just as Euclidean geometry is the appropriate mathematical representation of passive force.

4. Mathematics and the Physical World: The Case of Dead Force

At this point I would like to turn back to the question that originally started us off, the question about the reality of dead force as a genuine infinitesimal magnitude in nature.

Let us return to passages A and B from the SD, where Leibniz introduces the distinction between living and dead forces. The logic of the argument seems to be something like this. We can distinguish between the body in actual motion, and what Leibniz calls the "solicitation" to motion, the infinitesimal increment of additional motion that, over time, will result in the acceleration of the body along the tube. This is a distinction within the mathematical formalism for dealing with motion. Leibniz seems to argue from this distinction in the realm of *motion* to there being a corresponding distinction with respect to the *force* that causes motion: "Hinc Vis quoque duplex [...]."[41] (GM VI, 238) The argument thus seems to be a direct inference from the mathematical representation of two varieties of *motion*, to a mathematical representation of the two varieties of active *force*. Or so it would seem.

But the text is actually more complicated than that. In the manuscript the text goes directly from passage A to passage B.[42] But in the version printed in the *Acta eruditorum*, Leibniz adds the following sentence at the very end of passage A: "[...] quanquam non ideo velim haec Entia Mathematica reapse sic reperiri in natura, sed tantum ad accuratas aestimationes abstractione animi faciendas prodesse."[43] (GM VI, 238) This changes the argu-

41 "From this it follows that force is also twofold." (AG 121)

42 On this, see the textual note to lines 162–65 in Leibniz, 1982, 12. One can also view the original ms. page (LH 35, 9, 4, 1v) on the website for Reihe VIII of the Akademie edition, available at: http://leibniz.bbaw.de/ritter/Scans/lhxxxv_09.html, at image LH035,09!04_001+va-08295.jpg.

43 "Nevertheless, I wouldn't want to claim on these grounds that these mathematical entities are really found in nature, but I only wish to advance them for making careful calculations through mental abstraction." (AG 121)

ment considerably. After originally penning the discussion of the disinction between living and dead force in such straightforwardly mathematical terms, Leibniz seems to have stepped back and distanced the mathematical representation from the physical reality. In another passage, written some years later, in 1716, Leibniz does something similar with respect to infinitesimals. In a letter to Samuel Masson he writes:

> Le Calcul infinitesimal est utile, quand il s'agit d'appliquer la Mathematique à la Physique, cependant ce n'est point par là que je pretends rendre compte de la nature des choses. Car je considere les quantités infinitesimales comme des fictions utiles.[44] (GP VI, 629)

Again, the point seems to be that nature is one thing, and its mathematical representation is another.

Leibniz does not go into much detail here, in the case of dead force, about the relation between the physical reality of dead force and the mathematical representation in terms of infinitesimals. But it is illuminating to put this case along side of the cases of geometrical extension and passive force, and motion and living force, to which Leibniz devoted more explicit consideration. Leibniz is quite clear about the way in which the geometrical notion of continuous extension applies to physical reality is only approximate. Concrete things are not continuous, nor do they have exact shapes. But while extension is not literally *in* bodies as the Cartesians hold, geometrical extension is an ideal notion that *applies to* real things. Leibniz is also quite clear that there is a radical difference between living force, the ability to accomplish an effect, and motion, the change of place. But even though it is of an altogether different nature, the conservation of living force can be expressed mathematically by the expression 'mv^2.' And as with geometrical extension, even though the mathematical notions of motion and speed are not really in bodies in the same way as living force is, those mathematical notions apply to bodies. Now, Leibniz does not offer detailed arguments for distinguishing dead force from its mathematical representation as an infinitesimal. But these few passages cited above give us good reason to believe that he would have taken the same broad attitude toward the relation between dead force and its mathematical representation. Matters are a bit more delicate here since infinitesimals are in an important sense fictional, as he suggests to Masson, and not even real ideal things, as are

44 "The infinitesimal calculus is useful with respect to the application of mathematics to physics; however, that is not how I claim to account for the nature of things. For I consider infinitesimal quantities to be useful fictions." (AG 230)

geometrical extension and mathematical motion. But even so, the point stands. Dead force should not be identified with its mathematical representation, and the reality of dead force should not be taken to entail the reality of infinitesimals. At the same time, just as other mathematical concepts are useful in the description of physical phenomena ("Le Calcul infinitesimal est utile, quand il s'agit d'appliquer la Mathematique à la Physique [...]")[45] (GP VI, 629), so infinitesimals are useful in talking about dead force.

5. Mathematical Representations and Mathematical Physics

With this I think that we have resolved the question that we originally posed about the dead force and infinitesimals. Mathematics can represent physical phenomena without being identified with it; even a good mathematical representation is going to have features that don't correspond to the features of concrete reality, and vice versa. But a general question remains: if Leibniz wants to distance the physical reality of force, active and passive, living and dead from its mathematical representation as sharply as he does, why bother with the mathematical representation at all? What's the point?

The question comes up in a letter to de Volder on 20 June 1703. De Volder is a Cartesian, and it is quite natural that such questions would come up in the context of an exchange over the foundations of their two rather different conceptions of physics. In discussing the Cartesian ontology on which the only things in bodies are their broadly geometrical properties, Leibniz notes:

> Quae differunt, debent aliquo differre seu in se assignabilem habere diversitatem, mirumque est manifestissimum hoc axioma cum tot aliis ab hominibus adhibitum non fuisse. Sed vulgo homines imaginationi satisfacere contenti rationes non curant, hinc tot monstra introducta contra veram philosophiam. Scilicet non nisi incompletas abstractasque adhibuere notiones sive mathematicas, quas cogitatio sustinet sed quas nudas non agnoscit natura, ut temporis, item spatii seu extensi pure mathematici, massae mere passivae, motus mathematice sumti etc. [...].[46] (GP II, 249)

[45] "The infinitesimal calculus is useful with respect to the application of mathematics to physics [...]" (AG 230)

[46] "Things that differ ought to differ in some way, that is, have an intrinsic difference that can be designated; it is amazing that people have not made use of this most obvious axiom, along with so many others. But content to satisfy the imagination, people don't usually at-

Here Leibniz chides de Volder, and through de Volder the whole Cartesian school for applying mathematical notions to nature in an uncritical way and imposing onto the natural world features of the mathematical tools used to represent nature. But, at the same time, Leibniz does not reject the mathematicization of nature. He writes: "At in phaenomenis sive aggregatis mutatio omnis nova a concursu derivatur secundum praescriptas partim ex metaphysica partim ex Geometria Leges, abstractionibus enim opus est ut res scientifice explicentur."[47] (GP II, 252) And a bit later in the same letter: "[...] quo [i.e. abstractione] definire liceat in phaenomenis quid cuique parti massae sit ascribendum, cunctaque distingui et rationibus explicari possint, quae res necessario abstractiones postulat."[48] (GP II, 253)

Leibniz's view here is that to express things mathematically is to express them abstractly, and that such abstraction is necessary in order to be able to express the general truths of physics in a systematic way. Statics deals with the laws of dead force; dynamics deals with the laws of living force. And if we use mathematical representations of these notions, then we can express in a rigorous way the relations between the two, the fact that living force arises from an infinite repetition of dead forces. This is not to say that dead force is literally an infinitesimal, any more than living force is literally mv^2 or impenetrability is literally extension. But representing them in that way allows us to treat them and their relations in a systematic and rational way, to state the laws that they observe and the relations between them in an exact way. And as long as we don't try to impose every feature of the

tend to reasons, and from this, many monstrosities have been introduced, contrary to the true philosophy. Indeed, they have made use only of notions that are incomplete and abstract or mathematical, notions which thought supports but which nature doesn't know, taken by themselves. Take, for example, the notion of time, likewise space or purely mathematical extension, the notion of purely passive mass [*massa*], of motion considered mathematically, etc." (AG 174–75)

47 "But in phenomena or aggregates, all new change derives from the collision of bodies in accordance with laws prescribed, in part, by metaphysics and, in part, by geometry, for abstractions are needed to explain things in an orderly way." (AG 177) The Latin 'scientifice' doesn't mean the same as the English 'scientifically;' the concept of science as a general concept which embraces all of the particular systematic and empirical studies of nature and, by extension, a way of approaching investigation systematically doesn't emerge until much later. It can only mean here a "in the manner of a *scientia*," that is, in an organized and systematic way.

48 "In the phenomena we can define through abstraction whatever we want to ascribe to each part of mass, and everything can be distinguished and explained rationally, something that necessarily requires abstractions." (AG 178)

mathematics that we use to represent reality onto that reality, then we shouldn't get into trouble.

In this way mathematics and its representation of the physical world can play an important role in our understanding of the physical world. Leibniz's point in distinguishing mathematics from the physical world is not to reject the mathematical representation of physical magnitudes, but simply for us to understand what is going on when we mathematize nature, and what the role of mathematics is in the understanding of nature. His opponents are the Cartesians who have tried to make nature mathematical in a literal sense, to make the physical world over into a physical instantiation of mathematical concepts. He wants to reject this, but in doing so, he wants only to restore mathematics to its proper place in the enterprise, and not to reject it altogether. In this way we can embrace the mathematical representation of dead force in terms of infinitesimals, without having to say that there are real infinitesimals in nature.

Abbreviations

Editions including several volumes are quoted by volume and page if not mentioned otherwise. In addition the numbers of chapters and articles might be given thereafter.

A	Gottfried Wilhelm Leibniz, *Sämtliche Schriften und Briefe*, (Ed.) Akademien der Wissenschaften, Darmstadt and Berlin: Akademie Verlag 1923-. (also referred to as *Akademieausgabe* resp. *Akademie edition*; quoted by series, vol., page)
AE	*Acta eruditorum*, Leipzig.
AG	Gottfried Wilhelm Leibniz: *Philosophical Essays*, (Eds.) Roger Ariew, Daniel Garber. Indianapolis: Hackett 1989.
AT	René Descartes, *Œuvres*, (Eds.) Charles Adam & Paul Tannery, Paris: Vrin 1996.
C	Gottfried Wilhelm Leibniz, *Opuscules et fragments inédits*, (Ed.) Louis Couturat, Paris: Presses Universitaires de France 1903 (Reprint Hildesheim: Olms, 1961).
Cc2	*Catalogue critique des manuscrits de Leibniz*, Fascicule II (Mars 1672 – Novembre 1676), (Eds.) A. Rivaud et al., Poitiers 1914–1924 (Reprint Hildesheim: Olms 1986).
Child	Gottfried Wilhelm Leibniz, *The Early Mathematical Manuscripts of Leibniz*, (Ed. and Trans.) James Mark Child, Chicago: Open Court Publishing 1920 (Reprint, Mineola, NY: Dover Publishing Inc 2005).
Dutens	Gothofredi Guillelmi Leibnitii *Opera omnia*, (Ed.) Louis Dutens, Geneva: Fratres des Tournes 1768, 6 vols.
DLC	Gottfried Wilhelm Leibniz, *The Labyrinth of the Continuum. Writings on the Continuum Problem, 1672–1686*, (Trans., Ed. and Comm.) Richard Arthur, (= The Yale Leibniz), New Haven and London: Yale University Press 2001.
DQA	Gottfried Wilhelm Leibniz, *De quadratura arithmetica circuli ellipseos et hyperbolae cujus corollarium est trigonometria sine tabulis*, (Ed. and Comm.) Eberhard Knobloch (= Abhandlungen der Akademie der Wissenschaften in Göttingen, Math.-Naturwiss. Klasse, 3. Folge, N. 43), Göttingen: Vandenhoeck & Ruprecht 1993.
EW	Thomas Hobbes, *English Works*, (Ed.) William Molesworth, 11 Volumes, London: John Bohn (1–6); Longman (7–11) 1839–1845 (Reprint Aalen: Scientia Verlag 1962).

GP Gottfried Wilhelm Leibniz, *Die Philosophischen Schriften*, (Ed.) C. I. Gerhardt, 7 vols., Berlin: Weidmannsche Buchhandlung 1875–90 (Reprint Hildesheim: Olms, 1960).

GM Gottfried Wilhelm Leibniz, *Die Mathematischen Schriften*, (Ed.) C. I. Gerhardt, 7 vols., Berlin and Halle: Ascher and Schmidt 1849–63 (Reprint Hildesheim: Olms, 1971).

H&O Gottfried Wilhelm Leibniz, *Historia et Origo Calculi Differentialis*, (Ed.) C.I. Gerhardt, Hannover: Hahnsche Hofbuchhandlung 1846.

L Gottfried Wilhelm Leibniz, *Philosophical Papers and Letters*, (Ed. and trans.) Leroy E. Loemker, Dordrecht: D. Reidel 1969 (2nd ed.).

MPN Isaac Newton, *The Mathematical Papers of Isaac Newton*, (Eds.) Derek Thomas Whiteside with Michael Hoskin and Adolf Prag, 8 vols, Cambridge: Cambridge University Press 1967–81.

NE Gottfried Wilhelm Leibniz, *New Essays on Human Understanding*, (Ed. and Trans.) Peter Remnant and Jonathan Bennett, Cambridge: Cambridge University Press 1981.

OL Thomas Hobbes, *Opera philosophica quae latine scripsit, omnia in unum corpus nunc primum collecta*, (Ed.) William Molesworth, 5 Volumes, London: John Bohn 1839–1845 (Reprint Aalen: Scientia 1961 and 1966; Bristol: Thoemmes in 1999).

Bibliographical References

Adams, Robert M., 1994, *Leibniz: Determinist, Theist, Idealist,* New York: Oxford University Press.

Aiton, Eric J., 1984, "The Mathematical Basis of Leibniz's Theory of Planetary Motion," in: Albert Heinekamp (Ed.), *Leibniz's Dynamica* (=*Studia leibnitiana*, Sonderheft 13), 209–25.

Andersen, Kirsti, 1985, "Cavalieri's method of indivisibles," in: *Archive for History of Exact Sciences*, 31, 291–367.

Andersen, Kirsti, 1986, "The method of indivisibles: changing understandings," in: Albert Heinekamp (Ed.), *300 Jahre "Nova methodus" von G. W. Leibniz (1684–1984)*, Stuttgart: Steiner, 14–25.

Arnauld, Antoine, 1667, *Nouveaux éléments de géométrie*, Paris: Charles Savreux.

Arnauld, Antoine, 1781, *Œuvres de Messire Antoine Arnauld*, vol. 42, Paris: Sigismond d'Arnay.

Arthur, Richard T. W., 2001, "Leibniz on Infinite Number, Infinite Wholes and the Whole World: A Reply to Gregory Brown," in: *The Leibniz Review*, 11, 103–116.

Arthur, Richard, T. W., 2003, "The Enigma of Leibniz's Atomism," in: *Oxford Studies in Early Modern Philosophy*, 1, 183–227.

Arthur, Richard, T. W., 2008a, "Leibniz's Syncategorematic Infinitesimals, Smooth Infinitesimal Analysis, and Newton's Proposition 6," in: William Harper and Wayne C. Myrvold (Eds.), *Infinitesimals*, forthcoming.

Arthur Richard T. W., 2008b, "'A complete denial of the continuous?' Leibniz's philosophy of the continuum," submitted for a special edition of *Synthese* devoted to the Mathematics and Philosophy of the Continuum, forthcoming.

Arthur, Richard, T. W., 2008c, "Actual Infinitesimals in Leibniz's Early Thought," in: Mark Kulstad and Moegens Laerke (Eds.), The Young Leibniz (=*Studia Leibnitiana*, Sonderheft), forthcoming.

Baron, Margaret, 1987, *The Origins of the Infinitesimal Calculus*, New York: Dover 1987.

Bassler, O. Bradley, 1998a, "Leibniz on the Indefinite as Infinite," in: *The Review of Metaphysics*, 51, 849–874.

Bassler, O. Bradley, 1998b, "The Leibnizian Continuum in 1671," in: *Studia Leibnitiana*, 30, 1–23.

Bassler, O. Bradley, 1999, "Towards Paris: The Growth of Leibniz's Paris Mathematics out of the Pre-Paris Metaphysics," in: *Studia Leibnitiana*, 31, 160–180.

Bassler, O. Bradley, 2002, "Motion and Mind in the Balance: The Transformation of Leibniz's Early Philosophy," in: *Studia Leibnitiana*, 34, 221–231.

Beeley, Philip, 1996, *Kontinuität und Mechanismus. Zur Philosophie des jungen Leibniz in ihrem ideengeschichtlichen Kontext*, Stuttgart: Steiner.

Beeley, Philip, 2004, "A Philosophical Apprenticeship: Leibniz's Early Correspondence with the Secretary of the Royal Society, Henry Oldenburg," in: Paul Lodge (Ed.), *Leibniz and his Correspondents*, Cambridge: Cambridge University Press 2004, 47–73.

Belaval, Yvon, 1960, *Leibniz critique de Descartes*, Paris: Gallimard.

Bell, J. L., 1998, *A Primer of Infinitesimal Analysis*, Cambridge: Cambridge University Press.

Bernoulli, Jacob, 1691, "Specimen alterum calculi differentialis," in: *Acta eruditorum*, 282–290.

Bernoulli, Jacob, 1701, [Anonymous review], "Responsio ad Cl. Viri Bernhardi Nieuwentiit Considerationes Secundas circa Calculi Differentialis principia editas Basileae, literis Joh. Conradi a Mechel MDCC," in: *Acta Eruditorum*, January, 28–29.

Bernoulli, Jacob, 1993, *Der Briefwechsel von Jacob Bernoulli*, (Eds.)André Weil with Clifford Truesdell and Fritz Nagel, Basel, Boston, Berlin: Birkhäuser.

Bernoulli, Johann I, 1691/1692, *De calculo differentialis*, University Library Basel, manuscript Ms L I a 6.

Bernoulli, Johann I, 1697, "Principia calculi exponentialium seu percurrentium," in: *Acta Eruditorum*, March, 125–133.

Bernoulli, Johann I, *Correspondence of Johann Bernoulli and Christian Wolff*, University Library Basel, manuscript Ms L I a 671.

Bernstein, Howard R., 1980, "Conatus, Hobbes, and the Young Leibniz," in: *Studies in History and Philosophy of Science*, 11, 25–37.

Bertoloni Meli, Domenico, 1993, *Equivalence and Priority: Newton versus Leibniz, Including Leibniz's Unpublished Manuscripts on the* Principia, Oxford: Clarendon Press.

Blay, Michel, 1986, "Deux moments de la critique du calcul infinitesimal: Michel Rolle et George Berkeley," in: *Revue d'Histoire des Sciences*, 39, 223–253.

Blay, Michel, 1992, *La Naissance de la mécanique analytique. La Science du mouvement au tournant des XVIIe et XVIIIe siècles*, Paris: Presses Universitaires de France.

Bos, Henk, 1972, "Christiaan Huygens," in: *Dictionary of Scientific Biography*, Charles C. Gillispie (Ed.), vol. VI, New York: Scribner's Sons, 597–613.

Bos, Henk, 1974, "Differentials, Higher-Order Differentials and the Derivative in the Leibnizian Calculus," in: *Archive for History of Exact Sciences*, 14, 1–90.

Bos, Henk, 1980, "Huygens and Mathematics," in: *Studies on Christiaan Huygens*, Lisse: Swets and Zeitlinger, 126–146.

Boyer, Carl B, 1959, *The History of the Calculus and its Conceptual Development*, New York: Dover.

Breger, Herbert, 1984, "Elastizität als Strukturprinzip der Materie bei Leibniz," in: Albert Heinekamp (Ed.), *Leibniz's Dynamica* (=*Studia Leibnitiana*, Sonderheft, 13), 112–21.

Breger, Herbert, 1986, "Weyl, Leibniz, und das Kontinuum," in: *Studia Leibnitiana Supplementa*, 26, 316–330.

Breger, Herbert, 1990a, "Das Kontinuum bei Leibniz," in: Antonio Lamarra (Ed.), *L'Infinito in Leibniz: Problemi e Terminologia*, Roma et al.: Ed. dell'Ateneo, 53–67.

Breger, Herbert, 1990b, "Know-how in der Mathematik," in: Detlef Spalt (Ed.), *Rechnen mit dem Unendlichen*, Basel: Birkhäuser, 43–57.

Breger, Herbert, 1992, "Le Continu chez Leibniz," in: *Le Labyrinthe du Continu*, Jean-Michel Salanskis and Hourya Sinaceur (Eds.), Paris: Springer France, 76–84.

Breger, Herbert, 1994: "The Mysteries of Adaequare: A Vindication of Fermat," in: *Archive for History of Exact Sciences*, 46, 193–219.

Breger, Herbert, 1999, "Analysis und Beweis," in: *Internationale Zeitschrift für Philosophie*, 95–116.

Brown, Gregory, 1984, "'*Quod ostendendum susceperamus*': What Did Leibniz Undertake to Show in the *Brevis Demonstratio*?," in: Albert Heinekamp (Ed.), *Leibniz's Dynamica* (=*Studia Leibnitiana*, Sonderheft 13), Stuttgart: Franz Steiner, 122–37.

Cajori, Florian, 1928–29, *A History of Mathematical Notations*, 2 vols, Chicago: Open Court.

Cantor, Moritz, 1901, *Vorlesungen über Geschichte der Mathematik*, vol. 3, 2nd edition, Leipzig: Teubner.

Cavalieri, Bonaventura, 1635, *Geometria indivisibilibus continuorum nova quadam ratione promota*, Bologna: Clemens Ferronius.

Cavalieri, Bonaventura, 1647, *Exercitationes Geometricæ Sex*, Bologna: Montius.

Cavalieri, Bonaventura, 1653, *Geometria indivisibilibus continuorum nova quadam ratione promota*, 2nd Ed., Bologna: Typographia de Duciis.

Costabel, Pierre, 1973, *Leibniz and Dynamics*, Ithaca, NY: Cornell University Press.

Couturat, Louis, 1901, *La logique de Leibniz d'après des documents inédits*, Paris: Alcan.

Dascal, Marcelo, 1987, *Leibniz: Language, Signs and Thought*, Amsterdam, Philadelphia: Benjamin.

De Gandt, François, 1989, "Les indivisibles de Torricelli," in: François De Gandt (Ed.), *L'Œuvre de Torricelli: science Galiléenne et nouvelle géométrie*, Nice: Les Belles Lettres, 151–206.

De Gandt, François, 1992a, "Cavalieri's Indivisibles and Euclid's Canons," in: Roger Ariew and Peter Barker (Eds.), *Revolution and Continuity. Essays in the History and Philosophy of Early Modern Science*, Washington D.C.: Catholic University of America Press, 157–182.

De Gandt, François, 1992b, "L'évolution de la théorie des indivisibles et l'apport de Torricelli," in: Massimo Bucciantini and Maurizio Torrini (Eds.), *Geometria e atomismo nella scuola galileiana*, Florence: Leo S. Olschki, 103–118.

Debus, Allen G., 1970, *Science and Education in the Seventeenth Century: The Webster-Ward Debate*, London: Macdonald, New York: American Elsevier.

Dijksterhuis, Eduard J., 1987, *Archimedes*, Princeton: Princeton University Press.

Duchesneau, François, 1993, *Leibniz et la méthode de la science*, Paris: Presses Universitaires de France.

Duchesneau, François, 1994, *La Dynamique de Leibniz*, Paris: Vrin.

Duchesneau, François, 1999, "Leibniz's theoretical shift in the *Phoranomus* and *Dynamica de potentia*," in: *Perspectives on Science*, 6, 77–109.

Duchesneau, François, 2006, "Leibniz et la méthode des hypothèses," in: François Duchesneau & Jérémie Griard (Eds.), *Leibniz selon les Nouveaux essais sur l'entendement humain*, Paris: Vrin; Montréal: Bellarmin, 113–127.

Earman, John, 1975, "Infinities, Infinitesimals, and Indivisibles: The Leibnizian Labyrinth," in: *Studia Leibnitiana*, 7, 236–251.

Euler, Leonhard, 1755, *Institutiones calculi differentialis cum eius usu in analysi finitorum ac doctrina serierum*, Petersburg: Academia Imperialis Scientiarum Petropolitanae (=Leonard Euler. *Opera omnia*, vol. I, 10: *Institutiones calculi differentialis*, Gerhard Kowalewski (Ed.), Leipzig, Berlin: Teubner 1913)

Feingold, Mordechai, 1997, "The Mathematical Sciences and New Philosophies," in: Nicholas Tyacke (Ed.), *The History of the University of Oxford*, vol. IV: Seventeenth-Century Oxford, Oxford: Oxford University Press, 359–448.

Fermat, Pierre de, 1891, *Œuvres*, vol. 1, Adam Tannery and Charles Henry (Eds.), Paris: Gauthier-Villars.

Festa, Egidio, 1992, "Quelques aspects de la controverse sur les indivisibles," in: Massimo Bucciantini and Maurizio Torrini (Eds.), *Geometria e atomismo nella scuola galileiana*, Florence: Leo S. Olschki, 193–207.

Flood, Raymond and John Fauvel, "John Wallis," in: John Fauvel, Raymond Flood, Robin Wilson (Eds.), *Oxford Figures: 800 Years of the Mathematical Sciences*, Oxford: Oxford University Press 2000, 97–115.

Galilei, Galileo, 1890–1909, *Le Opere di Galileo Galilei*, Antonio Favaro (Ed.), 20 vols., Florence: Tip. di G. Barbèra.

Galilei, Galileo, 1974, *Two New Sciences*, Stillman Drake (Ed. and trans.), Madison, WI: University of Wisconsin Press.

Garber, Daniel, 1995, "Leibniz: Physics and Philosophy," in: Nicholas Jolley (Ed.), *The Cambridge Companion to Leibniz*, Cambridge: Cambridge University Press, 270–352.

Gerhardt, Carl Immanuel, 1891, "Leibniz und Pascal," in: *Sitzungsberichte der Kgl. Preuß. Akad. d. Wiss. zu Berlin*, 1053–1068.

Giusti, Enrico, 1980, *Bonaventura Cavalieri and the Theory of Indivisibles*, Bologna: Edizioni Cremonese.

Goldenbaum, Ursula, 1999, "Die *Commentatiuncula de judice* als Leibnizens erste philosophische Auseinandersetzung mit Spinoza nebst der Mitteilung über ein neuaufgefundenes Leibnizstück. Beilage: Leibniz' Marginalien zu Spinozas *Tractatus theologico-politicus* im Exemplar der Bibliotheca Boineburgica in Erfurt, also zu datieren auf 1670–71," in: Hartmut Rudolph, et al. (Eds.) *Labora diligenter* (=*Studia Leibnitiana Sonderheft*, 29). Steiner: Wiesbaden 1999, 61–127.

Goldenbaum, Ursula, 2002a, "Leibniz' Philosophie des Geistes als Gegenentwurf zu Hobbes' Philosophie des Körpers", in: Hans Poser together with Christoph Asmuth, Ursula Goldenbaum, Wencho Li (Eds.), *Nihil sine ratione. Mensch, Natur und Technik im Wirken von G. W. Leibniz*. Nachtragsband zum VII. Internationalen Leibniz-Kongress, Hannover, 204–210.

Goldenbaum, Ursula, 2002b, "All you need is love, love. Leibniz' Vermittlung von Hobbes' Naturrecht und christlicher Nächstenliebe als Grundlage seiner Definition der Gerechtigkeit," in: Günter Abel, Hans-Jürgen Engfer, Christoph Hubig (Eds.), *Neuzeitliches Denken. Festschrift für Hans Poser zum 65. Geburtstag*, De Gruyter: Berlin, New York, 209–231.

Goldenbaum, Ursula, 2008, "It's Love! Leibniz' Foundation of Natural Law as the Outcome of His Struggle with Hobbes' and Spinoza's Naturalism," in: Mark Kulstad, Moegens Laerke (Eds.), *The Young Leibniz*, Studia Leibnitiana, Sonderheft, forthcoming.

Grosholz, Emily, 1992, "Was Leibniz a Mathematical Revolutionary?" in: Donald Gillies (Ed.), *Revolutions in Mathematics*, Oxford: Clarendon Press, 117–133.

Gueroult, Martial, 1967, *Leibniz. Dynamique et métaphysique*, Paris: Aubier-Montaigne.

Guicciardini, Niccolò, 2002, "Analysis and synthesis in Newton's mathematical work," in: I. Bernard Cohen and George E. Smith (Eds.), *The Cambridge Companion to Newton*, Cambridge: Cambridge University Press, 308–328.

Guicciardini, Niccolò, 2003, "Conceptualism and contextualism in the recent historiography of Newton's *Principia*," in: *Historia Mathematica*, 30, 407–431.

Hall, A. Rupert, 1980, *Philosophers at War. The Quarrel between Newton and Leibniz*, Cambridge: Cambridge University Press.

Hannequin, Arthur, 1908, "La première philosophie de Leibniz," in: Arthur Hannequin, *Études des sciences et d'histoire de la philosophie*, vol. 2, 17–225.

Hartkopf, Werner, 1992, *Die Berliner Akademie der Wissenschaften. Ihre Mitglieder und Preisträger 1700–1990*, Berlin: Akademie-Verlag.

Hartz, Glenn A., and Jan A. Cover, 1988, "Space and Time in the Leibnizian Metaphysic," in: *Noûs*, 22, 493–519.

Hermann, Jacob, 1700, *Responsio ad Clarissimi Viri Bernh. Nieuwentiit Considerationes Secundas circa calculi differentialis principia editas* Basileae, Literis Johannis Conradi a Mechel.

Heytesbury, Guillaume, 1994, *Sophismata asinina. Une introduction aux disputes médiévales*, Fabienne Pironet, (Ed.), (= Collection Sic et Non), Paris: Vrin.

Hobbes, Thomas, 1651, *Leviathan; Or the Matter, Forme and Power of a Commonwealth Ecclesiastical and Civil*, London: Andrew Crooke.

Hobbes, Thomas, 1655, *Elementorum Philosophiæ Sectio Prima: De Corpore*, Londini: Andrea Crook sub signo Draconis.

Hobbes, Thomas, 1656, *Elements of Philosophy, the First Section: Concerning Body*, London: Andrew Crooke.

Hobbes, Thomas, 1656, *Six Lessons to the Professors of the Mathematiques, one of geometry, the other of astronomy: in the chaires set up by the noble and learned Sir Henry Savile, in the University of Oxford*, London: F. M. for Andrew Crooke.

Hobbes, Thomas, 1668, *Opera Philosophica, quae in Latine scripsit, omnia*, Amstelodami apud Ioannem Blaev.

Hobbes, Thomas, 1672, *Lux mathematica, Excussa Collisionibus Johannis Wallisii Theol. Doctoris, Geometriae in celeberrima Academia Oxoniensi Professoris Publici, et Thomae Hobbesii Malmesburiensis*, London: John Crooke for William Crooke.

Hofmann, Joseph E., 1974, *Leibniz in Paris 1672–1676. His Growth to Mathematical Maturity*, Cambridge, New York: Cambridge University Press.

Hotson, Howard, 2000, *Johann Heinrich Alsted, 1588–1638: between Renaissance, Reformation, and Universal Reform*, Oxford: Clarendon Press, New York: Oxford University Press.

Huygens, Christiaan, 1673, *Horologium oscillatorium*, Paris: Muguet.

Huygens, Christiaan, 1888–1950, *Œuvres*, 22 vol., The Hague: Nijhoff.

Ishiguro, Hidé, 1990, *Leibniz's Philosophy of Logic and Language* (2nd ed.), Cambridge: Cambridge University Press.

Jesseph, Douglas M., 1998, "Leibniz of the Foundations of the Calculus: The Question of the Reality of Infinitesimal Magnitudes," in: *Perspectives on Science*, 6, 6–40.

Jesseph, Douglas M., 1999, *Squaring the Circle: The War between Hobbes and Wallis*, Chicago, London: University of Chicago Press.

Jesseph, Douglas, 2003, "Hobbes, Galileo, and the Book of Nature," in: *Perspectives on Science*, 12, 191–211.

Jesseph, Douglas, 2006, "Hobbesian Mechanics," in: *Oxford Studies in Early Modern Philosophy*, 3, 119–152.

Kabitz, Willy, 1909, *Die Philosophie des jungen Leibniz. Untersuchungen zur Entwicklungsgeschichte seines Systems*, Hildesheim, New York: Olms 1974.

Kangro, Hans, 1969, "Der Begriff der physikalischen Größe, insbesondere der action motrice, bei Leibniz," in: *Studia Leibnitiana Supplementa*, II, 133–149.

Keisler, Jerome, 2002, *Elementary Calculus: An Approach using infinitesimals*; online edition: http://www.math.wisc.edu/~keisler/calc.html.

Kepler, Johannes, 1615, *Nova Stereometria doliorum vinariorum, in primis Austriaci, figurae omnium aptisimae; et usus in eo virgae cubicae compendiosissimus et plane singularis*, Linz: Johannes Plank (=Johannes Kepler, *Gesammelte Werke*, vol. IX: *Mathematische Schriften*, Franz Hammer (Ed.), München: C.H. Beck 1960, 5–133)

Knobloch, Eberhard, 1990, "L'infini dans les mathématiques de Leibniz," in: Antonio Lamarra (Ed.), *L'Infinito in Leibniz: Problemi e Terminologia*, Roma et al.: Ed. dell'Ateneo, 33–52.

Knobloch, Eberhard, 1993, "Les courbes analytiques simples chez Leibniz", in: *Sciences et techniques en perspective*, 26, 74–96.

Knobloch, Eberhard, 1994, "The Infinite in Leibniz's Mathematics – The Historiographical Method of Comprehension in Context," in: Kostas Gavroglu, Jean Christianidis, Efthymios Nicolaidis (Eds.), *Trends in the Historiography of Science*, *Boston Studies in the Philosophy of Science*, vol. 151, Dordrecht: Kluwer, 265–278.

Knobloch, Eberhard, 1999, "Galileo and Leibniz: Different Approaches to Infinity," in: *Archive for History of Exact Sciences*, 54, 87–99.

Knobloch, Eberhard, 2000, "Archimedes, Kepler and Guldin: the Role of Proof and Analogy," in: Rüdiger Thiele (Ed.), *Mathesis. Festschrift zum siebzigsten Geburtstag von Matthias Schramm*, Berlin-Diepholz: Verlag für Geschichte der Naturwissenschaften und der Technik, 82–100.

Knobloch, Eberhard, 2002, "Leibniz's Rigorous Foundation of Infinitesimal Geometry by Means of Riemannian Sums," in: *Synthese*, 133, 59–73.

Knobloch, Eberhard, 2005, "Mathesis perennis: Mathematics in Ancient, Renaissance, and Modern Times," in: *Studies in the History of Natural Sciences*, 24 Suppl., 10–22. (Reprint in: *The American Mathematical Monthly*, 113 (2006), 352–365).

Knobloch, Eberhard, 2006a, "Beyond Cartesian limits: Leibniz's passage from algebraic to 'transcendental' mathematics," in: *Historia Mathematica*, 33, 113–131.

Knobloch, Eberhard, 2006b, "La généralité dans les mathématiques leibniziennes," in: Herbert Breger, Jürgen Herbst, Sven Erdner (Eds.), *Einheit in der Vielheit*, VIII. Internationaler Leibniz-Kongress, Hanover: Hartmann, 382–389.

Koyré, Alexander, 1973, "Bonaventura Cavalieri et la géométrie des continus," in: *Études d'histoire de la pensée scientifique*, Paris: Gallimard, 334–361.

Laeven, Augustinus Hubertus, 1986, *De "Acta Eruditorum" onder redactie van Otto Mencke*, Amsterdam & Maarssen: APA.

Lange, Marc, 2005, "How Can Instantaneous Velocity Fulfill Its Causal Role?," in: *Philosophical Review*, 114, 433–68.

Laugwitz, Detlef, 1983, "Die Nichtstandard-Analysis: Eine Wiederaufnahme der Ideen und Methoden von Leibniz und Euler," in: *Leonhard Euler 1707–1783. Beiträge zu Leben und Werk*. Gedenkband des Kantons Basel-Stadt, Basel: Birkhäuser, 185–198.

Laugwitz, Detlef, 1986, *Zahlen und Kontinuum*, Darmstadt: Wissenschaftliche Buchgesellschaft.

Lavine, Shaughan, 1994, *Understanding the Infinite*, Cambridge: Harvard.

Leibniz, Gottfried Wilhelm, 1671, *Theoria Motus Abstracti; seu Rationes motuum universales, à sensu & Phaenomenis independentes*, London: J. Martyn.

Leibniz, Gottfried Wilhelm, 1682, "De vera proportione circuli ad quadratum circumscriptum in Numeris rationalibus a Gothofredo Gulielmo Leibnitio expressa," in: *Acta Eruditorum*, February, 41–46.

Leibniz, Gottfried Wilhelm, 1684, "Nova Methodus pro maximis et minimis, itemque tangentibus, quae nec fractas, nec irrationales quantitates moratur, et singulare pro illis calculi genus, per G. G. L.," in: *Acta Eruditorum*, October, 467–473.

Leibniz, Gottfried Wilhelm, 1692, "De la chainette," in: *Journal des Sçavans*, N. 13, March 31, 147–153.

Leibniz, Gottfried Wilhelm, 1695, "G. G. L. Responsio ad nonnullas difficultates, a Dn. Bernardo Nieuwentijt circa methodum differentialem seu infinitesimalem motas," in: *Acta Eruditorum*, July, 310–316.

Leibniz, Gottfried Wilhelm, 1696, Review of John Wallis, *Opera mathematica*, Volumes I and II, in: *Acta eruditorum*, June, 249–259.

Leibniz, Gottfried Wilhelm, 1701, "Mémoire de M.G.G. Leibniz touchant son sentiment sur le calcul différentiel," in: *Journal de Trevoux*, Novembre, 270–272 (= GM V, 350).

Leibniz, Gottfried Wilhelm, 1710, "Monitum de Characteribus Algebraicis," in: *Miscellanea Berolinensia ad incrementum scientarum ex scriptis Societati Regiae Scientarum exhibitis edita*, vol. 1, Berlin: Papen.

Leibniz, Gottfried Wilhelm, 1875, "Scientarum diversos gradus nostra imbecillitas facit ...," in: C. I. Gerhardt (Ed.), "Zum zweihundertjährigen Jubiläum der Entdeckung des Algorithmus der höheren Analysis durch Leibniz," in: *Mon.-ber. Kön. Preuss. Akad. Wiss. Berlin*, 28 Oct., 588–608.

Leibniz, Gottfried Wilhelm, 1982, *Specimen dynamicum*, Hans Günter Dosch, Glenn W. Most, Enno Rudolph, Jörg Aichelin (Eds. and trans.), Hamburg: Meiner.

Leibniz, Gottfried Wilhelm, 1989, *La naissance du calcul différentiel: 26 articles des "Acta eruditorum"*, Marc Parmentier (Ed. and trans.), Paris: Vrin.

Leibniz, Gottfried Wilhelm, 1991, *Phoranomus seu De potentia et legibus naturæ, Rome, juillet 1689*, André Robinet (Ed.), in: *Physis*, 38, 429–541; 797–885.

Leibniz, Gottfried Wilhelm, 1994, *La Réforme de la dynamique. De corporum concursu (1678) et autres textes inédits*, Michel Fichant (Trans., ed. and comm.), Paris: Vrin.

Leibniz, Gottfried Wilhelm, 2004, *Quadrature arithmétique du cercle, de l'ellipse et de l'hyperbole*, Marc Parmentier (Trans. and Ed.)/Latin text Eberhard Knobloch (Ed.), Paris: J. Vrin.

Lessing, Gotthold Ephraim, 2005, *Philosophical and Theological Writings*, (Ed.) H. B. Nisbet, Cambridge, New York: Cambridge University Press.

Levey, Samuel, 1998, "Leibniz on Mathematics and the Actually Infinite Division of Matter," in: *The Philosophical Review*, 107, 49–96.

Levey, Samuel, 2003, "The Interval of Motion in Leibniz's *Pacidius Philalethi*," in: *Noûs*, 37, 371–416.

Levey, Samuel, 2005, "Leibniz on Precise Shapes and the Corporeal World," in: Donald Rutherford and Jan A. Cover (Eds.), *Leibniz: Nature and Freedom*, Oxford: Oxford University Press, 69–94.

L'Hôpital, Guilleaume de, 1696, *Analyse des infiniment petits*, Paris: L'Imprimerie Royale.

Mahnke, Dietrich, 1926, "Neue Einblicke in die Entdeckungsgeschichte der höheren Analysis," in: *Abhandlungen der Preußischen Akademie der Wissenschaften*, Jahrgang 1925, phys.-math. Klasse, Nr. 1, Berlin, 1–64.

Malet, Antoni, 1996, *From Indivisibles to Infinitesimals. Studies on Seventeenth-Century Mathematizations of Infinitely Small Quantities*, Barcelona: Universitat Autònoma de Barcelona.

Malet, Antoni, 1997, "Barrow, Wallis, and the Remaking of Seventeenth Century Indivisibles," in: *Centaurus*, 39, 67–92.

Mancosu, Paulo, 1996, *Philosophy of Mathematics and Mathematical Practice in the Seventeenth Century*, Oxford, New York: Oxford University Press.

Marchi, Peggy, 1974, "The Controversy between Leibniz and Bernoulli on the Nature of Logarithms of Negative Numbers," in: *Studia Leibnitiana Supplementa, 13* (=Akten des II. Internationalen Leibniz-Kongresses), Albert Heinekamp (Ed.), 67–75.

Mercator, Nicholas, 1668, *Logarithmotechnia*, London: William Godbid for Moses Pitt (Reprinted in 1975, Hildesheim, New York: Olms).

Müller, Kurt/Gisela Krönert, 1969, *Leben und Werk von Gottfried Wilhelm Leibniz. Eine Chronik*, Frankfurt a.M.: Klostermann.

Mycielski, Jan, 1981, "Analysis without Actual Infinity," in: *Journal of Symbolic Logic*, 46, 625–633.

Nagel, Fritz, 1991, "A Catalog of the Works of Jacob Hermann (1678–1733)," in: *Historia mathematica*, 18, 36–54.

Nagel, Fritz, 2005, "Jacob Hermann. Skizze einer Biographie," in: Fritz Nagel, Andreas Verdun (Eds.), "*Geschickte Leute, die was praestiren können ...*". *Gelehrte aus Basel an der St. Petersburger Akademie der Wissenschaften des 18. Jahrhunderts*, Aachen: Shaker, 55–75.

Nagel, Fritz, 2007, "Nicolaus Cusanus. Mathematicus theologus. Unendlichkeitsdenken und Infinitesimalmathematik," in: *Trierer Cusanus Lecture 13*, Trier: Paulinus.

Nauenberg, Michael, 2003, "Kepler's Area Law in the *Principia*: Filling in Some Details in Newton's Proof of Proposition 1," in: *Historia Mathematica*, 30, 441–456.

Newton, Isaac, 1964, *The Mathematical Works of Isaac Newton*, Derek Thomas Whiteside (Eds.), New York and London: Johnson Reprint.

Newton, Isaac, 1999, *The Principia: Mathematical Principles of Natural Philosophy*, I. Bernard Cohen and Anne Whitman (Trans. and Eds.), Berkeley: University of California Press.

Nieuwentijt, Bernhard, 1694, *Considerationes circa Analyseos ad quantitates infinite parvas applicatae Principia, & calculi differentialis usum in resolvendis problematibus Geometricis*, Amsterdam: Wolters.

Nieuwentijt, Bernhard, 1695, *Analysis infinitorum, seu Curvilineorum proprietates ex polygonarum natura deductae*, Amsterdam: Wolters.

Nieuwentijt, Bernhard, 1696, *Considerationes secundae circa calculi differentialis principia et responsio ad Virum Nobilissimum G. G. Leibnitium*, Amsterdam: Wolters.

Oldenburg, Henry, 1965–86, *The Correspondence of Henry Oldenburg*, A. Rupert Hall and Marie Boas Hall (Eds. and trans.), 13 vols., Madison, Wisconsin: University of Wisconsin Press; London: Mansell; London and Philadelphia: Taylor and Francis.

Pascal, Blaise, 1659, *Lettres de Dettonville*, Paris: Guillaume Desprez (Reprinted in 1966, London: Dawsons of Pall Mall).

Pascal, Blaise, 1904–1914, *Oeuvres*, 14 Vols., Paris: Hachette and cie. (Reprint: Vaduz: Kraus 1965)

Pascal, Blaise, 1980, *Œuvres complètes*, Louis Lafuma (Ed.), Paris: du Seuil.

Pasini, Enrico, 1985–1986, *La nozione di infinitesimo in Leibniz; tra matematica e metafisica.* Tesi di Laurea in Storia della Filosofia Moderna, Universita' degli studi di Torino.

Pasini, Enrico, 1993, *Il reale e l'immaginario. La fondazione del calcolo infinitesimale nel pensiero di Leibniz*, Milano, Torino: Edizioni Sonda.

Pell, John, 2005, *John Pell (1611–1685) and his Correspondence with Sir Charles Cavendish. The Mental World of an Early Modern Mathematician*, 2005, Noel Malcolm and Jacqueline Stedall (Eds.), Oxford: Oxford University Press.

Perkins, Franklin, 2004, *Leibniz and China. A Commerce of Light*, Cambridge: Cambridge University Press.

Probst, Siegmund, 1997, *Die mathematische Kontroverse zwischen Thomas Hobbes und John Wallis*, Hanover: private imprint (= Dissertation Univ. Regensburg 1994).

Probst, Siegmund, 2006a, "Differenzen, Folgen und Reihen bei Leibniz (1672–1676)," in: Magdalena Hykšová, Ulrich Reich (Eds.), *Wanderschaft in der Mathematik.* Tagung zur Geschichte der Mathematik in Rummelsberg bei Nürnberg (4. 5. bis 8. 5. 2005), Augsburg: Rauner, 164–173.

Probst, Siegmund, 2006b, "Zur Datierung von Leibniz' Entdeckung der Kreisreihe," Herbert Breger, Jürgen Herbst, Sven Erdner (Eds.), *Einheit in der Vielheit*, VIII. Internationaler Leibniz-Kongress, Hanover, 813–817.

Ranea, Alberto Guillermo, 1989, "The *a priori* method and the *actio* concept revisited: dynamics and metaphysics in an unpublished controversy between Leibniz and Denis Papin," in: *Studia Leibnitiana*, 21, 42–68.

Rigaud, Stephen J. (Ed.), 1841, *Correspondence of Scientific Men*, Vol. I, Oxford: Oxford University Press.

Roberval, Gilles Personne de, 1736a, "Epistola ad Evangelistam Torricellium," in: *Ouvrages de mathématique*, Amsterdam: Pierre Mortier (= *Mémoires de l'Académie Royale des Sciences* [...] *avant son renouvellement en 1699*, Vol. 3).

Roberval, Gilles Personne de, 1736b, "Traité des indivisibles," in: *Ouvrages de mathématique*, Amsterdam: Pierre Mortier (= *Mémoires de l'Académie Royale des Sciences* [...] *avant son renouvellement en 1699*, Vol. 3).

Robinet, André, 1986, *Architectonique disjonctive, automates systémiques et idéalité transcendentale dans l'œuvre de Leibniz*, Paris: Vrin.

Robinet, André, 1991, *L'Empire Leibnizien, La conquête de la chaire de mathématiques de l'université de Padoue, Jacob Hermann et Nicolas Bernoulli*, Trieste: Edizioni LINT.

Robinson, Abraham, 1966, *Non-standard analysis*, Amsterdam: North Holland.

Rosen, Gideon, 2006, "Problems in the History of Fictionalism," in: Mark E. Kalderon (Ed.), *Fictionalism in Metaphysics*, Oxford, New York: Oxford University Press; Oxford: Clarendon Press, 14–64.

Ross, George MacDonald, 2007, "Leibniz's Debt to Hobbes," in: Pauline Phemister and Stuart Brown (Eds.), *Leibniz and the English-Speaking World*, Dordrecht: Springer, 19–33.

Russell, Bertrand, 1905, "On Denoting," in: *Mind* 14 (56), 479–93.

Russell, Bertrand, 1919, *Introduction to Mathematical Philosophy*, London: Routledge.

Rutherford, Donald, 1995, *Leibniz and the Rational Order of Nature*, New York: Cambridge University Press.

Rutherford, Donald, 2004, "Idealism Declined: Leibniz and Christian Wolff," in: Paul Lodge, (Ed.), *Leibniz and His Correspondents*, Cambridge: Cambridge University Press, 214–37.

Rutherford, Donald, 2005, "Leibniz on Spontaneity," in: Donald Rutherford and Jan A. Cover, (Eds.), *Leibniz: Nature and Freedom*, New York: Oxford University Press, 156–80.

Rutherford, Donald, 2008, "Leibniz as Idealist," in: *Oxford Studies in Early Modern Philosophy*, 4, 141–90.

Schmieden, Curt/Laugwitz, Detlef, 1958, "Eine Erweiterung der Infinitesimalrechnung," in: *Mathematische Zeitschrift*, 69, 1–39.

Scholtz, Lucie, 1934, *Die exakte Grundlegung der Infinitesimalrechnung bei Leibniz* (part of a dissertation), Marburg, Görlitz: Kretschmer.

Schuhmann, Karl, 2005, "Leibniz' Briefe an Hobbes," in: *Studia Leibnitiana*, XXXVII, 147–160.

Sleigh, Robert C., 1990, *Leibniz and Arnauld: A Commentary on their Correspondence*, New Haven: Yale University Press.

Smith, Sheldon, 2003, "Are Instantaneous Velocities Real and Really Instantaneous?: An Argument for the Affirmative," in: *Studies in the History and Philosophy of Modern Physics*, 34, 261–80.

Stedall, Jacqueline, 2002, *A Discourse Concerning Algebra: English Algebra to 1685*, Oxford: Oxford University Press.

Suisky, Dieter, 2006, "The foundation of Euler's mechanics using non-standard analysis," Power Point file of an unpublished paper.

Tönnies, Ferdinand, 1887, "Leibniz und Hobbes," in: *Philosophische Monatshefte* 23, 557–573 (=Ferdinand Tönnies, Studien zur Philosophie und Gesellschaftslehre im 17. Jahrhundert, E. G. Jacoby (Ed.), Frommann-Holzboog: Stuttgart-Bad Cannstatt 1975, 151–167).

Torricelli, Evangelista, 1919, "De solido hyperbolico acuto," in: *Opere di Evangelista Torricelli*, Gino Loria and Giuseppe Vassura (Eds.), Vol. I, Part 1, Faenza: Montanari.

Van Fraassen, Bas, 1980, *The Scientific Image*, Oxford: Clarendon Press.

Walker, Evelyn, 1932, *A Study of the Traité des Indivisibles of Gilles Personne de Roberval*, New York: Columbia University Press.

Wallis, John, 1655a, *De sectionibus conicis*, Oxford: Leonard Lichfield for Thomas Robinson.

Wallis, John, 1655b, *Arithmetica infinitorum*, Oxford: Leonard Lichfield for Thomas Robinson.

Wallis, John, 1656, *Due Correction for Mr. Hobbes; or Schoole Discipline, for not Saying his Lessons right. In Answer to his Six Lessons, Directed to the Professors of Mathematics*, Oxford: Printed by L. Lichfield, printer to the University for T. Robinson.

Wallis, John, 1659, *Tractatus duo. Prior, de cycloide et corporibus inde genitis. Posterior, epistolaris: inqua agitur, de cissoide, et corporibus inde genitis*, Oxford: Lichfield.

Wallis, John, 1670, *Mechanica: sive de motu, tractatus geometricus*, part II, London: William Godbid for Moses Pitt.

Wallis, John, 1672, "Binae methodi tangentium," In: *Philosophical Transactions*, No. 81, 25 March, 4010–4016.

Wallis, John, 1685, *A Treatise of Algebra, both Historical and Practical. Shewing, the Original, Progess, and Advancement thereof, from Time to Time; and by what Steps it hath attained to the Heighth at which it now is*, London: John Playford for Richard Davies.

Wallis, John, 1695, *Opera mathematica*, Vol. I, Oxford: at the Sheldonian Theater (Reprinted in 1972, Hildesheim, New York: Olms).

Wallis, John, 2003, *The Correspondence of John Wallis* (1616–1703), Philip Beeley and Christoph J. Scriba (Eds.), Vol. I, Oxford: Oxford University Press.

Wallis, John, 2004, *The Arithmetic of Infinitesimals*, Jacqueline A. Stedall (Trans.), New York: Springer.

Ward, Seth, *Vindiciae Academiarum, containing some briefe Animadversions upon Mr Websters Book, stiled, The Examination of Academies*, Oxford: Leonard Lichfield for Thomas Robinson 1654.

Webster, Charles, 1975, *The Great Instauration. Science, Medicine and Reform 1626–1660*, London: Duckworth.

Webster, John, 1654, *Academiarum Examen, or the Examination of the Academies*, London: Giles Calvert.

Whiteside, Derek, 1960/1962, "Patterns of Mathematical Thought in the later Seventeenth Century," in: *Archive for History of Exact Sciences*, 1, 179–388.

Wilson, Catherine, 1997, "Motion, Sensation, and the Infinite: The Lasting Impression of Hobbes on Leibniz," in: *British Journal for the History of Philosophy*, 5, 339–351.

Wolff, Christian, 1730, *Philosophia prima sive Ontologia, methodo scientifica pertracta, qua omnis cognitionis humanae principia continentur*, Francofurti & Lipsiae: Renger.

Yoder, Joella, 1988, *Unrolling Time*, Cambridge: Cambridge University Press, 1988.

Zeuthen, Hieronymus G., 1903, *Geschichte der Mathematik im XVI. und XVII. Jahrhundert*, Leipzig: Teubner.

Index of Persons

The dates of birth and death are added only to those persons who are subject of this book. The index includes only persons mentioned in the main text of contributions.

Affiliations of the Authors

RICHARD T. W. ARTHUR, McMaster University, Hamilton, Canada
O. BRADLEY BASSLER, The University of Georgia, Athens, USA
PHILIP BEELEY, Linacre College, Oxford, UK
HERBERT BREGER, Editorial Center of the Academy of Sciences at Göttingen, Hannover, Germany
FRANÇOIS DUCHESNEAU, Université de Montréal, Canada
DANIEL GARBER, Princeton University, USA
URSULA GOLDENBAUM, Emory University, Atlanta, USA
EMILY GROSHOLZ, Pennsylvania State University, USA
DOUGLAS JESSEPH, North Carolina State University, USA
EBERHARD KNOBLOCH, Technische Universität Berlin / Berlin-Brandenburg Academy of Sciences and Humanities, Berlin, Germany
SAMUEL LEVEY, Dartmouth College, Hanover, USA
FRITZ NAGEL, Bernoulli-Archiv, Basel, Switzerland
SIEGMUND PROBST, Editorial Center of the Academy of Sciences at Göttingen, Hannover, Germany
DONALD RUTHERFORD, University of California, San Diego, USA

Bei Fragen zur Produktsicherheit wenden Sie sich bitte an:
If you have any questions regarding product safety,
please contact:

Walter de Gruyter GmbH
Genthiner Straße 13
10785 Berlin
productsafety@degruyterbrill.com